512.74 Rib
Ribenboim.
Catalan's conjecture.

**The Lorette Wilmot Library
Nazareth College of Rochester**

D1595913

CATALAN'S CONJECTURE

Are 8 and 9 the Only Consecutive Powers?

CATALAN'S CONJECTURE

*Are 8 and 9
the Only Consecutive Powers?*

PAULO RIBENBOIM
Queen's University
Kingston, Ontario

ACADEMIC PRESS
Harcourt, Brace and Company, Publishers

Boston San Diego New York
London Sydney Tokyo Toronto

This book is printed on acid-free paper

Copyright © 1994 by Academic Press, Inc.
All rights reserved
No part of this publication may be reproduced or
transmitted in any form or by any means, electronic
or mechanical, including photocopy, recording, or
any information storage and retrieval system, without
permission in writing from the publisher.

ACADEMIC PRESS, INC.
525 B Street, Suite 1900, 92101-4495

United Kingdom Edition published by
ACADEMIC PRESS LIMITED
24-28 Oval Road, London NW1 7DX

Library of Congress Cataloging-in-Publication Data
Ribenboim, Paulo.
    Catalan's conjecture: are 8 and 9 the only consecutive powers? /
  Paulo Ribenboim.
       p. cm.
    Includes bibliographical references and indexes.
    ISBN 0-12-587170-8
    1. Consecutive powers (Algebra)   I. Title.
QA161.E95R53   1994
512'.74--dc20   93-23740
                                       CIP

Printed in the United States of America

94 95 96 97 BC 9 8 7 6 5 4 3 2 1

Dedication

A Henri, Charles et Gaston—tes couloirs, salles et bibliothèque m'ont toujours bien accueilli.

A Eugène (Catalan) qui a suscité la curiosité et Michel (Waldschmidt)—avec enthousiasme tu m'as fait goûter le vin de la connaissance.

Aux voix lointaines et admirables de las Scandinavie, passées et présentes, Carl (Størmer), Trygve (Nagell) et Kustaa (Inkeri).

Celui qui multiplie les questions, répondit-elle, augmente ses problèmes. Réjouis-toi, mon doux cadavre, de tout ce qui t'est donné, et ne pose pas de questions qui ne comportent pas de réponses.

La Dame et le Colporteur
S. J. Agnon
(Nobel Prize in Literature)

Contents

Preface xi

Reader's Guide xiii

Introduction 1

History 5

Part P PRELIMINARIES 9

    1. Binomials and Cyclotomic Polynomials 9
    2. The Cyclotomic Field 27
    3. The Pythagorean Equation, Special Cases of Fermat's Last Theorem and Related Equations 31
    4. Continued Fractions 55
    5. The Equations $EX^2 - DY^2 = \pm C$ 60

Part A SPECIAL CASES 67

    1. Preliminary Lemmas 67
    2. The Sequence of Squares or Cubes 69
    3. The Equation $X^m - Y^2 = 1$ 78
    4. The Result of Størmer on Fermat's Equation 80
    5. The Attempts to Solve $X^2 - Y^n = 1$ 89
    6. The Equation $X^2 - Y^n = 1$, $n \geq 3$ 92

7. The Equations $X^3 - Y^n = 1$ and $X^m - Y^3 = 1$, with $m, n \geq 3$ 96
8. The Equation $\dfrac{X^n - 1}{X - 1} = Y^m$ 110
9. The Sequence of Powers of 2 or 3 124
10. Interlude 127
11. The Equation $2X^n - 1 = Z^2$ 129
12. π and Gravé's Problem 132
13. A Problem of Fermat on Pythagorean Triangles and the Equation $2X^4 - Y^4 = Z^2$ 144
14. The Equations $X^4 \pm 2^m Y^4 = \pm Z^2$ and $X^4 \pm Y^4 = 2^m Z^2$ 164
15. Representation of Integers by Binary Cubic Forms 177
16. Some Quartic Equations 192

Part B DIVISIBILITY CONDITIONS 201

1. Getting the Consecutive Powers 8 and 9 201
2. The Theorem of Cassels and First Consequences 204
3. Prime Factors of Solutions of Catalan's Equation 214
4. The Theorem of Hyyrö 216
5. The Theorems of Inkeri 219

Part C ANALYTICAL METHODS 241

1. Some General Theorems for Diophantine Equations 241

I. The Equation $X^m - Y^n = 1$ 247
2. Upper Bounds for the Number and Size of Solutions ... 247
3. Lower Bounds for Solutions 256
4. Algorithm to Determine the Eventual Solutions 266

II. The Equation $a^U - b^V = 1$ 269
5. What Will Be Discussed 269
6. Finiteness of the Number of Solutions 271
7. Algorithm to Determine the Eventual Solutions 282
8. The Largest Prime Factor of Values of Quadratic Polynomials 283
9. Effective Results 288

III. The Equation $X^U - Y^V = 1$ 298

Contents ix

    10. The Theorem of Tijdeman 298
    11. A Density Result...................................... 308

Appendix 1. Catalan's Equation in Other Domains 313

    (A) Catalan's Equation over Number Fields 313
    (B) Catalan's Equation over Fields $K(t)$ and
         Domains $K[t]$ 314
    (C) Catalan's Equation Over Function Fields of
         Projective Varieties................................... 318

Appendix 2. Powerful Numbers 319

    (A) Distribution of Powerful Numbers..................... 320
    (B) Additive Problems 322
    (C) Difference Problems.................................. 323

Bibliography 331
    Index of Names 359
    Subject Index... 363

Preface

"Yet another book And all about consecutive powers ... I only know of 8 and 9."

"Same with me".

"Then it must be a very short book. Just the statement that 8 and 9 are the only consecutive powers and its proof. But, as I open the book, I see pages and pages of contorted arguments and not even this simple statement. What is this all about?"

"Be generous. We are not so smart. We only know that any consecutive powers are smaller than some fixed number T". After much thought, my enemy said:

"This sounds excellent to me".

"But, my friend, why don't you check all numbers below T and clear up this question at once? Your book would be a list of consecutive powers and perhaps you would only find 8 and 9". (Dear enemy, I wanted to say, but I am polite.)

"We are not smart enough, I told you already. The value of T which we are able to determine is greater than the number of stars in the Universe."

He cut me off, saying:

"Nobody ever counted the stars in the Universe, and I want only 8 and 9," he said with a slight irritation. I was visibly in the defensive. Despite a deep inner feeling of my weakness, I decided to counterattack.

"Let me explain what this book is about. Catalan said that 8 and 9 may well be the only consecutive powers. Since no one has found others, mathematicians are trying to prove the statement. It is known, for instance, that if one of the powers is a square, then 8 and 9 are the only consecutive powers. Also for many exponents m, n, it is known that a mth power and a nth power cannot be consecutive. Tijdeman showed that all consecutive powers must be below a definite number T. Most of the proofs are clever and appeal to all kinds of methods, from elementary number theory, to algebraic numbers, to diophantine approximation.

I personally find this very beautiful and instructive. I like to imagine that I am leading my readers to a destination which I don't quite know how to reach. But, as we go, we contemplate a charming landscape – here and there flat, somewhere else high peaks and plenty of flowers to be picked up. It is really not so important to arrive."

And after a pause, I began another discourse.

"And suppose that someday, someone could really establish that 8 and 9 are the only consecutive powers. Will mathematics or the world change? Will the birds in spring begin singing:

> 8 and 9, 9 and 8,
> Powers we are
> And consecutive too.
> Like the couples
> We will be
> In this bright sunshine.

Friends, it is what you encounter in your journey that makes it memorable. At arrival, we are no more than tired heroes. Pitiful".

P. Ribenboim

Reader's Guide

This book begins with two short sections. In the Introduction, you will become acquainted with the easily explained, but deep, problem of consecutive powers, first for fixed exponents, then for arbitrary exponents, and finally for fixed bases. Some variants of these questions are also spelled out. Simple-minded as these problems might seem, they have resisted mathematicians' attempts to solve them. The section on the history of these efforts gives you a feeling of the most important partial achievements as they happened. You will thus know what to expect later in the book, when studying the proofs in detail.

I found it convenient to gather in the Preliminaries results, usually of great independent interest, that will be invoked in the sequel. Then come the three main parts that constitute the body of the book. Part A contains only the study of special cases, and the methods are elementary throughout. Part B, devoted to divisibility conditions, appeals also to facts from the theory of algebraic numbers. In Part C, I begin with a short section on diophantine approximation and other powerful results of general nature on diophantine equations, for which no proofs are given. These strong theorems are then applied to reach the best-known results to date.

Finally, the Appendix goes somewhat beyond the main scope. I hope that you will reach it, as it contains a discussion of interesting variations and of powerful numbers.

Wishing that your study of this book be pleasant, I did not spare details in proofs and have explicitly quoted all theorems used in the steps of the proofs.

Paulo Ribenboim

Acknowledgements

To Queen's University, which provided funds to type a first version of this book in Chiwriter. The delay that was caused by the decision to retype it in TeX gave an opportunity to incorporate newer results and thereby improve the quality of the book. The final typing was done expertly by Diane Berezowski in Ottawa.

To Gary Walsh, who read my manuscript spotting mistakes ... and found Diane (no mistake).

To Kustaa Inkeri, who so patiently and generously checked the text and gave only good advice.

INTRODUCTION

 (a) If I write the sequence of squares and cubes of integers in increasing order

$$1\ 4\ 8\ 9\ 16\ 25\ 27\ 36\ 49\ 64\ 81\ 100\cdots$$

say

$$z_1 < z_2 < z_3 < z_4 < \cdots < z_n < z_{n+1} < \cdots$$

many questions may be asked. For example:

I) Are there consecutive integers in this sequence? Of course, yes: 8 and 9. Are there others? How many? Only finitely many?

Examining a list of squares and cubes up to 1,000,000, no other example is found. Is this always true? Or will there be, so to say by accident, other consecutive squares and cubes?

If a search with a computer is pushed further, it may be observed that the differences appear to become larger (though not monotonically); that is, squares and cubes appear more sparsely. Yet, from this experimental observation, it should not be hastily concluded that 8 and 9 are the only consecutive cube and square.

A second question is the following:

II) Given k (now $k \geq 2$), how many indices n are there such that $z_{n+1} - z_n \leq k$? Only finitely many?

It is also possible to consider other sequences involving powers:

(b) The sequence $z_1 < z_2 < z_3 \cdots$ of all proper powers of integers: squares, cubes, 5th powers, 7th powers, etc.

I may ask the same questions (I) and (II) for the sequence (b). Note that (a) is a subsequence of (b). Thus, if 8, 9 are the only consecutive integers in (b), they are also the only consecutive integers in (a); similarly, if (b) has only finitely many consecutive terms with difference at most k (where $k \geq 2$), then the same is true for the sequence (a).

The following question (which is of no interest for the sequence (a)), may now be asked:

III) Are there three or more consecutive powers? How many? Only finitely many?

(c) If $a, b \geq 2$, $a \neq b$, consider the sequence $z_1 < z_2 < z_3 < \cdots$ of all proper powers of a or b.

For example, if $a = 2$, $b = 3$:

$$4 \; 8 \; 9 \; 16 \; 27 \; 32 \; 64 \; 81 \; 128 \; 243 \; 256 \; \cdots$$

(d) Let $E = \{p_1, \cdots, p_r\}$, where $r \geq 2$ and each p_i is a prime number; let $E^\times$ be the set of all natural numbers, all of whose prime factors are in E:

$$E^\times : z_1 < z_2 < z_3 < \cdots$$

Introduction

The sequence (c) is of course a subsequence of one of type (d). It is interesting to ask the same questions (I) and (II) for each of the sequences (c) and (d).

This book deals with these problems. At various occasions, I will (on purpose!) sidetrack and discuss other more or less related questions which I find interesting in their own right and enrich this study.

HISTORY

I shall outline here very briefly some of the main historic developments in the study of the problem.

In Dickson's useful *History of the Theory of Numbers*, Volume II, it is stated that the first mention of this problem is in a question asked by Phillippe de Vitry: Is there any integer $m \geq 2$ such that $3^m \pm 1$ is a power of 2? This was solved by Levi ben Gerson (alias Leo Hebraeus), who lived in Spain from 1288 to 1344. He showed that if $3^m \pm 1 = 2^n$ then $m = 2$, $n = 3$, so these numbers are 9 and 8.

Using the method of infinite descent, which had been invented by Fermat, Euler showed in 1738 that if the difference between a square and a cube of rational numbers is ± 1, then these numbers are 9 and 8.

In a letter of 1844, published in Crelle's Journal, Catalan wrote:

> *Je vous prie, Monsieur, de vouloir bien énoncer dans votre recueil, le théorème suivant, que je crois vrai, bien que je n'aie pas encore réussi à le démontrer complètement; d'autres*

> *seront peut-être plus heureux. Deux nombres entiers consécutifs, autres que 8 et 9, ne peuvent être des puissances exactes; autrement dit: l'equation $x^m - y^n = 1$, dans laquelle les inconnues sont entières et positives, n'admet qu'une seule solution.*

This assertion is now called "Catalan's conjecture." In other words, he proposed to prove that the equation $X^U - Y^V = 1$, in four unknown quantities, two of which are in the exponent, has only the solution $x = 3$, $u = 2$, $y = 2$, $v = 3$ in natural numbers greater than 1. The only results of Catalan on this equation are simple observations, which are in his *Mélanges Mathématiques*, XV, published much later in 1885. Among the various statements, Catalan asserted, without proof, that if $x^y - y^x = 1$ then $x = 2$, $y = 3$ — but this is a rather simple exercise to prove.

Shortly after Catalan's conjecture was formulated, in 1850 Lebesgue used Gaussian integers to show that the equation $X^m - Y^2 = 1$ has no solution in positive integers x, y when $m > 2$.

The subsequent progress also concerned Catalan's equation with special small exponents. Thus, Nagell considered in 1921 the equations $X^3 - Y^n = 1$, $X^m - Y^3 = 1$, showing that they have no solution in positive integers when $n > 2$, $m > 2$.

In 1932, Selberg proved that $X^4 - Y^n = 1$ has no solution in positive integers when $n > 1$. Finally, in 1964, Chao Ko proved the stronger result that $X^2 - Y^n = 1$ does not have a solution in positive integers when $n > 1$. 120 years were required to establish this special case of Catalan's conjecture. Here it should be added that in 1976 Chein used results of Størmer and Nagell to give a much shorter proof of Chao Ko's result.

In 1953 and 1961, Cassels dealt with the equation $X^p - Y^q = 1$, where p, q are odd primes. He showed that if x, y are positive integers, solutions of the equation, then p divides

y and q divides x.

This was at once used by Mąkowski (and also by Hyyrö) to show that there are no three consecutive proper powers.

Hyyrö sharpened Cassels's result in 1964, indicating various congruences satisfied by x, $y > 1$, p, q if $x^p - y^q = 1$. In the same year, and again in 1989 and 1991, Inkeri drew further conclusions with a method involving the class numbers of $\mathbb{Q}(\sqrt{-p})$, $\mathbb{Q}(\sqrt{-q})$ and certain congruences. In this way, Inkeri established Catalan's conjecture for many pairs of exponents (p, q).

The equation $a^U - b^V = 1$, where $2 \leq a, b$, $a \neq b$, was studied in 1951 by LeVeque, who showed that it has at most one solution, or two solutions in positive integers when $a = 3$, $b = 2$.

The equation $X^m - Y^n = 1$, where m, $n \geq 2$, $\max\{m, n\} \geq 3$, has at most finitely many solutions, as follows from a theorem of Siegel (1929).

In 1964, Hyyrö determined an upper bound for the number of positive solutions of $X^m - Y^n = 1$. Hyyrö also gave lower bounds for the positive solutions.

Baker's famous results on effective bounds for solutions of certain types of diophantine equations are applicable to the equation $X^m - Y^n = 1$, where m, $n \geq 3$. Thus, for every pair (m, n) as above, there exists an effectively computable positive integer $C(m, n)$ such that if $x^m - y^n = 1$ then x, $y < C(m, n)$.

In 1976, Tijdeman used the results of Baker and proved the most important theorem to date about Catalan's equation:

There exists an effectively computatable positive constant C, such that if m, $n \geq 2$, x, $y \geq 1$ and $x^m - y^n = 1$, then x, y, m, $n < C$.

The value of C, estimated by Langevin, is still too big. This prevents deciding by numerical computations whether there are in fact, consecutive proper powers other than 8 and 9. Recent computational activity is aimed at narrowing the interval where consecutive powers might be found.

Part P

PRELIMINARIES

For the convenience of the reader, I assemble in this part many results not dealing directly with the problem of consecutive powers, but which will be needed in the sequel in various proofs. Even though they have independent interest, the reader may wish to go directly to Part A and return to the preliminaries only when needed.

Unless the contrary is stated, the numbers appearing in the equations are positive integers.

Briefly, a power of an integer is meant to be a proper power, that is, an integer a^n with $n \geq 2$, $a \geq 2$.

1. Binomials and Cyclotomic Polynomials

For every prime p and non-zero integer a, denote by $v_p(a)$ the *p-adic value* of a:

$$v_p(a) = e \text{ when } p^e | a, \text{ but } p^{e+1} \nmid a.$$

It is convenient to define $v_p(0) = \infty$.

For each rational number $\frac{a}{b}$ (with $b \neq 0$, a and b integers), define $v_p\left(\frac{a}{b}\right) = v_p(a) - v_p(b)$. The mapping $v_p : \mathbb{Q} \to \mathbb{Z} \cup \{\infty\}$ is the *p-adic valuation*.

Here is a list of properties of the p-adic valuation.

(P1.1). *If $a, b, b_1, \ldots, b_k$ are integers, then:*

i) $v_p(ab) = v_p(a) + v_p(b)$.
ii) $v_p(a+b) \geq \min\{v_p(a), v_p(b)\}$.
iii) *If* $v_p(a) < v_p(b_1), \ldots, v_p(b_k)$, *then*

$$v_p(a + b_1 + \ldots + b_k) = v_p(a).$$

iv) *If $a \neq b$, $p \neq 2$, $p \nmid ab$ and $v_p(a-b) = e \geq 1$, then $v_p(a^{p^r} - b^{p^r}) = e + r$ for every $r \geq 1$.*
v) *If $a \neq b$, $2 \nmid ab$ and $v_2(a-b) = e \geq 2$, then $v_2(a^{2^r} - b^{2^r}) = e + r$ for every $r \geq 1$.*
vi) *If p is any prime and $p | a^p - b^p$, then $p^2 | a^p - b^p$.*

Proof: The proofs are very simple. As an illustration, I give the proofs of (iv), (v) and (vi).

Proof of (iv): It suffices to show that $v_p(a^p - b^p) = e + 1$ and then repeat the argument.

By hypothesis, $a = b + kp^e$, where $p \nmid k$. Then

$$a^p = b^p + \binom{p}{1} b^{p-1} kp^e + \binom{p}{2} b^{p-2} k^2 p^{2e} + \ldots + k^p p^{pe}.$$

Since p divides $\binom{p}{j}$ for $j = 1, \ldots, p-1$, then

$$v_p\left[\binom{p}{j} b^{p-j} k^j p^{je}\right] \geq 1 + je.$$

From $v_p(k^p p^{pe}) = pe$ and (iii), it follows that $v_p(a^p - b^p) = e+1$.

Proof of (v): As in (iv), it suffices to show that $v_2(a^2 - b^2) = e + 1$. By hypothesis $a = b + 2^e k$ with $e \geq 2$ and a, b, k odd.

Then $a^2 = b^2 + 2^{e+1}k + 2^{2e}k^2$; since $e+1 < 2e$, then by (iii), $v_2(a^2 - b^2) = e + 1$.

Proof of (vi): By hypothesis, $a \equiv a^p \equiv b^p \equiv b \pmod{p}$; raising to the pth power, $a^p \equiv b^p \pmod{p^2}$, so $p^2 | a^p - b^p$. ∎

Special cases of the following basic result were given by Euler:

(P1.2). *Let $n > 1$ and let x, y be non-zero relatively prime integers. Then:*

i) $\frac{x^n - y^n}{x - y} = k(x - y) + ny^{n-1}$, *where*

$$k = (x-y)^{n-2} + \binom{n}{1} y(x-y)^{n-3} + \ldots + \binom{n}{n-2} y^{n-2}.$$

In particular, if $n = p$ is a prime, then $k = (x-y)^{p-2} + uyp$, where u is an integer.

ii) $\gcd\left(x - y, \frac{x^n - y^n}{x-y}\right) = \gcd(x - y, n)$.

iii) *If m is a positive integer, $m|x - y$ and $m \nmid n$, then $m \nmid \frac{x^n - y^n}{x-y}$.*

iv) *If p is an odd prime and $p|x - y$, then*

$$v_p\left(\frac{x^n - y^n}{x - y}\right) = v_p(n).$$

In particular, $p^2 \nmid \frac{x^p - y^p}{x-y}$.

v) *If $4|x - y$, then $v_2\left(\frac{x^n - y^n}{x-y}\right) = v_2(n)$, whereas if $2|x - y$ but $4 \nmid x - y$, then*

$$v_2\left(\frac{x^n - y^n}{x - y}\right) \geq v_2(n) + 1.$$

vi) *If n is odd, then $\frac{x^n - y^n}{x-y}$ is odd.*

vii) *If n is odd and $0 < i$, then*

$$\gcd\left(\frac{x^n - y^n}{x - y}, x^{2^i n} + y^{2^i n}\right) = 1.$$

viii) *If p is an odd prime and $x > y \geq 1$, then*

$$\frac{x^p + y^p}{x + y} \geq p, \text{ and if } \frac{x^p + y^p}{x + y} = p, \text{ then } p = 3,$$
$$x = 2, \ y = 1.$$

ix) *If p is an odd prime and $x > y \geq 1$, then*

$$\frac{x^p - y^p}{x - y} > p.$$

Proof:

i) $\dfrac{x^n - y^n}{x - y} = \dfrac{[(x-y) + y]^n - y^n}{x - y}$

$= (x-y)^{n-1} + \binom{n}{1} y(x-y)^{n-2} + \binom{n}{2} y^2 (x-y)^{n-3}$

$+ \ldots + \binom{n}{n-2} y^{n-2}(x-y) + n y^{n-1}$

$= k(x-y) + n y^{n-1},$

where k has the value indicated. Clearly, if $n = p$ is a prime, then $k = (x-y)^{p-2} + uyp$, with u an integer.

ii) Since $\gcd(x, y) = 1$, the statement follows at once from (i).

iii) This follows immediately from (ii).

iv) Let $n = p^e m$, with $p \nmid m$, $e \geq 0$, so $e = v_p(n)$. Let $x_1 = x^m$, $y_1 = y^m$. By (iii) $v_p\left(\frac{x^m - y^m}{x - y}\right) = 0$, so $v_p(x_1 - y_1) = v_p(x - y) \geq 1$. By (P1.1) $v_p\left(\frac{x_1^{p^e} - y_1^{p^e}}{x_1 - y_1}\right) = e$, hence $v_p\left(\frac{x^n - y^n}{x - y}\right) = e = v_p(n)$.

Binomials and Cyclotomic Polynomials

v) The proof is similar. Let $n = 2^e m$, with m odd, $e \geq 0$, and $x_1 = x^m$, $y_1 = y^m$. By (iii), $v_2(x_1 - y_1) = v_2(x-y) \geq 1$. If $v_2(x-y) \geq 2$, then by (P1.1) $v_2\left(\frac{x_1^{2^e} - y_1^{2^e}}{x_1 - y_1}\right) = e = v_2(n)$.
If $v_2(x-y) = v_2(x_1 - y_1) = 1$, then $x_1 \equiv -y_1 \pmod 4$, so $v_2(x_1 + y_1) \geq 2$. If $e = 1$, then $v_2\left(\frac{x_1^2 - y_1^2}{x_1 - y_1}\right) \geq 2$. If $e \geq 2$, let $x_2 = x_1^2, y_2 = y_1^2$ so $v_2\left(\frac{x_1^{2^e} - y_1^{2^e}}{x_1 - y_1}\right) = v_2\left(\frac{x_2^{2^{e-1}} - y_2^{2^{e-1}}}{x_2 - y_2}\right) + v_2\left(\frac{x_2 - y_2}{x_1 - y_1}\right) \geq (e-1) + 2 = e+1$, by the first part of the proof.

vi) If $x \not\equiv y \pmod 2$, then $x^n - y^n$ is odd, and so is $\frac{x^n - y^n}{x-y}$.
If $x \equiv y \pmod 2$, since x, y are relatively prime, they are odd. It follows that $\frac{x^n - y^n}{x-y} = x^{n-1} + x^{n-2}y + \ldots + xy^{n-2} + y^{n-1}$ (sum of an odd number n of odd summands) is odd.

vii) Let p be a prime, $e \geq 1$, such that $p^e \left| \frac{x^n - y^n}{x-y} \right.$, $p^e \left| x^{2^i n} + y^{2^i n} \right.$. By (vi), $p \neq 2$. Since $p^e | x^n - y^n$ then $x^{2^i n} + y^{2^i n} \equiv ex^{2^i n} \pmod{p^e}$, so p^e divides $2x^{2^i n}$. But $p \nmid x$, otherwise $p|y$, which is a contradiction. So $p^e / 2$, thus $p = 2$, which has been excluded.

viii) Let $x \geq 2$ and $p \geq 3$, then
$$\frac{x^p + y^p}{x+y} = (x^{p-1} - x^{p-2}y) + \ldots + (x^2 y^{p-3} - xy^{p-2}) + y^{p-1}$$
$$\geq x^{p-2}(x-y) + y^{p-1} \geq 2^{p-2} + 1 \geq p.$$
If $\frac{x^p + y^p}{x+y} = p$ then $2^{p-2} + 1 = p$, so $p = 3$, $x = 2$, and $y = 1$.

ix) $\frac{x^p - y^p}{x-y} = x^{p-2} + x^{p-2}y + \ldots + xy^{p-1} + y^{p-2} \geq 2^{p-1} > p$. ∎

Now I shall collect some facts about cyclotomic polynomials.

For every $n \geq 1$, let
$$\zeta_n = \cos\frac{2\pi}{n} + i\sin\frac{2\pi}{n}$$

be a primitive n^{th} root of 1. Then the primitive n^{th} roots of 1 are precisely the powers ζ_n^j, where $1 \leq j < n$, $gcd(j,n) = 1$.

The nth *cyclotomic polynomial* is

$$\Phi_n(X) = \prod_{gcd(j,n)=1} (X - \zeta_n^j). \tag{1.1}$$

It is a monic polynomial with coefficients in $\mathbb{Z}$ and degree equal to $\varphi(n)$ (the *totient* of n). If n is odd, then $\Phi_{2n}(X) = \Phi_n(-X)$.

As is well known,

$$X^n - 1 = \prod_{d|n} \Phi_d(X). \tag{1.2}$$

Hence, if $m|n$, $m \neq n$, then

$$X^n - 1 = (X^m - 1) \prod_{\substack{d|n \\ d \nmid m}} \Phi_d(X). \tag{1.3}$$

Thus, for example, $\Phi_2(X) = X+1$, $\Phi_4(X) = X^2+1$, $\Phi_{2^a}(X) = X^{2^{a-1}} + 1$. More generally, for every prime power $p^e (e \geq 1)$:

$$\Phi_{p^e}(X) = \frac{X^{p^e} - 1}{X^{p^{e-1}} - 1}$$
$$= X^{p^{e-1}(p-1)} + X^{p^{e-1}(p-2)} + \ldots + X^{p^{e-1}} + 1.$$

Hence for every $e \geq 1$, $\Phi_{p^e}(1) = p$. Moreover, if p is odd then $\Phi_{p^e}(a)$ is odd, for every integer a.

If p is a prime, $p \nmid m$, $e \geq 1$, then

$$\Phi_{p^e m}(X) = \frac{\Phi_m(X^{p^e})}{\Phi_m(X^{p^{e-1}})} \tag{1.4}$$

and

$$X^{p^e m} - 1 = \prod_{i=0}^{e} \Phi_{p^i}(X^m). \tag{1.5}$$

It follows by induction on the number of distinct prime factors of n that if n is odd, $n \geq 3$, then $\Phi_n(a)$ is odd for every integer a.

If $p|m$, then
$$\Phi_{pm}(x) = \Phi_m(X^p). \tag{1.6}$$

Let μ denote the *Möbius function* (defined for every $n \geq 1$):
$$\mu(n) = \begin{cases} (-1)^r & \text{if } n \text{ is the product of } r \text{ distinct primes} \\ 0 & \text{if } n \text{ is not square-free.} \end{cases}$$

In particular, $\mu(1) = 1$.

With the Möbius function it is possible to express the cyclotomic polynomials as follows:
$$\Phi_n(X) = \prod_{d|n}(X^d - 1)^{\mu\left(\frac{n}{d}\right)}. \tag{1.7}$$

It is useful to consider also the corresponding homogenized polynomials
$$\Phi_n(X, Y) = Y^{\varphi(n)} \Phi_n\left(\frac{X}{Y}\right). \tag{1.8}$$

Note that if n is odd, then
$$\Phi_{2n}(X, Y) = \Phi_n(X, -Y).$$

The following relations hold:
$$X^n - Y^n = \prod_{d|n} \Phi_d(X, Y), \tag{1.9}$$

and if $m|n$, $m \neq n$, then
$$X^n - Y^n = (X^m - Y^m) \prod_{\substack{d|n \\ d \nmid m}} \Phi_d(X, Y). \tag{1.10}$$

If p is a prime, $p \nmid m$, $e \geq 1$, then

$$\Phi_{p^e m}(X,Y) = \frac{\Phi_m(X^{p^e}, Y^{p^e})}{\Phi_m(X^{p^{e-1}}, Y^{p^{e-1}})} \qquad (1.11)$$

and also

$$X^{p^e m} - Y^{p^e m} = \prod_{i=0}^{e} \Phi_{p^i}(X^m, Y^m). \qquad (1.12)$$

If $p|m$, then

$$\Phi_{pm}(X,Y) = \Phi_m(X^p, Y^p) \qquad (1.13)$$

and

$$\Phi_n(X,Y) = \prod_{d|n}(X^d - Y^d)^{\mu(\frac{n}{d})}. \qquad (1.14)$$

If $1 \leq b < a$, by (1.14), $\Phi_n(a,b) > 0$. If n is odd, $n \geq 3$, and a, b are not both even, then $\Phi_n(a,b)$ is odd.

Also, if n is odd, $n \geq 3$, and a, b are not both even, then $\Phi_{2n}(a,b) = \Phi_n(-a,b)$ is odd.

If a, b are distinct positive relatively prime integers and $n \geq 1$, the prime p is said to be a *primitive factor* of $a^n \pm b^n$ if p divides $a^n \pm b^n$, but p does not divide $a^m \pm b^m$, for every m, $1 \leq m < n$.

Thus, if $p = 2$ is a primitive factor of $a^n \pm b^n$ then a, b have the same parity, so $n = 1$.

If $p|a^n - b^n$, $\gcd(a,b) = 1$, then $p \nmid ab$; let b' be an integer such that $bb' \equiv 1 \pmod{p}$; then $(ab')^n \equiv 1 \pmod{p}$, so the order of $ab' \bmod p$ divides n, as well as $p - 1$.

(P1.3). *With the above notations, the following statements are equivalent:*

1) *p is a primitive factor of $a^n - b^n$.*
2) *$p|a^n - b^n$, but $p \nmid a^m - b^m$ for every proper divisor m of n.*
3) *$p|\Phi_n(a,b)$, but $p \nmid \Phi_m(a,b)$ for every m such that $1 \leq m < n$.*

Binomials and Cyclotomic Polynomials   17

4) $p|\Phi_n(a,b)$, but $p \nmid \Phi_m(a,b)$ for every proper divisor m of n.
5) $\text{ord}\,(ab',\bmod p) = n$.

Proof: The implications $(1) \Rightarrow (2)$ and $(3) \Rightarrow (4)$ are trivial.

The equivalences $(1) \Leftrightarrow (3)$ and $(2) \Leftrightarrow (4)$ follow from the expressions
$$a^m - b^m = \prod_{p|m} \Phi_d(a,b),$$
for every $m = 1, \ldots, n$. Noting that $(ab')^d \equiv 1 \pmod{p}$ if and only if $p|a^d - b^d$, then (5) is visibly equivalent to (1) and also to (2). ∎

It follows from the above condition (2) that if $p|a^n - b^n$, then there exists an integer d dividing n, such that p is a primitive factor of $a^d - b^d$.

(P1.4). *Let a,b be distinct positive relatively prime integers, $n \geq 1$ and p be an odd prime. Then the following conditions are equivalent:*

1) *p is a primitive factor of $a^n - b^n$.*
2) *$p|\Phi_n(a,b)$ and $p \equiv 1 \pmod{n}$.*
3) *$p|\Phi_n(a,b)$ and $p \nmid n$.*

Proof: $(1) \Rightarrow (2)$. By hypothesis $p|a^n - b^n$, but $p \nmid a^m - b^m$ for every m, $1 \leq m < n$. By (P1.3), $p|\Phi_n(a,b)$ and the order of ab' mod p is n, hence $n|p - 1$, so $p \equiv 1 \pmod{n}$.

$(2) \Rightarrow (3)$. Since $n < p$, $p \nmid n$.

$(3) \Rightarrow (1)$. Clearly $p|a^n - b^n$. If there exists a proper divisor m of n such that $p|a^m - b^m$, from $a^n - b^n = \Phi_n(a,b)(a^m - b^m) \prod_{\substack{d|n, d\nmid m \\ d \neq n}} \Phi_d(a,b)$, it follows that p divides $\frac{a^n - b^n}{a^m - b^m}$. By (P1.2) part (iv),
$$v_p\left(\frac{n}{m}\right) = v_p\left(\frac{a^n - b^n}{a^m - b^m}\right) \geq v_p(\Phi_n(a,b)) \geq 1,$$

so $p|n$, which is a contradiction. ∎

(P1.5). *Let p be an odd prime. Then p is a primitive factor of $a^n + b^n$ if and only if p is a primitive factor of $a^{2n} - b^{2n}$.*

Proof: Let p be a primitive factor of $a^n + b^n$, so $p|a^{2n} - b^{2n}$. If p is not a primitive factor of $a^{2n} - b^{2n}$, there exists k, $1 \leq k < 2n$, k dividing $2n$, such that $p|a^k - b^k$. If $k = 2h$, then $h < n$ and $n \nmid a^n + b^n$, hence $p|a^n - b^n$, which is absurd. Thus k is odd. From $a^n \equiv -b^n \pmod{p}$, then $a^{kn} \equiv -b^{kn} \pmod{p}$, thus $p|a$ and $p|b$, which is a contradiction.

Conversely, if p is a primitive factor of $a^{2n} - b^{2n} = (a^n - b^n)(a^n + b^n)$, then $p \nmid a^n - b^n$, so $p|a^n + b^n$. If there exists h, $1 \leq h < n$, such that $p|a^h + b^h$, then $p|a^{2h} - b^{2h}$ with $2h < 2n$, which is impossible. ∎

For every integer $n \geq 2$, let $P[n]$ denote the largest prime factor of n.

(P1.6). *Let $a > b \geq 1$, $\gcd(a,b) = 1$ and let $n \geq 2$. Let p be a primitive factor of $a^f - b^f$ such that $p|\Phi_n(a,b)$. Then:*

i) *There exists $j \geq 0$ such that $n = fp^j$ with $p \nmid f$.*
ii) *If $j > 0$, then $p = P[n]$.*
iii) *If $j > 0$ and $p^2|\Phi_n(a,b)$, then $n = p = 2$.*
iv) *$\gcd(\Phi_n(a,b), n) = 1$ or $P[n]$.*

Proof: i) By (1.9), $\Phi_n(a,b)$ divides $a^n - b^n$; then $p|a^n - b^n$, hence $f|n$ by (P1.3). Since $p|a^{p-1} - b^{p-1}$, again $f|p-1$, so $f < p$. Let $n = fp^j w$ with $j \geq 0$, $p \nmid fw$. Write $r = fp^j$. By (P1.2)
$$\frac{a^n - b^n}{a^r - b^r} \equiv wb^{w-1} \pmod{a^r - b^r}.$$

Since $p|a^r - b^r$ (because $f|r$, then
$$\frac{a^n - b^n}{a^r - b^r} \equiv wb^{w-1} \pmod{p}.$$

If $n < m$ then by (1.9), $\Phi_n(a,b)$ divides $\frac{a^n - b^n}{a^r - b^r}$. Since $p \nmid b$, (because $\gcd(a,b) = 1$), then $p|w$, which is absurd. So $n = fp^j$.

ii) From $f < p$, if $j > 0$, then $p = P[n]$.

iii) Let $j \geq 1$ and $s = fp^{j-1}$, so $n = ps$. Then

$$\frac{a^n - b^n}{a^s - b^s} = \frac{[(a^s - b^s) + b^s]^p - b^{sp}}{a^s - b^s} = pb^{s(p-1)}$$
$$+ \binom{p}{2}(a^s - b^s)b^{s(p-2)}$$
$$+ \binom{p}{3}(a^s - b^s)^2 b^{s(p-3)} + \ldots + (a^s - b^s)^{p-1}.$$

If $p \geq 3$, since $p | a^s - b^s$, then

$$\frac{a^n - b^n}{a^s - b^s} \equiv p \pmod{p^2}.$$

On the other hand, by (1.10), $\Phi_n(a,b)$ divides $\frac{a^n - b^n}{a^s - b^s}$, hence $p^2 | \Phi_n(a,b)$. Thus, if $p^2 | \Phi_n(a,b)$, then necessarily $p = 2$. So $f \leq p - 1$ implies $f = 1$ and $n = 2^j$, so $\Phi_n(a,b) \not\equiv 0 \pmod{r}$, which is absurd. This shows that $j = 1$ and $n = 2$.

iv) Assume that there exists a prime p dividing $\gcd(\Phi_n(a,b), n)$. By (i) and (ii), $p = P[n]$. By (iii), if $p^2 | \gcd(\Phi_n(a,b), n)$, then $n = p = 2$, so $p^2 \nmid n$. This shows the assertion. ∎

The following very interesting theorem was proved by Bang (1886) in a particular case. In 1892, Zsigmondy proved the stronger version presented here. It was rediscovered by Birkhoff and Vandiver (1904) and by various other mathematicians, like Dickson (1905), Carmichael (1913), Kanold (1950), Artin (1955), Hering (1974), Lüneburg (1981) and maybe others.

(P1.7). *Let $a > b \geq 1$, $gcd(a,b) = 1$, $n \geq 1$.*

i) $a^n - b^n$ *has a primitive factor, with the following exceptions:*

   a) $n = 1$, $a - b = 1$.
   b) $n = 2$, $a + b$ *a power of* 2.
   c) $n = 6$, $a = 2$, $b = 1$.

ii) $a^n + b^n$ *has a primitive factor, with the following exception:* $n = 3$, $a = 2$, $b = 1$.

Proof: i) It is clear that in cases (a), (b), (c), $a^n - b^n$ does not have a primitive factor. If $n = 1$ and $a - b$ does not have a primitive factor, then $a - b = 1$.

Let $n = 2$ and assume that $a^2 - b^2$ does not have a primitive factor. From $a^2 - b^2 = (a+b)(a-b)$ and $gcd(a+b, a-b) = 1$ or 2, if p is an odd prime dividing $a+b$, then p divides $a^2 - b^2$. But p is not a primitive factor, so $p | a - b$, hence p divides a and b, which is absurd. This shows that $a + b$ is a power of 2.

Now let $n \geq 3$ and assume again that $a^n - b^n$ does not have a primitive factor. Let $p = P[n]$ and $v_p(\Phi_n(a,b)) = j \geq 0$. Define

$$\Phi_n^*(a,b) = \frac{\Phi_n(a,b)}{p^j}.$$

1°) Assume that $\Phi_n^*(a,b) = 1$.

Let $\zeta_1, \zeta_2, \ldots, \zeta_{\varphi(n)}$ be the primitive n^{th} roots of 1. From

$$|a - \zeta_i b| = b\left|\frac{a}{b} - \zeta_i\right| > b\left(\frac{a}{b} - 1\right) = a - b$$

and a previous remark,

$$\Phi_n(a,b) = |\Phi_n(a,b)| = \prod_{i=1}^{\varphi(n)} |a - \zeta_i b| > (a-b)^{\varphi(n)} \geq 1 = \Phi_n^*(a,b).$$

So $j \geq 1$ and $p | \Phi_n(a,b)$, hence p divides $a^n - b^n$; so p is a primitive factor of $a^f - b^f$, where f divides n. By (P1.6), $gcd(n, \Phi_n(a,b)) = p$ and also $p^2 \nmid \Phi_n(a,b)$.

Binomials and Cyclotomic Polynomials

In conclusion, $\Phi_n(a,b) = p$, because $\Phi_n^*(a,b) = 1$. Moreover, from $p|n$, it follows that $p-1$ divides $\varphi(n)$. This implies in turn that $p = \Phi_n(a,b) \geq (a-b)^{\varphi(n)} \geq (a-b)^{p-1}$, hence $a-b=1$.

If $p^2|n$ let $n = pm$, then $p-1 \leq \varphi(m)$ and by (1.13)

$$p = \Phi_n(a,b) = \Phi_m(a^p - b^p) > (a^p - b^p)^{\varphi(m)} \geq (a^p - b^p)^{p-1},$$

because $p|m$. Thus $a^p - b^p = 1$, which is not compatible with $a-b=1$.

Thus, from (P1.6), $n = pf$, $p \nmid f$, where p is a primitive factor of $a^f - b^f$. Note also that $f | p-1$, so $f < p$. From $\varphi(n) = (p-1)\varphi(f)$ it follows that

$$p(a^p - b^p) > p(a^f - b^f) \geq \Phi_n(a,b)\Phi_f(a,b)$$
$$= \Phi_f(a^p - b^p) > (a^p - b^p)^{\varphi(f)}$$

using (1.11). Therefore $p > (a^p - b^p)^{\varphi(f)-1}$, hence necessarily $\varphi(f) = 1$, thus $f = 1$ or $f = 2$, so $n = p$ or $n = 2p$.

If $n = p$, then $p = \Phi_p(a,b) = a^{p-1} + a^{p-2} + \ldots ab^{p-2} + b^{p-1} = \frac{a^p - b^p}{a-b} = a^p - b^p$, and this is absurd because $a - b = 1$.

If $n = 2p$, from $3 \leq p$ it follows from (1.11) that

$$p = \Phi_{2p}(a,b) = \frac{a^p + b^p}{a+b}.$$

By (P1.2), necessarily $a = 2$, $b = 1$ and $p = 3$, so $n = 6$.

2°) Assume that $a^n - b^n$ does not have a primitive factor. It suffices to show that $\Phi_n^*(a,b) = 1$ and the result follows from (1°).

Let p be a prime dividing $\Phi_n(a,b)$, so $p|a^n - b^n$. Then there exists f, dividing n, $1 \leq f < n$, such that p is a primitive factor of $a^f - b^f$. By (P1.6), $p = P[n]$ and $\Phi_n(a,b) = p^j$ with $j \geq 1$. Hence $\Phi_n^*(a,b) = 1$.

ii) If $n = 3$, $a = 2$, $b = 1$, then $a^n + b^n = 2^3 + 1$ has no primitive factor.

Conversely, if $n = 1$ then $a + b \geq 2$, so there is a primitive factor.

If $n = 2$ and $a^2 + b^2$ does not have a primitive factor, then $a^2 + b^2 = 2^k$ (with $k \geq 2$). Indeed, if p is an odd prime dividing $a^2 + b^2$, then $p | a + b$, so $p | a^2 - b^2$, hence, $p | 2a^2$; it follows that $p | a$ and also $p | b$, which is absurd.

From $a^2 + b^2 = 2^k$ ($k \geq 2$), $gcd(a,b) = 1$, it follows that a, b are odd, hence $a^2 + b^2 \equiv 2 \pmod{4}$, which is a contradiction, proving that $a^2 + b^2$ has a primitive factor.

If $n \geq 3$, it follows from (i) that $a^{2n} - b^{2n}$ has a primitive factor p with the only exception $n = 3$, $a = 2$, $b = 1$. If $p = 2$ then a, b are odd, so $2 | a + b$, which is not compatible with 2 being a primitive factor of $a^n - b^n$.

By (P1.5), $a^n + b^n$ has a primitive factor, with the exception indicated. ∎

It follows from this theorem and (P1.3) that if $a \geq 2$, then each number in the sequence

$$\Phi_3(a), \ \Phi_4(a), \ \Phi_5(a), \ \Phi_6(a), \ \Phi_7(a) \ldots$$

(with $\Phi_6(a)$ deleted when $a = 2$) has a prime factor which is not a factor of any of the preceding numbers.

The following results will also be needed later in this book.

(P1.8). Let $1 \leq m < n$, and $a > b \geq 1$, with $gcd(a,b) = 1$. If $gcd(\Phi_m(a,b), \Phi_n(a,b)) \neq 1$, then $P[n] = gcd(\Phi_m(a,b), \Phi_n(a,b))$.

Proof: If $n = 2$, then $m = 1$. If $gcd(a - b, a + b) \neq 1$, then $gcd(a - b, a + b) = 2$.

Now assume $n \geq 3$.

Let p be a prime and $e \geq 1$ be such that $p^e | \Phi_m(a,b)$, $p^e | \Phi_n(a,b)$. Then $p | a^m - b^m$, $p | a^n - b^n$, so p is not a primitive factor of $a^n - b^n$. By (P1.4) $p | n$, and by (P1.6) $p =$

$P[n]$, $\Phi_n(a,b) = pc$, $p \nmid c$, so $e = 1$. Since p was an arbitrary common divisor of $\Phi_m(a,b)$, $\Phi_n(a,b)$, this proves that $P[n] = gcd(\Phi_m(a,b), \Phi_n(a,b))$. ∎

(P1.9). *Let p be any prime, let $0 \leq i < j$ and let $a \geq b > 1$, $gcd(a,b) = 1$. Then*

$$gcd(\Phi_{p^i}(a,b), \Phi_{p^j}(a,b)) = \begin{cases} 1 & \text{if } p \nmid a - b \\ p & \text{if } p | a - b. \end{cases}$$

Proof: By (P1.8), if $d = gcd(\Phi_{p^i}(a,b), \Phi_{p^j}(a,b)) \neq 1$ then $d = p$.

Assume first that $p \neq 2$.

If $p | a - b$, then $a^{p^{j-1}} \equiv a \equiv b \equiv b^{p^{j-1}} \pmod{p}$ so by (P1.2), part (iv), p divides $\Phi_{p^j}(a,b) = \frac{a^{p^j} - b^{p^j}}{a^{p^{j-1}} - b^{p^{j-1}}}$. Similarly, p divides $\Phi_{p^i}(a,b)$.

Finally, if $p \nmid a - b$, then $a^{p^j} \equiv a \not\equiv b \equiv b^{p^j} \pmod{p}$, so $p \nmid a^{p^j} - b^{p^j}$ and a fortiori $p \nmid \Phi_{p^j}(a,b)$. Thus, $gcd(\Phi_{p^i}(a,b), \Phi_{p^j}(a,b)) = 1$.

If $p = 2$, then $\Phi_1(a,b) = a - b$ and $\Phi_{2^k}(a,b) = a^{2^{k-1}} + b^{2^{k-1}}$ (for $k \geq 1$). So if $a \equiv b \pmod{2}$, then 2 divides $gcd(\Phi_{2^i}(a,b), \Phi_{2^j}(a,b))$, and conversely. ∎

In §B5 I will use a classical result of Gauss (1801). The proof requires some well-known facts about Gaussian sums, which I recall now.

Let p be an odd prime, and let

$$p^* = (-1)^{\frac{p-1}{2}} p = \begin{cases} p & \text{if } p \equiv 1 \pmod{4} \\ -p & \text{if } p \equiv 3 \pmod{4}. \end{cases} \tag{1.15}$$

Let $R^+ = \{a | 1 \leq a \leq p-1, \left(\frac{a}{p}\right) = 1\}$ (the set of quadratic residues modulo p) and $R^- = \{b | 1 \leq b \leq p-1, \left(\frac{b}{p}\right) = -1\}$ (the set of non-quadratic residues mod p).

Observe that if $p > 3$, then $\sum_{a \in R^+} a \equiv 0 \pmod{p}$ and $\sum_{b \in R^-} b \equiv 0 \pmod{p}$. Indeed, there exists k, $1 < k \leq p-1$, such that $\left(\frac{k}{p}\right) = 1$. Note that $R^+ = \{c | 1 \leq c \leq p-1$ and there exists $a \in R^+$ such that $c \equiv ka \pmod{p}\}$. Then $k\left(\sum_{a \in R^+} a\right) \equiv \sum_{a \in R^+} a$ $\pmod{p}$, hence $(k-1)\left(\sum_{a \in R^+} a\right) \equiv 0 \pmod{p}$, and therefore $\sum_{a \in R^+} a \equiv 0 \pmod{p}$. From $\sum_{j=1}^{p-1} j \equiv 0 \pmod{p}$ it follows that $\sum_{b \in R^-} b \equiv 0 \pmod{p}$.

Let $\zeta = \cos\frac{2\pi}{p} + i\sin\frac{2\pi}{p}$. The value of the *Gaussian sum*

$$\tau = \sum_{m=1}^{p-1} \left(\frac{m}{p}\right)\zeta^m = \sum_{a \in R^+} \zeta^a - \sum_{b \in R^-} \zeta^b \qquad (1.16)$$

is known to satisfy $\tau^2 = p^*$.

Let

$$\begin{cases} A(X) = \prod_{a \in R^+} (X - \zeta^a) \\ B(X) = \prod_{b \in R^-} (X - \zeta^b), \end{cases} \qquad (1.17)$$

so

$$4\frac{X^p - 1}{X - 1} = 4A(X)B(X) = [A(X) + B(X)]^2 - [A(X) - B(X)]^2. \qquad (1.18)$$

If $p = 3$, $4\frac{X^3-1}{X-1} = (2X+1)^2 + 3$. More generally:

(**P1.10**). *If $p > 3$ is a prime, there exist polynomials F, $G \in \mathbb{Z}[X]$ such that*

$$4\frac{X^p - 1}{X - 1} = F(X)^2 - p^*G(X)^2$$

Moreover, $\deg(F) = \frac{p-1}{2}$, $\deg(G) = \frac{p-3}{2}$ and

$$F(X) = (-X)^{\frac{p-1}{2}} F\left(\frac{1}{X}\right), \qquad G(X) = X^{\frac{p-1}{2}} G\left(\frac{1}{X}\right),$$
$$G(X) = X + a_2 X^2 + \ldots + a_{\frac{p-5}{2}} X^{\frac{p-5}{2}} + X^{\frac{p-3}{2}}.$$

Proof: Let

$$\begin{cases} F(X) = A(X) + B(X) \\ G(X) = -\frac{\tau}{p^*}[A(X) - B(X)]; \end{cases}$$

thus $\deg(F) = \frac{p-1}{2}$, $\deg(G) = \frac{p-3}{2}$. Moreover the leading coefficient of $G(X)$ is

$$-\frac{\tau}{p^*}\left(-\sum \zeta^a + \sum \zeta^b\right) = -\frac{\tau}{p^*}(-\tau) = 1.$$

Also, it follows at once that $4\frac{X^p-1}{X-1} = F(X)^2 - p^*G(X)^2$, and it suffices to show that $F(X)$, $G(X) \in \mathbb{Z}[X]$.

First, I prove that $F(X)$, $G(X) \in \mathbb{Q}[X]$, that is, the coefficients of $F(X)$, $G(X)$ are invariant by all automorphisms σ of $\mathbb{Q}(\zeta)|\mathbb{Q}$. These automorphisms are $\sigma_1, \ldots, \sigma_{p-1}$, where $\sigma_i(\zeta) = \zeta^i$ and σ_i leaves fixed each rational number.

Since $\sigma_i(\zeta^a) = \zeta^{ia}$, if $\left(\frac{i}{p}\right) = 1$, then applying σ_i to each factor: $\sigma_i(A(X)) = \prod_a(X - \zeta^{ia}) = \prod_a(X - \zeta^a) = A(X)$ because ia runs with a through the set of all quadratic residues. In the same way, $\sigma_i(B(X)) = B(X)$ and $\sigma_i(\tau) = \tau$. On the other hand, if $\left(\frac{i}{p}\right) = -1$, then $\left(\frac{a}{b}\right) = 1$ implies $\left(\frac{ja}{p}\right) = -1$, hence $\sigma_j(A(X)) = B(X)$, $\sigma_j(B(X)) = A(X)$ and also $\sigma_j(\tau) = \left(\frac{j}{p}\right)\tau = -\tau$. Therefore $\sigma_i(F(X)) = F(X)$ and $\sigma_i(G(X)) = G(X)$ for every $i = 1, 2, \ldots, p-1$, showing that $F(X)$, $G(X) \in \mathbb{Q}[X]$.

Thus, the coefficients of $F(X) = A(X) + B(X)$ are in $\mathbb{Q} \cap \mathbb{Z}[\zeta] = \mathbb{Z}$. From $p^*G(X)^2 = F(X)^2 - 4\frac{X^p-1}{X-1} \in \mathbb{Z}[X]$, it follows that $G(X) \in \mathbb{Z}[X]$. Indeed, if $m > 1$ is the least common multiple of the denominators of the coefficients of $G(X)$, then $\frac{p}{m^2} \in \mathbb{Z}$, which is impossible. Thus

$$(-X)^{\frac{p-1}{2}} A\left(\frac{1}{X}\right) = \prod_a (-1 + \zeta^a X) = (\zeta^{\Sigma a}) \prod_a (X - \zeta^{-a})$$

$$= \begin{cases} A(X) & \text{when } p \equiv 1 \pmod 4, \text{ that is } \left(\frac{-1}{p}\right) = 1 \\ B(X) & \text{when } p \equiv -1 \pmod 4, \text{ that is } \left(\frac{-1}{p}\right) = -1, \end{cases}$$

using a previous observation about the sum of quadratic residues. Similarly

$$(-X)^{\frac{p-1}{2}} B\left(\frac{1}{X}\right) = \begin{cases} B(X) & \text{when } p \equiv 1 \pmod 4 \\ A(X) & \text{when } p \equiv -1 \pmod 4. \end{cases}$$

Therefore $(-X)^{\frac{p-1}{2}} F\left(\frac{1}{X}\right) = F(X)$ in all cases. Similarly, $X^{\frac{p-1}{2}} G\left(\frac{1}{X}\right) = G(X)$. Finally, if

$$G(X) = a_0 + a_1 X + \ldots + a_{\frac{p-5}{2}} X^{\frac{p-5}{2}} + X^{\frac{p-3}{2}},$$

then $a_0 = 0$ because $\deg(G) = \frac{p-3}{2}$, and $a_1 = a_{\frac{p-3}{2}} = 1$. ∎

The following identity will be required in §B5. It was first given by Lagrange (1741) in his *Leçons sur le Calcul des Fonctions*.

(P1.11). *Let X, Y be indeterminates, let $n \geq 1$. Then*

$$X^n + Y^n - (X+Y)^n = \sum_{i=1}^\infty (-1)^i \frac{n}{i} \binom{n-i-1}{i-1} (XY)^i (X+Y)^{n-2i}.$$

The Cyclotomic Field

Proof: If $n = 1$ or 2, it is true. Proceeding by induction

$$X^{n+1} + Y^{n+1} = (X^n + Y^n)(X + Y) - XY(X^{n-1} + Y^{n-1})$$
$$= (X+Y)^{n+1} + \sum_{i=1}^{\infty}(-1)^i \frac{n}{i}\binom{n-i-1}{i-1}(XY)^i(X+Y)^{n+1-2i}$$
$$- (XY)(X+Y)^{n-1}$$
$$- \sum_{i=1}^{\infty}(-1)^i \frac{n-1}{i}\binom{n-2-i}{i-1}(XY)^{i+1}(X+Y)^{n-1-2i}$$
$$= (X+Y)^{n+1} + \sum_{i=1}^{\infty}(-1)^i c_i (XY)^i (X+Y)^{n+1-2i},$$

where $c_1 = n + 1$, and if $i \geq 2$, then

$$c_i = \frac{n}{i}\binom{n-1-i}{i-1} + \frac{n-1}{i-1}\binom{n-i-1}{i-2} = \frac{n+1}{i}\binom{n-i}{i-1},$$

as may easily be verified. This concludes the proof. ∎

2. The Cyclotomic Field

Let p be an odd prime, and let

$$\zeta = \zeta_p = \cos\frac{2\pi}{p} + i\sin\frac{2\pi}{p}$$

be a primitive p^{th} root of 1.

$\Phi_p(\zeta) = 0$ and Φ_p is an irreducible polynomial. Let $K = \mathbb{Q}(\zeta)$ be the *cyclotomic field* associated to p. Its elements are of the form $\sum_{i=0}^{p-2} a_i \zeta^i$, where each $a_i \in \mathbb{Q}$.

The elements of the form $\sum_{i=0}^{n-2} a_i \zeta^i$ with each $a_i \in \mathbb{Z}$ are the *cyclotomic integers*. The set $A = \mathbb{Z}[\zeta]$ of cyclotomic integers is a ring, having field of quotients equal to K.

If $\alpha, \beta \in K$, α *divides* β if there exists $\gamma \in A$ such that $\alpha\gamma = \beta$; the notation $\alpha|\beta$ is also used. If $\alpha, \beta \in K\backslash\{0\}$ and $\alpha|\beta$, $\beta|\alpha$, then α, β are said to be *associated*; this is denoted by $\alpha \sim \beta$.

I shall indicate here some facts about cyclotomic fields which will be required later; no attempt is made to mention all important concepts and aspects of the theory.

A *fractional ideal* of K is a non-empty subset I, such that:

1) If $\alpha, \beta \in I$, then $\alpha - \beta \in I$.
2) If $\alpha \in A$, $\beta \in I$, then $\alpha\beta \in I$.
3) There exists $\alpha \in A$, $a \neq 0$, such that $\alpha\beta \in A$ for every $\beta \in I$.

Every fractional ideal contained in A is called an *integral ideal*, or simply an ideal.

If $\alpha \in K$, the set $A\alpha = \{\beta\alpha | \beta \in A\}$ is a fractional ideal, called the *principal fractional ideal* generated by α; it is usually denoted by (α). In particular, (0) is the *zero ideal*, $A = (1)$ is the *unit ideal*.

If $\alpha, \beta \in K$, and if I is a non-zero fractional ideal, the *congruence modulo* I is defined as follows: $\alpha \equiv \beta \pmod{I}$ when $\alpha - \beta \in I$. If $I = (\gamma)$ (with $\gamma \neq K$, $\gamma \neq 0$), the notation $\alpha \equiv \beta \pmod{\gamma}$ will be used.

If I is an integral ideal, $I \neq (0)$, then the *residue ring* A/I is a finite ring.

If I, J are fractional ideals, define

$$IJ = \Big\{ \sum_{i=1}^{n} \alpha_i\beta_i \big| n \geq 0,\ a_i \in I,\ \beta_i \in J \text{ for every } i \Big\}.$$

Then IJ is a fractional ideal. Clearly $I(0) = (0)$, $I(1) = I$ for every I. The product is commutative and associative: $IJ = JI$, $(II')I'' = I(I'I'')$ for any fractional ideals I, I', I'', J.

The fractional ideal I *divides* the fractional ideal J if there exists an integral ideal I' such that $II' = J$. Then (α) divides (β) exactly when α divides β.

The Cyclotomic Field

An integral ideal P is a *prime ideal* when $P \neq (0)$, $P \neq (1)$ and the only integral ideals dividing P are P and (1).

The fundamental theorem of Kummer is the following

(P2.1). *Every non-zero integral ideal is, in a unique way, the product of prime ideals.*

From this theorem, it is possible to define, in the usual way, the *greatest common divisor*, $\gcd(I, J)$, and the *least common multiple*, $\mathrm{lcm}(I, J)$, of non-zero integral ideals, I, J.

The integral ideals I, J are *relatively prime* when $\gcd(I, J) = (1)$, or equivalently, $1 = \alpha + \beta$, where $\alpha \in I$, $\beta \in J$. The ideal I divides J if and only if $J \subseteq I$. In particular, $I | (\alpha)$ exactly when $\alpha \in I$.

If I is any fractional ideal, then $I \cap \mathbb{Z} = \mathbb{Z}r$, for some $r \in \mathbb{Q}$; if $I \subseteq A$, then $r \in \mathbb{Z}$. If P is a prime ideal, then $P \cap \mathbb{Z} = \mathbb{Z}p$, where p is a prime number.

Now I describe the decomposition of the principal ideals $(q) = Aq$ (where q is any prime number) as a product of prime ideals.

1°) $Ap = (\lambda)^{p-1}$, where $\lambda = 1 - \zeta$; note that (λ) is a prime ideal. The residue ring $A/(\lambda)$ is the field $\mathbb{F}_p$, with p elements.

If $\alpha \in A \setminus (\lambda)$, then Fermat's little theorem holds:

$$\alpha^{p-1} \equiv 1 \pmod{\lambda}.$$

Also, if $\alpha \equiv \beta \pmod{\lambda^e}$, then $\alpha^p \equiv \beta^p \pmod{\lambda^{e+1}}$, for $e \geq 1$.

2°) Let q be a prime different from p. Let f be the order of q modulo p, so $f | p - 1$; write $p - 1 = fs$. Then

$$(q) = Q_1 \ldots Q_s$$

where $Q_1, \ldots, Q_s$ are distinct prime ideals and the residue rings A/Q_i are isomorphic to the field $\mathbb{F}_{q^f}$ with q^f elements.

In particular, (q) is a prime ideal if and only if q is a primitive root modulo p.

An element $\epsilon \in A$ is a *unit* if there exists $\epsilon' \in A$ such that $\epsilon \epsilon' = 1$. The set U of units is a multiplicative subgroup of $K\backslash\{0\}$.

If $\alpha, \beta \in K\backslash\{0\}$, $(\alpha) = (\beta)$ if and only if $\alpha = \epsilon\beta$, where ϵ is a unit.

Among the units of K, there are the roots of unity, which are precisely $\pm 1, \pm \zeta, \pm \zeta^2, \ldots, \pm \zeta^{p-1}$. Kummer showed:

(P2.2). *Every unit ϵ of K is of the form $\epsilon = \pm \zeta^j \eta$, where $0 \leq j \leq p-1$, and η is a real unit (i.e., $\eta = \bar{\eta}$, its complex conjugate).*

The elements
$$\frac{1-\zeta^k}{1-\zeta} = 1 + \zeta + \ldots + \zeta^{k-1}$$

(for $k = 1, 2, \ldots, p-1$) are units.

Similarly, the elements
$$\delta_k = \sqrt{\frac{1-\zeta^k}{1-\zeta} \cdot \frac{1-\zeta^{-k}}{1-\zeta^{-1}}}$$

(for $k = 1, 2, \ldots, \frac{p-1}{2}$) are real positive units.

Two non-zero fractional ideals I, J are said to be *equivalent* (denoted by $I \equiv J$) when there exists a non-zero principal fractional ideal (α) such that $I = (\alpha)J$. Denote by $[I]$ the equivalence class of I.

The operation $[I] \cdot [J] = [IJ]$ is well defined and makes the set of equivalence classes of non-zero fractional ideals into an abelian group. It is the *ideal class group* of K, denoted by $C\ell(K)$.

Kummer proved:

(P2.3). *The ideal class group is finite.*

The number of elements of $C\ell(K)$ is called the *class number* of K and denoted by h_p.

It follows at once:

(P2.4). *If $gcd(k, h_p) = 1$, and if I is a fractional ideal such that I^k is a principal fractional ideal, then so is I.*

K contains a maximal real subfield, namely $K^+ = K \cap \mathbb{R}$. Its elements are of the form $\sum_{i=0}^{\frac{p-3}{2}} a_i(\zeta + \zeta^{-1})^i$, with $a_i \in \mathbb{Q}$.

All the above concepts may also be introduced for the field K^+. In particular, the class number of K^+ is finite and denoted by h_p^+.

Kummer proved:

(P2.5). h_p^+ *divides* h_p.

Thus, it is possible to write

$$h_p = h_p^- \cdot h_p^+,$$

where h_p^- is a natural number.

Kummer gave formulas to express h_p^-, h_p^+ in terms of various numbers attached to the field K, but I shall not require these facts.

3. The Pythagorean Equation, Special Cases of Fermat's Last Theorem and Related Equations

In this section, I shall describe the solutions of the Pythagorean equation and show that Fermat's equations

$$X^n + Y^n = Z^n$$

with $n = 3, 4$, have only trivial solutions. I shall also discuss some related equations, which will be needed in the sequel.

For obvious reasons, the equation

$$X^2 + Y^2 = Z^2 \tag{3.1}$$

is called the *Pythagorean equation*.

A solution in integers (x, y, z) is non-trivial when x, y, z are non-zero. It suffices to determine the solutions with x, y, z positive. If (x, y, z) is a solution, and $d = gcd(x, y, z)$, then $\left(\frac{x}{d}, \frac{y}{d}, \frac{z}{d}\right)$ is a solution in relatively prime integers. Thus, it suffices to determine the solutions (x, y, z) with $gcd(x, y, z) = 1$, which are called the *primitive* solutions. In this case, x, y, z are actually pairwise relatively prime.

Note also that x, y cannot both be odd, otherwise $z^2 = x^2 + y^2 \equiv 1 + 1 \equiv 2 \pmod 4$, and this is impossible. Similarly, in a primitive solution, x, y cannot both be even, otherwise z is even, and the solution would not be primitive. Thus, it may be agreed that x is odd, y even, and hence z is odd.

The following result may be traced back to Diophantus but may have been known beforehand, at least in part.

(P3.1). Let a, b be relatively prime integers, not both odd, $a > b \geq 1$, and let

$$\begin{cases} x = a^2 - b^2 \\ y = 2ab \\ z = a^2 + b^2. \end{cases} \quad (3.2)$$

Then (x, y, z) is a primitive solution of the Pythagorean equation, in relatively prime positive integers, with y even. Every such solution may be obtained from a unique couple (a, b) of the type indicated by the relations (3.2).

Proof: Given a, b as said, let x, y, z be defined as in the statement. Then

$$x^2 + y^2 = (a^2 - b^2)^2 + (2ab)^2 = (a^2 + b^2)^2 = z^2.$$

Also y is even, $gcd(x, y, z) = 1$, because if d divides x, y, z, then d divides $2a^2$ and $2b^2$. But since x is odd, so is d; hence d divides a, b, which implies that $d = 1$.

The Pythagorean Equation

Different pairs (a, b) and (a', b') cannot give rise to the same triple, otherwise

$$a^2 - b^2 = x = a'^2 - b'^2$$
$$a^2 + b^2 = z = a'^2 + b'^2;$$

hence it follows immediately that $a = a'$, $b = b'$.

Now let (x, y, z) be a primitive solution of the Pythagorean equation. Then

$$x^2 = z^2 - y^2 = (z - y)(z + y).$$

But $gcd(z-y, z+y) = 1$, because $z-y$ is odd and $gcd(y, z) = 1$.

By the fundamental theorem of unique factorization, there exist positive relatively prime integers t, u such that

$$\begin{cases} z + y = t^2 \\ z - y = u^2; \end{cases}$$

then $t > u$ and t, u are odd. Let a, b be such that $2a = t + u$, $2b = t - u$, so $a > b \geq 1$.

Hence $t = a + b$, $u = a - b$, and

$$\begin{cases} x = tu = (a+b)(a-b) = a^2 - b^2 \\ y = \frac{t^2 - u^2}{2} = \frac{(a+b)^2 - (a-b)^2}{2} = 2ab \\ z = \frac{t^2 + u^2}{2} = \frac{(a+b)^2 + (a-b)^2}{2} = a^2 + b^2. \end{cases}$$

Finally, I note that $gcd(a, b) = 1$, because $gcd(t, u) = 1$, and a, b are not both odd, since x is odd. This concludes the proof. ∎

A *Pythagorean triangle* is a right-angled triangle whose sides are measured in integers a, b, c; if c is the size of the hypotenuse, then $a^2 + b^2 = c^2$ and (a, b, c) is a solution of equation (2.1).

Fermat considered the following problem:

Can the area of a Pythagorean triangle be the square of an integer?

Assume that a, b, c are the lengths of the sides of the triangle, c being the length of the hypotenuse; hence $a^2 + b^2 = c^2$. If the area is the square of an integer h, then $\frac{1}{2}ab = h^2$. Therefore

$$\begin{cases} c^2 + 4h^2 = (a+b)^2 \\ c^2 - 4h^2 = (a-b)^2. \end{cases}$$

Hence,

$$c^4 - 16h^4 = (a^2 - b^2)^2.$$

So $(c, 2h, |a^2 - b^2|)$ would be a solution in positive integers of the equation

$$X^4 - Y^4 = Z^2. \tag{3.3}$$

But Fermat showed with his famous method of descent:

(P3.2). *The equation $X^4 - Y^4 = Z^2$ has no solution in nonzero integers.*

Proof: If the statement is false, let (x, y, z) be a triple of positive integers, solution of (3.3), with smallest possible x. Then $gcd(x, y) = 1$, because if a prime p divides x, y then p^4 divides z^2, so p^2 divides z; letting $x = px'$, $y = py'$, $z = p^2z'$, then $x'^4 - y'^4 = z'^2$, with $0 < x' < x$, which is contrary to the choice of x minimal.

From $z^2 = x^4 - y^4 = (x^2 - y^2)(x^2 + y^2)$ and $gcd(x^2 - y^2, x^2 + y^2) = 1$ or 2, as easily seen, two cases are then possible.

<u>Case 1</u>. $gcd(x^2 - y^2, x^2 + y^2) = 1$.

The Pythagorean Equation

By the unique factorization of integers, there exist positive relatively prime integers s, t, such that
$$\begin{cases} x^2 + y^2 = s^2 \\ x^2 - y^2 = t^2. \end{cases}$$

Since $2x^2 = s^2 + t^2$, then s, t have the same parity, so they are odd. Thus, there exist positive relatively prime integers u, v such that
$$\begin{cases} u = \frac{s+t}{2} \\ v = \frac{s-t}{2}. \end{cases}$$

Then $uv = \frac{s^2-t^2}{4} = \frac{y^2}{2}$, so $y^2 = 2uv$.

Since $\gcd(u, v) = 1$, by the unique factorization of integers, there exist positive integers ℓ, m such that
$$\begin{cases} u = 2\ell^2 \\ v = m^2 \end{cases} \quad \text{or} \quad \begin{cases} u = \ell^2 \\ v = 2m^2. \end{cases}$$

But
$$u^2 + v^2 = \frac{(s+t)^2 + (s-t)^2}{4} = \frac{s^2 + t^2}{2} = x^2.$$

By (P3.1), there exist positive relatively prime integers a, b such that
$$\begin{cases} 2\ell^2 = u = 2ab \\ m^2 = v = a^2 - b^2 \\ x = a^2 + b^2 \end{cases} \quad \text{or} \quad \begin{cases} \ell^2 = u = a^2 - b^2 \\ 2m^2 = v = 2ab \\ x = a^2 + b^2. \end{cases}$$

Hence $\ell^2 = ab$, respectively $m^2 = ab$. By the unique factorization, there exist positive relatively prime integers c, d, such that
$$\begin{cases} a = c^2 \\ b = d^2, \end{cases}$$

and so $m^2 = c^4 - d^4$, respectively $\ell^2 = c^4 - d^4$.

Note that $0 < c \le a < x$, so by the minimal choice of x, (c,d,m), respectively (c,d,ℓ) could not be a solution of equation (3.3), and this is a contradiction.

Case 2. $gcd(x^2 - y^2, x^2 + y^2) = 2$.

Now x, y are odd and z is even. Since $y^4 + z^2 = x^4$, by (P3.1) there exist positive relatively prime integers a, b such that
$$\begin{cases} y^2 = a^2 - b^2 \\ z = 2ab \\ x^2 = a^2 + b^2. \end{cases}$$

Then $x^2 y^2 = a^4 - b^4$ with $0 < a < x$, and this is contrary to the choice of x as minimal possible. ∎

In particular, the equation
$$X^4 + Y^4 = Z^4 \tag{3.4}$$
has no solution in non-zero integers (x, y, z), otherwise (z, y, x^2) would be a solution of (3.3).

The equation
$$X^4 + Y^4 = Z^2 \tag{3.5}$$
may be treated in a similar way:

(P3.3). *The equation $X^4 + Y^4 = Z^2$ has no solution in non-zero integers.*

Proof: Assume that (x, y, z) is a solution in positive integers for which z is minimal possible. As in the proof of (P3.2), $gcd(x, y) = 1$. Since $(x^2)^2 + (y^2)^2 = z^2$, for example y is even, x is odd and by (P3.1) there exist positive relatively prime

The Pythagorean Equation

integers a, b such that

$$\begin{cases} x^2 = a^2 - b^2 \\ y^2 = 2ab \\ z = a^2 + b^2. \end{cases}$$

Since x is odd, then $x^2 \equiv 1 \pmod{4}$, so a is odd and b is even. From $x^2 + b^2 = a^2$, by (P3.1) there exist positive relatively prime integers c, d, such that

$$\begin{cases} x = c^2 - d^2 \\ b = 2cd \\ a = c^2 + d^2. \end{cases}$$

Then $y^2 = 4cd(c^2 + d^2)$. But $c, d, c^2 + d^2$ are pairwise relatively prime positive integers and their product is a square. By the unique factorization, there exist positive relatively prime integers ℓ, m, p such that

$$\begin{cases} c = \ell^2 \\ d = m^2 \\ c^2 + d^2 = p^2. \end{cases}$$

Hence
$$\ell^4 + m^4 = p^2,$$
where
$$z = a^2 + b^2 > (c^2 + d^2)^2 = p^4 \geq p.$$

By the minimality of z, (ℓ, m, p) could not be a solution of (3.5), and this is a contradiction. ∎

(P3.4). *The only solutions in non-zero integers of*

$$2X^4 + 2Y^4 = Z^2 \qquad (3.6)$$

and of

$$X^4 + Y^4 = 2Z^2 \qquad (3.7)$$

with $gcd(x,y) = 1$ are $x^2 = y^2 = 1$ and $z^2 = 4$ or 1, respectively.

Proof: If $2x^4 + 2y^4 = z^2$, with non-zero integers x, y, such that $gcd(x, y) = 1$, then

$$4(x^4 - y^4)^2 = 4(x^8 - 2x^4y^4 + y^8)$$
$$= 4[(x^4 + y^4)^2 - 4x^4y^4] = z^4 - 16x^4y^4.$$

From (P3.2), $x^4 - y^4 = 0$, thus $x^2 = y^2 = 1$ and $z^2 = 4$.

Similarly, if $x^4 + y^4 = 2z^2$, with $gcd(x, y) = 1$, then $2x^4 + 2y^4 = (2z)^2$, hence by the above, $x^2 = y^2 = z^2 = 1$. ∎

In §A13, we will require an explicit description of the solutions of

$$X^2 + 2Y^2 = Z^2. \qquad (3.8)$$

A triple of positive integers (x, y, z) is a solution of (3.8) if $x^2 + 2y^2 = z^2$. It is a primitive solution if $gcd(x, y) = 1$. If (x, y, z) is a primitive solution, then x, z are odd and y is even. Indeed, if x is even, then so is z, hence $4|2y^2$, so y would be even, contrary to the assumption. Thus x is odd and so is z. Finally

$$2y^2 = z^2 - x^2 \equiv 0 \pmod 4,$$

hence y is even.

(P3.5). *There is a bijection between the set $\{(a, b) | a, b \text{ are pairwise relatively prime positive integers and } b \text{ is odd}\}$ and the set*

The Pythagorean Equation

of all primitive solutions (x, y, z) of $X^2 + 2Y^2 = Z^2$, which is given by $(a, b) \mapsto (x, y, z)$ with

$$\begin{cases} x = \begin{cases} 2a^2 - b^2 & \text{if } 2a^2 > b^2 \\ b^2 - 2a^2 & \text{if } 2a^2 < b^2 \end{cases} \\ y = 2ab \\ z = 2a^2 + b^2. \end{cases}$$

Proof: Let (a, b) be given, where $a, b \geq 1$, b is odd, $gcd(a, b) = 1$, and define x, y, z as indicated. Then

$$x^2 + 2y^2 = (2a^2 - b^2)^2 + 8a^2b^2 = (2a^2 + b^2)^2 = z^2.$$

Moreover, $gcd(x, y) = 1$, because if p is a prime and $p|x$, $p|y$, then necessarily $p \neq 2$ (since x is odd), hence $p|z$; so $p|4a^2$, $p|2b^2$, thus $p|a$ and $p|b$, which is impossible.

Now I show that the mapping is surjective. Let (x, y, z) be a primitive solution of (3.8). Then $z - x$, $z + x$ are even and $\frac{z+x}{2}, \frac{z-x}{2}$ are not both even (because x, z are odd); let $y = 2y_1$, then $2y_1^2 = \frac{z+x}{2} \cdot \frac{z-x}{2}$, so $\frac{z+x}{2}, \frac{z-x}{2}$ are both odd. Moreover, $gcd\left(\frac{z+x}{2}, \frac{z-x}{2}\right) = 1$, because if p is a prime dividing $\frac{z+x}{2}$ and $\frac{z-x}{2}$ then $p \neq 2$ and $p|z$, $p|x$; therefore $p|2y^2$, so $p|y$, which is contrary to the assumption that $gcd(x, y) = 1$. It follows that either

(a) $\begin{cases} \frac{z+x}{2} = 2a^2 \\ \frac{z-x}{2} = b^2 \end{cases}$ or (b) $\begin{cases} \frac{z-x}{2} = 2a^2 \\ \frac{z+x}{2} = b^2 \end{cases}$

where a, b are relatively prime positive integers and b is odd. In case (a), $2a^2 > b^2$ and in case (b), $2a^2 < b^2$. Then

$$\begin{cases} x = \begin{cases} 2a^2 - b^2 & \text{if } 2a^2 > b^2 \\ b^2 - 2a & \text{if } 2a^2 < b^2 \end{cases} \\ z = 2a^2 + b^2. \end{cases}$$

This shows that the mapping is surjective.

Finally, if (a,b) and (a',b') give rise to the same primitive solution (x,y,z), then $2a^2 + b^2 = 2a'^2 + b'^2$. If $2a^2 - b^2 = b'^2 - 2a'^2$, then $2b^2 = 4a'^2$, which is impossible. Thus $2a^2 - b^2 = 2a'^2 - b'^2$ and from this it follows at once that $(a,b) = (a',b')$. ∎

The next result will be needed in §A8.

Fermat stated the following fact (see *Oeuvres*, II, p. 441). A proof was given by Genocchi in 1883.

(P3.6). *The only solutions in non-zero integers of the system*

$$\begin{cases} 1 + X = 2Y^2 \\ 1 + X^2 = 2Z^2 \end{cases}$$

are $(X, Y, Z) = (1, \pm 1, \pm 1)$ *and* $(7, \pm 2, \pm 5)$.

Proof: If x, y, z are non-zero integers, solutions of the given system, then $x = 2y^2 - 1$, hence

$$1 + (2y^2 - 1)^2 = 2z^2,$$

so

$$y^4 + (y^2 - 1)^2 = z^2.$$

If $y = \pm 1$, then $x = 1$, $z = \pm 1$.

Now assume $y^2 \neq 1$. If y is odd, by (P3.1) there exist positive integers m, n, $\gcd(m, n) = 1$, such that

$$\begin{cases} y^2 = m^2 - n^2 \\ y^2 - 1 = 2mn \\ z = m^2 + n^2. \end{cases}$$

Since $y^2 \equiv 1 \pmod{4}$, then m is odd and n is even.

The Pythagorean Equation

Let f, g be defined by

$$\begin{cases} f = m+n \\ g = m-n, \end{cases}$$

so f, g are odd, $gcd(f, g) = 1$ and $m = \frac{f+g}{2}$, $n = \frac{f-g}{2}$. Then

$$\begin{cases} y^2 = m^2 - n^2 = fg \\ y^2 - 1 = 2mn = \frac{f^2 - g^2}{2}. \end{cases}$$

But $gcd(f, g) = 1$, so there exist integers $d, e \neq 0$, such that $f = d^2, g = e^2$.

From $\frac{f^2 - g^2}{2} = fg - 1$, it follows that

$$(f+g)^2 = f^2 + 2fg + g^2 = 2f^2 + 2 = 2d^4 + 2.$$

It follows from (P3.4) that $f = d^2 = 1$, hence $(1+g)^2 = 4$ and $g = 1$, hence $m = 1$, $n = 0$ and $y^2 = 1$, a case which was discarded.

Now let y be even, $y \neq 0$. Again, by (P3.1) there exist integers m, n, $gcd(m, n) = 1$, such that

$$\begin{cases} y^2 - 1 = m^2 - n^2 \\ y^2 = 2mn \\ z = m^2 + n^2. \end{cases}$$

But $y^2 - 1 \equiv -1 \pmod{4}$, so m is even and n is odd. It follows that there exist integers f, g such that $gcd(f, g) = 1$ and

$$\begin{cases} m = 2f^2 \\ n = g^2, \end{cases}$$

so g is odd. From this,
$$4f^2g^2 - 1 = 4f^4 - g^4,$$
hence
$$(2f^2 + g^2)^2 = 4f^4 + 4f^2g^2 + g^4 = 1 + 8f^4,$$
and
$$(2f^2 + g^2 - 1)(2f^2 + g^2 + 1) = 8f^4.$$

But $gcd(2f^2 + g^2 - 1,\ 2f^2 + g^2 + 1) = 2$, so there exist integers r, s, with $gcd(r, s) = 1$, such that
$$\begin{cases} 2f^2 + g^2 - 1 = 2r^4 \\ 2f^2 + g^2 + 1 = 4s^4, \end{cases}$$
and r is odd, or
$$\begin{cases} 2f^2 + g^2 - 1 = 4r^4 \\ 2f^2 + g^2 + 1 = 2s^4, \end{cases}$$
and s is odd.

This gives, by subtraction,
$$2 = 4s^4 - 2r^4,$$
so
$$1 = 2s^4 - r^4$$
or respectively
$$1 = s^4 - 2r^4.$$

In the first case, from $r^4 + 1 = 2s^4$, by (P3.4), $r^2 = s^2 = 1$, therefore $2f^2 + g^2 = 3$, hence $f^2 = g^2 = 1$ and $m = 2$, $n = 1$, $y^2 = 4$, and finally $x = 7$.

The Pythagorean Equation

In the second case, from
$$2r^4 = s^4 - 1 = (s^2 - 1)(s^2 + 1),$$
it follows that there exist integers t, u, with $gcd(t, u) = 1$, such that
$$\begin{cases} s^2 - 1 = t^4 \\ s^2 + 1 = 2u^4 \end{cases} \quad \text{or} \quad \begin{cases} s^2 - 1 = 2t^4 \\ s^2 + 1 = u^4. \end{cases}$$
However, the relations $s^2 - 1 = t^4$, respectively $s^2 + 1 = u^4$, are impossible. So this case cannot happen. ∎

In 1972, Inkeri proved the following extension:
If p is a prime, $e \geq 0$, y is odd, $x \geq 1$ and
$$\begin{cases} x + 1 = 2^e p y^2 \\ x^2 + 1 = 2z^2, \end{cases}$$
then $x = 1$ or 7, $p = 2$, and $e = 0$.

Now I shall study the cubic Fermat equation:
$$X^3 + Y^3 = Z^3. \tag{3.9}$$

In his book on algebra (1770), Euler gave a proof (with some details missing) that this equation has only the trivial solution. But I shall present here the proof of Gauss, found in his private papers after his death. The proof involves calculations in the field $K = \mathbb{Q}(\omega) = \mathbb{Q}(\sqrt{-3})$, where
$$\omega = \frac{-1 + \sqrt{-3}}{2}, \quad \omega^2 = \frac{-1 - \sqrt{-3}}{2}$$
are the primitive cubic roots of 1. Note that $1 + \omega + \omega^2 = 0$.

For the convenience of the reader, I shall state some specific facts about the cyclotomic field $K = \mathbb{Q}(\omega)$, which supplement the general results indicated in §P2.

The ring of integers of $K = \mathbb{Q}(\omega)$ is $A = \mathbb{Z}[\omega]$; that is, the integers are of the form

$$\frac{a + b\sqrt{-3}}{2}$$

where a, b are rational integers and $a \equiv b \pmod{2}$. The ring A is a unique factorization domain, so every ideal is principal. It has the following units: $\pm 1, \pm \omega, \pm \omega^2$, and no other. So, the units are the powers $(-\omega)^s$, for $0 \leq s \leq 5$.

Any pair of non-zero integers has a greatest common divisor, which is defined up to a unit of A. Moreover $\alpha, \beta \in A$ are associated elements if $\alpha = \beta \epsilon$, where ϵ is a unit; this is written as $\alpha \sim \beta$.

The decomposition of ordinary prime numbers p as a product of prime elements of the ring A is as follows:

1) $p = 3$ is ramified, that is, $3 = (-\omega^2)\lambda^2$, so $3 \sim \lambda^2$, where $\lambda = 1 - \omega = \frac{3-\sqrt{-3}}{2}$; λ is a prime element of A.

There are three residue classes of A modulo λ, namely the classes of $0, 1, -1$. The conjugate of λ is $\bar{\lambda} = 1 - \omega^2$ and the norm of λ is

$$N(\lambda) = \lambda \bar{\lambda} = (1 - \omega)(1 - \omega^2) = 1 - \omega - \omega^2 + 1 = 3.$$

2) If $p \equiv 1 \pmod{3}$ then p splits, that is, $p \sim \pi_1 \pi_2$ where π_1, π_2 are prime elements of A, which are not associated to each other. There are p^2 residue classes of A modulo p; more precisely, A/Ap is the product of two copies of the field $\mathbb{F}_p$, and $N(\pi_1) = N(\pi_2) = p$.

3) If $p \equiv -1 \pmod{3}$ then p is inert; that is, p is a prime element of A. Now, A/Ap is the field with p^2 elements and $N(p) = p^2$.

The facts (2), (3) above will not be needed in the next proof.

The Pythagorean Equation

If $\alpha, \beta, \gamma \in K$, $\gamma \neq 0$, I recall that the notation $\alpha \equiv \beta \pmod{\gamma}$ means that γ divides $\alpha - \beta$.

(P3.7). Lemma. *If $\alpha \in A$ and $\lambda = 1 - \omega$ does not divide α, then $\alpha^3 \equiv \pm 1 \pmod{\lambda^4}$.*

Proof: Since $\alpha \not\equiv 0 \pmod{\lambda}$ then $\alpha \equiv \pm 1 \pmod{\lambda}$.

First, I assume $\alpha \equiv 1 \pmod{\lambda}$, so $\alpha - 1 = \beta\lambda$ where $\beta \in A$. Then

$$\alpha - \omega = (\alpha - 1) + (1 - \omega) = \beta\lambda + \lambda = \lambda(\beta + 1)$$
$$\alpha - \omega^2 = (\alpha - \omega) + (\omega - \omega^2) = \lambda(\beta + 1) + \omega\lambda = \lambda(\beta - \omega^2).$$

Hence

$$\alpha^3 - 1 = (\alpha - 1)(\alpha - \omega)(\alpha - \omega^2) = \lambda^3 \beta(\beta + 1)(\beta - \omega^2).$$

But $1 - \omega^2 = (1+\omega)\lambda$, so $\omega^2 \equiv 1 \pmod{\lambda}$. Hence $\beta, \beta+1, \beta-\omega^2$ are in three different classes modulo λ, and at least one is a multiple of λ. Therefore $\alpha^3 \equiv 1 \pmod{\lambda^4}$.

If $\alpha \equiv -1 \pmod{\lambda}$, then $-\alpha^3 = (-\alpha)^3 \equiv 1 \pmod{\lambda^4}$, so $\alpha^3 \equiv -1 \pmod{\lambda^4}$. ∎

(P3.8). *The equation $X^3 + Y^3 = Z^3$ has no solution in algebraic integers of $K = \mathbb{Q}(\omega)$, all different from 0.*

Proof: It is equivalent to prove the same statement for the equation
$$X^3 + Y^3 + Z^3 = 0. \tag{3.10}$$

Assume that $\xi, \eta, \zeta \in A$ are non-zero and satisfy $\xi^3 + \eta^3 + \zeta^3 = 0$. If $\gcd(\xi, \eta, \zeta) = \delta$ then $\frac{\xi}{\delta}, \frac{\eta}{\delta}, \frac{\zeta}{\delta}$ satisfy the same equation and $\gcd\left(\frac{\xi}{\delta}, \frac{\eta}{\delta}, \frac{\zeta}{\delta}\right) = 1$. So it may be assumed without loss of generality that $\gcd(\xi, \eta, \zeta) = 1$ and therefore ξ, η, ζ are pairwise relatively prime. So λ cannot divide two of these elements ξ, η, ζ. Say, for example, that $\lambda \nmid \xi$, $\lambda \nmid \eta$.

Case 1. Assume that $\lambda \nmid \zeta$. As recalled in §P2,

$$\begin{cases} \xi^3 \equiv \pm 1 \pmod{\lambda^3} \\ \eta^3 \equiv \pm 1 \pmod{\lambda^3} \\ \zeta^3 \equiv \pm 1 \pmod{\lambda^3}, \end{cases}$$

(this is weaker than what was proved in Lemma (P3.7)). So $0 = \xi^3 + \eta^3 + \zeta^3 \equiv \pm 1 \pm 1 \pm 1 \pmod{\lambda^3}$. The eight combinations of signs give ± 1 or ± 3. These are not congruent to 0 modulo λ^3, since ± 1 are units, and ± 3 are associated with λ^2, hence not multiples of λ^3.

Case 2. Assume that $\lambda | \zeta$.

Let $\xi = \lambda^m \psi$, $\psi \in A$, where $m \geq 1$ and λ does not divide ψ.

The essential part of the proof consists in establishing the following assertion:

Let $n \geq 1$, and let ϵ be a unit of A. If there exist $\alpha, \beta, \gamma \in A$, pairwise relatively prime, not multiplies of λ, with $\alpha^3 + \beta^3 + \epsilon \lambda^{3n} \gamma^3 = 0$, then:

a) $n \geq 2$, and

b) there exists a unit ϵ_1 and $\alpha_1, \beta_1, \gamma_1 \in A$, pairwise relatively prime, not multiples of λ, such that $\alpha_1^3 + \beta_1^3 + \epsilon_1 \lambda^{3(n-1)} \gamma_1^3 = 0$.

The hypothesis is satisfied with $n = m$, $\epsilon = 1$, $\alpha = \xi, \beta = \eta$, and $\gamma = \psi$. By repeated application of the above assertion, it follows that there exists a unit ϵ', and $\alpha', \beta', \gamma' \in A$, not multiples of λ, such that $\alpha'^3 + \beta'^3 + \epsilon' \lambda^3 \gamma'^3 = 0$, and this contradicts (a) above.

First, I show that $n \geq 2$.

Indeed, $\lambda \nmid \alpha$ and $\lambda \nmid \beta$. So by Lemma (P3.7), $\alpha^3 \equiv \pm 1 \pmod{\lambda^4}$, $\beta^3 \equiv \pm 1 \pmod{\lambda^4}$ and $\pm 1 \pm 1 \equiv -\epsilon \lambda^{3n} \gamma^3 \pmod{}$

The Pythagorean Equation

λ^4), with $\lambda \nmid \gamma$. Since $\lambda | 3$ then $\lambda \nmid \pm 2$, so the left-hand side must be 0. From $\lambda \nmid \gamma$, it follows that $3n \geq 4$ so $n \geq 2$.

Now, I prove (b).

Note that

$$-\epsilon \lambda^{3n} \gamma^3 = \alpha^3 + \beta^3 = (\alpha + \beta)(\alpha + \omega\beta)(\alpha + \omega^2 \beta). \quad (3.11)$$

Since λ is a prime element dividing the right-hand side, it must divide one of the factors. But

$$\alpha + \beta \equiv \alpha + \omega\beta \equiv \alpha + \omega^2 \beta \pmod{\lambda}$$

because $\lambda = 1 - \omega$, $1 - \omega^2 = -\omega^2 \lambda$, so λ must divide all the three factors; hence

$$\frac{\alpha + \beta}{\lambda}, \frac{\alpha + \omega\beta}{\lambda}, \frac{\alpha + \omega^2 \beta}{\lambda} \in A$$

and

$$-\epsilon \lambda^{3(n-1)} \gamma^3 = \frac{\alpha + \beta}{\lambda} \cdot \frac{\alpha + \omega\beta}{\lambda} \cdot \frac{\alpha + \omega^2 \beta}{\lambda}.$$

Since $n \geq 2$, λ divides the right-hand side, hence at least one factor. It cannot divide two of the factors, otherwise two among $\alpha + \beta$, $\alpha + \omega\beta$, $\alpha + \omega^2 \beta$ are congruent modulo λ^2. I show now that this is not possible:

If

$$(\alpha + \beta) - (\alpha + \omega\beta) = \beta(1 - \omega) = \beta\lambda \equiv 0 \pmod{\lambda^2}$$

then $\lambda | \beta$, a contradiction.

If

$$(\alpha + \beta) - (\alpha + \omega^2 \beta) = \beta(1 - \omega^2) = -\beta\omega^2 \lambda \equiv 0 \pmod{\lambda^2}$$

then again $\lambda | \beta$, a contradiction. Finally, if

$$(\alpha + \omega\beta) - (\alpha + \omega^2 \beta) = \omega\beta(1 - \omega) = \omega\beta\lambda \equiv 0 \pmod{\lambda^2}$$

then again $\lambda|\beta$.

Assume for example that λ divides $\frac{\alpha+\beta}{\lambda}$ (the other cases are treated replacing β by $\omega\beta$ or $\omega^2\beta$). Then $\lambda^{3(n-1)}$ divides $\frac{\alpha+\beta}{\lambda}$. Therefore

$$\begin{cases} \alpha + \beta = \lambda^{3n-2}\kappa_1 \\ \alpha + \omega\beta = \lambda\kappa_2 \\ \alpha + \omega^2\beta = \lambda\kappa_3, \end{cases} \quad (3.12)$$

with $\kappa_1, \kappa_2, \kappa_3 \in A$, λ not dividing $\kappa_1, \kappa_2, \kappa_3$. Multiplying, it follows that

$$-\epsilon\gamma^3 = \kappa_1\kappa_2\kappa_3. \quad (3.13)$$

Note that $\kappa_1, \kappa_2, \kappa_3$ are pairwise relatively prime. For example, if $\delta \in A$ divides κ_1, κ_2, then δ divides $(\alpha + \beta) - (\alpha + \omega\beta) = \beta(1-\omega) = \beta\lambda$, and similarly when δ divides κ_1, κ_3 (or κ_2, κ_3), then δ divides $\beta\lambda$. But λ does not divide $\kappa_1, \kappa_2, \kappa_3$, so δ is not associated with λ; hence δ divides β, and therefore also α, which is a contradiction.

By the unique factorization in the ring A, it follows from (3.13) that the elements $\kappa_1, \kappa_2, \kappa_3$ are associated with cubes; i.e., there exist units $\tau_i \in A$ and elements $\mu_i \in A$ such that $\kappa_i = \tau_i\mu_i^3$ ($i = 1, 2, 3$). So

$$\begin{cases} \alpha + \beta = \lambda^{3n-2}\mu_1^3\tau_1 \\ \alpha + \omega\beta = \lambda\mu_2^3\tau_2 \\ \alpha + \omega^2\beta = \lambda\mu_3^3\tau_2. \end{cases} \quad (3.14)$$

Note again that μ_1, μ_2, μ_3 are pairwise relatively prime and λ does not divide μ_1, μ_2, μ_3. Thus

$$\begin{aligned} 0 &= (\alpha + \beta) + \omega(\alpha + \omega\beta) + \omega^2(\alpha + \omega^2\beta) \\ &= \lambda^{3n-2}\mu_1^3\tau_1 + \omega\lambda\mu_2^3\tau_2 + \omega^2\lambda\mu_3^3\tau_3 \end{aligned}$$

so, dividing by $\omega\lambda\tau_2$, this may be written as

$$\mu_2^3 + \tau\mu_3^3 + \tau'\lambda^{3(n-1)}\mu_1^3 = 0$$

where τ, τ' are units, $\mu_1, \mu_2, \mu_3 \in A$ are not zero, and $gcd(\mu_1, \mu_2, \mu_3) = 1$. If $\tau = 1$, the proof of (b) is complete. If $\tau = -1$, just replace μ_3 by $-\mu_3$ and this again shows (b).

To finish the proof, it will be shown that the unit τ cannot be equal to $\pm\omega$ or to $\pm\omega^2$. In fact, $\mu_2^3 + \tau\mu_3^3 \equiv 0 \pmod{\lambda^2}$. Since $\mu_2^3 \equiv 1 \pmod{\lambda^4}$, $\mu_3^3 \equiv \pm 1 \pmod{\lambda^4}$ then $\mu_2^3 + \tau\mu_3^3 \equiv \pm 1 \pm \tau \equiv 0 \pmod{\lambda^2}$.

However, $\pm 1 \pm \omega \not\equiv 0 \pmod{\lambda^2}$ and $\pm 1 \pm \omega^2 \not\equiv 0 \pmod{\lambda^2}$, so $\tau \neq \pm\omega, \pm\omega^2$, and thus the proof of (b) is now complete. As already explained, this suffices to prove the theorem.

∎

In particular, the equation $X^3 + Y^3 = Z^3$ has no solution in non-zero integers.

Now I shall indicate Legendre's result (1808, 1830) about the equations

$$X^3 + Y^3 = 2Z^3 \tag{3.15}$$

and

$$X^3 + Y^3 = 4Z^3. \tag{3.16}$$

However, I shall give a proof different from Legendre's and similar to the one for (P3.8) (see Mordell's book, p. 126), which is appropriate even to treat a wider class of equations. First, a simple lemma:

(P3.9). Lemma. *Let p be a prime such that $p \equiv 2$ or 5 (mod 9). Then there does not exist $\alpha \in A$ such that $\omega \equiv \alpha^3 \pmod{p}$.*

Proof: Since $p \equiv -1 \pmod{3}$, then A/Ap is the field with p^2 elements. If $\omega \equiv \alpha^3 \pmod{p}$, since $p \nmid \omega$ (which is a unit), then $p \nmid \alpha$. Hence $\alpha^{p^2-1} \equiv 1 \pmod{p}$. Writing $p = 3r - 1$ with $r \not\equiv 0 \pmod{3}$, then $p^2 - 1 = 3r(3r - 2)$, so $\omega^{3r(3r-2)} \equiv$

1 (mod p). But $\omega^3 = 1$, thus multiplying with ω^{2r}, it follows that $1 \equiv \omega^{2r}$ (mod p). Since $3 \nmid r$, then $\omega^{2r} = \omega$ or ω^2, hence p divides $1 - \omega = \lambda$ or p divides $1 - \omega^2 = \bar{\lambda}$, and hence also p divides λ; therefore $p = 3$, which is not true by hypothesis. ∎

(P3.10). *Let p be a prime such that $p \equiv 2$ or 5 (mod 9) and let $a = p$ or p^2. If ξ, η, ζ are non-zero elements of A and ϵ is a unit of A, such that*

$$\xi^3 + \eta^3 + a\epsilon\zeta^3 = 0, \qquad (3.17)$$

then $a = 2$, $\xi^3 = \eta^3 = \pm 1$, $\epsilon\zeta^3 = \pm 1$, $\xi\zeta^3 = \pm 1$.

Proof: Assume that there exist $\xi, \eta, \zeta, \epsilon$ as indicated. Among all possible $(\xi, \eta, \zeta, \epsilon)$ choose one for which $|N(\xi\eta\zeta\epsilon)| = |N(\xi\eta\zeta)|$ is minimal. Then $\gcd(\xi, \eta, \zeta) = 1$, because if a prime element $\pi \in A$ divides ξ, η, ζ, then $\left(\frac{\xi}{\pi}, \frac{\eta}{\pi}, \frac{\zeta}{\pi}, \epsilon\right)$ is still a solution of (3.17) and $\left|N\left(\frac{\xi}{\pi}\frac{\eta}{\pi}\frac{\zeta}{\pi}\right)\right| < |N(\xi\eta\zeta)|$, contrary to the hypothesis.

Then $\gcd(\xi, \zeta) = 1$ and $\gcd(\eta, \zeta) = 1$. Moreover, $\gcd(\xi, \eta) = 1$, because if a prime element π divides ξ, η, then π^3 divides $a\zeta^3$, but $a = p$ or p^2 where $p \neq 3$, so p is not ramified in $\mathbb{Q}(\omega)$. Hence $\pi^3 \nmid a$, so $\pi | \zeta$, which is not possible.

Let
$$\begin{cases} \alpha = \xi + \eta \\ \beta = \omega\xi + \omega^2\eta \\ \gamma = \omega^2\xi + \omega\eta. \end{cases}$$

Then $\alpha + \beta + \gamma = 0$ and from $1 + \omega + \omega^2 = 0$,

$$\alpha\beta\gamma = \xi^3 + \eta^3 = -\epsilon a\zeta^3.$$

Let $\delta = \gcd(\alpha, \beta, \gamma)$. I show that δ is a unit or $\delta = \lambda = 1 - \omega$. Indeed, if π is a prime element of A, $e \geq 1$, and π^e divides α, β, then π^e divides $\omega\alpha - \beta = \omega(1 - \omega)\eta$ and also

The Pythagorean Equation

$\omega^2\alpha - \beta = -\omega(1-\omega)\xi$; if $\pi \neq \lambda$ then π divides ξ, η, which is not true. So $\pi = \lambda$. Also if $e > 1$ then λ divides ξ, η, which is not true, thus $e = 1$.

Let $\alpha' = \frac{\alpha}{\delta}$, $\beta' = \frac{\beta}{\delta}$, $\gamma' = \frac{\gamma}{\delta}$, so $gcd(\alpha', \beta', \gamma') = 1$. Since $\alpha' + \beta' + \gamma' = 0$, then also $gcd(\alpha', \beta') = gcd(\alpha', \gamma') = gcd(\beta', \gamma') = 1$.

But δ^3 divides $\alpha\beta\gamma = -\epsilon a\zeta^3$. It follows that $\delta | \zeta$; indeed, this is clear if $\delta = 1$ and if $\delta = \lambda$, since $\lambda^3 \nmid a$ then $\lambda | \zeta$. Writing $\zeta' = \frac{\zeta}{\delta}$ then $\alpha'\beta'\gamma' = -\epsilon a \zeta'^3$. Since $p \equiv -1 \pmod{3}$, then p is a prime element of A; α', β', γ' are pairwise relatively prime, so p divides one and only one of these elements, say p divides γ'. It follows that

$$\begin{cases} \alpha' = \kappa_1 \xi_1^3 \\ \beta' = \kappa_2 \eta_1^3 \\ \gamma' = \kappa_3 a \zeta_1^3, \end{cases}$$

where $\kappa_1, \kappa_2, \kappa_3$ are units of A.

Then

$$\kappa_1 \xi_1^3 + \kappa_2 \eta_1^3 + \kappa_3 a \zeta_1^3 = \alpha' + \beta' + \gamma' = 0$$

and dividing by κ_1, the following relation holds:

$$\xi_1^3 + \mu \eta_1^3 + \nu a \zeta_1^3 = 0,$$

where μ, ν are units.

If $\mu = \pm 1$ then $\xi_1^3 + (\pm\eta_1)^3 + \nu a \zeta_1^3 = 0$. From the choice of $(\xi, \eta, \zeta, \epsilon)$ such that $|N(\xi\eta\zeta\epsilon)|$ is minimal, it follows that

$$\left|N(\xi\eta\zeta)\right|^3 \leq \left|N(\xi_1\eta_1\zeta_1)\right|^3 = \left|N\left(\frac{\alpha\beta\gamma}{\delta^3 a}\right)\right|$$
$$= \left|N\left(\frac{\zeta^3}{\delta^3}\right)\right| = \left|N\left(\frac{\zeta}{\delta}\right)\right|^3,$$

hence

$$\left|N(\delta\xi\eta)\right| \leq 1,$$

and necessarily
$$\left|N(\delta\xi\eta)\right| = 1.$$

So δ, ξ, η are units of the ring A, which implies $\xi^3 = \pm 1$, $\eta^3 = \pm 1$, hence $\xi^3 = \eta^3 = \pm 1$, and therefore $-\epsilon a \zeta^3 = \pm 2$. But 2 is a prime element of A, and a is not a unit. Thus ζ is a unit, so $\zeta^3 = \pm 1$, and this implies that $a = 2$, $\epsilon \zeta^3 = \pm 1$.

The proof will be concluded once it is shown that μ cannot be different from ± 1. Indeed, since p divides a, then $-\mu\eta_1^3 \equiv \xi_1^3 \pmod{p}$. From $\gcd(\xi_1, \eta_1) = 1$ it follows that p does not divide η_1 and therefore in the field A/Ap, $-\eta_1$ has an inverse θ; so $\mu \equiv (\theta\xi_1)^3 \pmod{p}$. But μ is a unit, $\mu \neq \pm 1$, so changing sign and/or raising to the square, $\omega \equiv \tau^3 \pmod{p}$, where $\tau \in A$. This is impossible, according to Lemma (P3.9). ∎

In particular, one obtains Legendre's results:
The only solution in non-zero integers of (3.15) is $x = y = z = \pm 1$, and Equation (3.16) has no solution in non-zero integers.

The proofs by Legendre and the one presented here are by descent or some equivalent method.

In 1977, McCallum studied the special case
$$X^3 \pm 1 = 2Z^3, \tag{3.18}$$

and gave a proof that the non-trivial solutions are $x = z = \pm 1$. His proof did not require the method of descent and was based on simple properties of the cubic field $\mathbb{Q}(\sqrt[3]{2})$.

In this field (see LeVeque, vol. II, p. 108–109), every unit $a - \sqrt[3]{2}b$ is equal, up to sign, to a power of the fundamental unit $-1 + \sqrt[3]{2}$, so $a - \sqrt[3]{2}b = \pm(-1 + \sqrt[3]{2})^n$, with n an integer not necessarily positive.

McCallum's proof:

Assume that a, b are non-zero integers such that $a^3 - 2b^3 = \pm 1$. In the field $K = \mathbb{Q}(\sqrt[3]{2})$ the following decomposition takes place:

$$\pm 1 = a^3 - 2b^3 = (a - \sqrt[3]{2}b)(a^2 + \sqrt[3]{2}ab + \sqrt[3]{4}b^2),$$

so $a - \sqrt[3]{2}b$ is a unit of K. By what was recalled, there exists an integer n such that

$$a - \sqrt[3]{2}b = \pm(-1 + \sqrt[3]{2})^n.$$

I show that $n > 0$. Indeed,

$$a^2 + \sqrt[3]{2}ab + \sqrt[3]{4}b^2 = \left(a + \frac{b}{\sqrt[3]{4}}\right)^2 + \frac{3}{2\sqrt[3]{2}}b^2$$
$$> 1.19 b^2 > 1.$$

On the other hand, from

$$\left|(a - \sqrt[3]{2}b)(a^2 + \sqrt[3]{2}ab + \sqrt[3]{4}b^2)\right| = 1,$$

it follows that $|a - \sqrt[3]{2}b| < 1$; also $|-1 + \sqrt[3]{2}| < 1$ hence $n > 0$.

Now I consider successively the cases that $n > 2$, $n = 2$, $n = 1$.

If $n > 2$ the coefficient of $\sqrt[3]{4}$ in $\pm(-1 + \sqrt[3]{2})^n$, is

$$\pm\left[(-1)^{n-2}\binom{n}{2} + (-1)^{n-5}2\binom{n}{5} + \ldots\right.$$
$$\left.\ldots + (-1)^{n-2-3k}2^k\binom{n}{3k+2} + \ldots\right] = 0.$$

I show that this is impossible, by considering congruences modulo 3. I note that

$$\binom{n}{3k+2} = \frac{n(n-1)}{(3k+1)(3k+2)}\binom{n-2}{3k},$$

so
$$\frac{1}{n(n-1)}\binom{n}{3k+2} = \frac{1}{(3k+1)(3k+2)}\binom{n-2}{3k}$$
$$\equiv -\binom{n-2}{3k} \pmod{3}.$$

Hence
$$\frac{1}{n(n-1)}(-1)^{n-2-3k}2^k\binom{n}{3k+2}$$
$$\equiv (-1)^{n-2-3k}(-1)^{k+1}\binom{n-2}{3k} \equiv (-1)^{n-1}\binom{n-2}{3k} \pmod{3}.$$

Therefore, I obtain the congruence
$$(-1)^{n-1}\sum_{k\geq 0}\binom{n-2}{3k} \equiv 0 \pmod{3}.$$

However, as I shall show now, for every $m \geq 1$,
$$S_0 = \sum_{k\geq 0}\binom{m}{3k} \not\equiv 0 \pmod{3}.$$

Indeed, let
$$S_1 = \sum_{k\geq 0}\binom{m}{3k+1} \quad \text{and} \quad S_2 = \sum_{k\geq 0}\binom{m}{3k+2}.$$

So
$$S_0 + S_1 + S_2 = \sum_{j\geq 0}\binom{m}{j} = (1+1)^m = 2^m \equiv (-1)^m \pmod{3}.$$

Now I observe that
$$S_2 = \sum_{k\geq 0}\binom{m}{3k+2} = \sum_{k\geq 0}\binom{m}{3k+1}\frac{m-3k-1}{3k+2}$$
$$\equiv -(m-1)S_1 \pmod{3},$$

and

$$S_1 = \sum_{k\geq 0} \binom{m}{3k+1} = \sum_{k\geq 0} \binom{m}{3k}\frac{m-3k}{3k+1} \equiv mS_0 \pmod{3}.$$

Thus

$$(-1)^m \equiv S_0 + S_1 + S_2 \equiv (1 + m - (m-1)m)S_0$$
$$\equiv (1 + 2m - m^2)S_0 \pmod{3}.$$

Hence $S_0 \not\equiv 0 \pmod{3}$, as I intended to show, arriving at a contradiction.

If $n = 2$, then

$$\pm(-1 + \sqrt[3]{2})^2 = \pm(1 - 2\sqrt[3]{2} + \sqrt[3]{4}) = a - \sqrt[3]{2}b,$$

which is not possible.

If $n = 1$, then

$$\pm(-1 + \sqrt[3]{2}) = a - \sqrt[3]{2}b$$

implies that $a = b = \pm 1$, and this concludes the proof. ∎

4. Continued Fractions

Lagrange showed the importance of continued fractions in the study of diophantine equations. In this section I recall some pertinent facts; for more details, see Perron (1913) and Hua (1982).

If $\xi_0, \xi_1, \ldots, \xi_n$ are positive real numbers let

$$[\xi_0, \xi_1, \ldots, \xi_n] = \zeta_0 + \cfrac{1}{\xi_1 + \cfrac{1}{\xi_2 + \cfrac{\cdots}{+\cfrac{1}{\xi_n}}}}.$$

The expression on the right is called a *continued fraction*, and $\xi_0, \xi_1, \ldots, \xi_n$ are its *partial quotients*.

If $\xi_i = a_i$ ($i = 0, 1, \ldots, n$) are integers, then $[a_0, a_1, \ldots, a_n]$ is equal to a rational number. Conversely, every rational number $\frac{c}{d}$ (with $c > 0$, $d > 0$ and $gcd(c,d) = 1$) has a continued fraction expansion

$$\frac{c}{d} = [a_0, a_1, \ldots, a_n]$$

which is unique (upon requiring that the last partial quotient is different from 1). It is obtained as follows:

$\frac{c}{d} = a_0 + \frac{c_1}{d}$, with a_0, c_1 integers, $0 \leq c_1 < d$; if $c_1 \neq 0$, then

$\frac{d}{c_1} = a_1 + \frac{c_2}{c_1}$, with a_1, c_2 integers, $0 \leq c_2 < c_1$; if $c_2 \neq 0$, then

$\frac{c_1}{c_2} = a_2 + \frac{c_3}{c_2}$, with a_2, c_3 integers, $0 \leq c_3 < c_2$; etc.

Since $d > c_1 > c_2 > c_3 > \ldots \geq 0$, the process must terminate; i.e., there exists $n \geq 1$, smallest such that $c_{n+1} = 0$. Then

$$\frac{c}{d} = [a_0, a_1, \ldots, a_n];$$

moreover, if $a_n = 1$ then also

$$\frac{c}{d} = [a_0, a_1, \ldots, a_{n-1} + 1],$$

so the last partial quotient is different from 1.

More generally, if $a_0, a_1, a_2, \ldots$ is an infinite sequence of positive integers, let

$$r_n = [a_0, a_1, \ldots, a_n] \quad \text{for every } n \geq 0.$$

Let A_n, B_n (for $n \geq -2$) be integers defined as follows:

$$A_{-2} = 0, \ A_{-1} = 1, \ A_0 = a_0, \ldots, A_n = a_n A_{n-1} + A_{n-2}$$

for $n \geq 0$, and

$$B_{-2} = 1, \ B_{-1} = 0, \ B_0 = 1, \ldots, B_n = a_n B_{n-1} + B_{n-2}$$

for $n \geq 0$.
Thus
$$A_0 < A_1 < A_2 < A_3 < \ldots,$$
$$1 = B_0 \leq B_1 < B_2 < B_3 < \ldots,$$

and the following facts are easy to prove:

(P4.1). *For every real number ξ and $n \geq 1$:*

$$[a_0, a_1, \ldots, a_{n-1}, \xi] = \frac{\xi A_{n-1} + A_{n-2}}{\xi B_{n-1} + B_{n-2}}.$$

(P4.2). $\quad r_n = \dfrac{A_n}{B_n}, \quad \gcd(A_n, B_n) = 1, \text{ for } n \geq 0.$

(P4.3). $A_n B_{n-1} - A_{n-1} B_n = (-1)^{n-1},$
$\quad A_n B_{n-2} - A_{n-2} B_n = (-1)^n a_n, \quad \text{for } n \geq 0.$

(P4.4). $\quad r_0 < r_2 < r_4 < \ldots; \ldots < r_5 < r_3 < r_1,$

(P4.5). *The following limits exist and are equal:*

$$\lim_{n \to \infty} r_{2n} = \lim_{n \to \infty} r_{2n-1} = \alpha.$$

The real number α is said to have the (infinite) continued fraction expansion

$$\alpha = [a_0, a_1, a_2, \ldots].$$

The rational number $r_n = \frac{A_n}{B_n}$ is called the nth *convergent* of α. Thus a convergent $r_n < \alpha$ if and only if n is even.

(P4.6). *Every infinite continued fraction defines an irrational number. Conversely, every irrational number has a unique infinite continued fraction expansion.*

The calculation of the continued fraction expansion may be performed using medians. The *median* of the rational numbers $\frac{a}{b}, \frac{c}{d}$, such that $\frac{a}{b} < \frac{c}{d}$ is defined to be $\frac{a+c}{b+d}$; clearly, $\frac{a}{b} < \frac{a+c}{b+d} < \frac{c}{d}$.

If $n \geq 0$, if $a_0, \ldots, a_n$ and the convergents $\frac{A_i}{B_i}$ ($0 \leq i \leq n$) of the continued fraction of α are known, then a_{n+1} is defined as follows:

If n is odd, a_{n+1} is the largest integer such that

$$\frac{A_{n-1}}{B_{n-1}} < \frac{A_{n-1} + A_n}{B_{n-1} + B_n} < \frac{A_{n-1} + 2A_n}{B_{n-1} + 2B_n}$$
$$< \ldots < \frac{A_{n-1} + a_{n+1}A_n}{B_{n-1} + a_{n+1}B_n} < \alpha;$$

note that each fraction is the median of the preceding fraction and $\frac{A_n}{B_n}$.

If n is even, a_{n+1} is defined in the same way, with opposite inequalities.

The continued fraction expansion of a number α provides approximations of α by rational numbers. In this respect, I note the inequalities:

(P4.7). $\quad \frac{1}{B_n(B_n + B_{n+1})} < \left| \alpha - \frac{A_n}{B_n} \right| < \frac{1}{B_n B_{n+1}} < \frac{1}{B_n^2}$

for every $n \geq 1$.

(P4.8). *For every $n \geq 1$, the convergent $\frac{A_n}{B_n}$ of α is the "best approximation" with denominator at most B_n, in the following sense:*

$$If \left| \alpha - \frac{a}{b} \right| < \left| \alpha - \frac{A_n}{B_n} \right|, \text{ then } b > B_n.$$

Conversely, if $\frac{a}{b}$ is a best approximation with denominator at most b, then $\frac{a}{b}$ is equal to a convergent of α.

It follows:

(P4.9). *If α is a real irrational number and $\left|\alpha - \frac{a}{b}\right| < \frac{1}{2b^2}$, then $\frac{a}{b}$ is a best approximation of α, hence $\frac{a}{b}$ is equal to a convergent of α.*

An infinite continued fraction $\alpha = [a_0, a_1, a_2, \ldots]$ is *periodic* if there exist $n_0 \geq 0$ and $k > 0$ such that $a_{n+k} = a_n$ for every $n \geq n_0$. It is then possible to consider the smallest n_0, k with the above property. If $n_0 = 0$, the continued fraction is said to be *purely periodic*. For periodic continued fractions, the customary notation is

$$\alpha = [a_0, \ldots a_{n_0-1}, \overline{a_{n_0}, a_{n_0+1}, \ldots, a_{n_0+k-1}}].$$

The sequence $(a_{n_0}, a_{n_0+1}, \ldots, a_{n_0+k-1})$ is called the *period*, the sequence $(a_0, \ldots, a_{n_0-1})$ is the *pre-period*, k is the *period length* and n_0 the *pre-period length*:

Euler proved (1737):

(P4.10). *If $\alpha = [a_0, \ldots, a_{n_0-1}, \overline{a_{n_0+1}, \ldots, a_{n_0+k-1}}]$ is an infinite periodic continued fraction, then α is a real quadratic irrational number; that is, α is an irrational number satisfying a quadratic relation $A\alpha^2 + B\alpha + C = 0$, with integers A, B, C and $A \neq 0$.*

The most important result about periodic continued fractions is the converse to the above statement. It was proved by Lagrange in 1770:

(P4.11). *The continued fraction expansion of every real quadratic irrational number α is periodic.*

Thus, for example,

$$\sqrt{2} = [1, 2, 2, \ldots] = [1, \bar{2}],$$

$$\frac{1 + \sqrt{5}}{2} = [1, 1, 1, \ldots] = [\bar{1}].$$

There are many classical results of Galois, Lagrange, and Legendre about the pre-period and period of continued fraction expansions of real quadratic irrational numbers, but they fall outside the scope of this book; the reader may wish to consult the sources already indicated. Just as an illustration, Lagrange showed:

(P4.12). *If $D > 0$ is an integer which is not a square, and if the length of the period of the continued fraction expansion of $\sqrt{D}$ is odd, then D is the sum of two squares.*

5. The Equations $EX^2 - DY^2 = \pm C$

Let D be a positive integer, not a square (I do not assume, however, that D is square-free). The equations $X^2 - DY^2 = \pm 1$ which were the object of many of Fermat's problems "have long borne the name of Pell, due to a confusion originating with Euler, [and] should have been designated as Fermat's equation" (see Dickson, 1920, vol. II, p. 341). I shall not enter here into historical considerations in relation to these very interesting equations, referring the reader to Heath (1885) for the early history and to Dickson's work for a thorough account up to 1920. Let it only be said that before Fermat, special cases were considered by Indian and Greek mathematicians, in problems connected with rational approximation to square roots and other quadratic problems (Baudhayana, Theon of Smyrna, Archimedes, Diophantus, Brahmegupta). Later, Alkarkhi and Bháscara Achárya solved special cases and gave methods to obtain new solutions from known solutions.

Fermat stated in February 1657, in a letter to Frénicle de Bessy with the "Second Défi aux Mathématiciens," that if $D > 0$ is not a square then the equation $X^2 - DY^2 = 1$ has an infinite number of solutions in integers (see a thoroughly documented discussion in the paper of Hoffmann, 1944). Brouncker and Wallis gave a method to find infinitely many solutions (but not necessarily all solutions). The problem and method of solution were thoroughly discussed in an extensive exchange of letters, involving Fermat, Frénicle, Digby, Wallis and Brouncker.

Euler contributed to the theory of these equations and Lagrange used continued fractions to give the first complete proof that if $D > 0$ is not a square then the equation $X^2 - DY^2 = 1$ has infinitely many solutions in integers. He also used his method to prove the very famous and important theorem that every real quadratic irrational number has an infinite periodic continued fraction expansion (see (P4.11)).

Legendre's *Théorie des Nombres* (1830) contains a classical exposition of the theory of the equations $X^2 - DY^2 = \pm 1$. A modern presentation is given, for example, by Mordell (1969).

In an article of 1977, Weil confronted the solution by Wallis and Brouncker, with Bachet's solution of the equation $aX + bY = c$, in the light of the method of continued fractions.

A thorough understanding of the theory is achieved by considering the units of real quadratic fields (see, e.g., Ribenboim 1972).

For the convenience of the reader, I shall summarize the facts about real quadratic fields which will be needed in the sequel.

Let $D > 1$ be a square-free integer. Denote by $K = \mathbb{Q}(\sqrt{D})$ the real quadratic field defined by $\sqrt{D}$. It consists of all real numbers of the form $r + s\sqrt{D}$, with $r, s \in \mathbb{Q}$. The ring of integers of K is denoted by A.

(P5.1). *If $D \equiv 2$ or $3 \pmod{4}$, then*

$$A = \{a+b\sqrt{D} \mid a,b \in \mathbb{Z}\} = \left\{\frac{a+b\sqrt{D}}{2} \mid a,b, \text{ are even integers}\right\}.$$

If $D \equiv 1 \pmod{4}$, then

$$A = \left\{\frac{a+b\sqrt{D}}{2} \mid a,b \in \mathbb{Z},\ a \equiv b \pmod{2}\right\}.$$

The units of K are the elements $\alpha \in A, \alpha \neq 0$, such that $\alpha^{-1} \in A$. Clearly, ± 1 are units. The set of all units may be easily described, as I indicate now.

If α is a unit, so are $-\alpha, \alpha^{-1}, -\alpha^{-1}$. If $\alpha \neq \pm 1$, then one of the four numbers above is greater than 1. It may be shown that $\alpha = \frac{a+b\sqrt{D}}{2} > 1$ if and only if $a > 0, b > 0$. If $\alpha' = \frac{a'+b'\sqrt{D}}{2}$ is also a unit such that $\alpha' > 1$, it may be shown that $\alpha' < \alpha$ if and only if $a' < a$. Hence among the units $\alpha > 1$ there is one which is the smallest. It is denoted by ϵ_D and called the *fundamental unit* of $K = \mathbb{Q}(\sqrt{D})$.

(P5.2). *Every unit of K is of the form*

$$\alpha = \pm \epsilon_D^k$$

where k is any integer.

The conjugate of the number $\alpha = r + s\sqrt{D} \in K$ is defined to be $\bar{\alpha} = r - s\sqrt{D}$. The norm of α is

$$N(\alpha) = \alpha \bar{\alpha} = r^2 - s^2 D \in \mathbb{Q}.$$

If $\alpha \in A$ then

$$N(\alpha) = \frac{a^2 - b^2 D}{4} \in \mathbb{Z},$$

because $a \equiv b \pmod{2}$, and if a, b are odd, then $D \equiv 1 \pmod 4$. Finally, if $\alpha \in A$, then α is a unit if and only if $N(\alpha) = \pm 1$.

It should be noted that the fundamental unit ϵ_D may have norm 1, or norm -1, depending on D.

If $\epsilon_D = \frac{t+u\sqrt{D}}{2}$, define the sequences of integers x_n, y_n (for $n \geq 1$) by the relation

$$\frac{x_n + y_n\sqrt{D}}{2} = \left(\frac{t+u\sqrt{D}}{2}\right)^n. \tag{5.1}$$

Since $\frac{x_n+y_n\sqrt{D}}{2} \in A$, then $x_n \equiv y_n \pmod 2$. If t, u are even then x_n, y_n are even for every $n \geq 1$. If t, u are odd, then $D \equiv 5 \pmod 8$, and x_n, y_n are even if and only if $3 | n$.

It is easy to see that

$$x_1 < x_2 < x_3 < \ldots \text{ and } y_1 < y_2 < y_3 < \ldots,$$

except for the case $D = 5$, in which $y_1 = y_2 = 1$.

Note that

$$N\left(\frac{x_n + y_n\sqrt{D}}{2}\right) = (N(\epsilon_D))^n.$$

The units of $\mathbb{Q}(\sqrt{D})$, $D > 0$, may be effectively computed in finitely many steps, using the continued fraction expansion of $\sqrt{D}$, as indicated by Lagrange in 1770.

Let k be the period and let $\frac{A_i}{B_i}$ be the ith convergent (for $i \geq 0$) of the continued fraction expansion of $\sqrt{D}$. Then:

(P5.3). *If k is even, then*

$$\begin{cases} x_n = A_{nk-1} \\ y_n = B_{nk-1} \end{cases} (for\ n \geq 1).$$

If k is odd, then

$$\begin{cases} x_n = A_{2nk-1} \\ y_n = B_{2nk-1} \end{cases} (for\ n \geq 1).$$

After these preliminaries, I indicate how to obtain all the solutions of some equations of the type

$$EX^2 - DY^2 = C \tag{5.2}$$

where $E, D > 0$ are not squares and C is any non-zero integer.

Writing $E = r^2 E_1$, $D = s^2 D_1$, with $r, s \geq 1$ and E_1, D_1 square-free integers, if (x, y) is a solution of (5.2), then (rx, sy) is a solution of $E_1 X^2 - D_1 Y^2 = C$. Thus, it suffices to determine the solutions of the above equation. In other words, there is no loss of generality to assume that E, D are square-free. Moreover, it is clearly sufficient to determine the solutions (x, y) with $x > 0$, $y > 0$.

Denote by $S_{E,D,C}$, or simply by S_C, the set of all solutions (x, y) of (5.2), with $x > 0$, $y > 0$.

As before, let $\epsilon_D = \frac{t+u\sqrt{D}}{2}$ be the fundamental unit of $K = \mathbb{Q}(\sqrt{D})$, and for every $n \geq 1$, let x_n, y_n be defined by the relation (5.1).

$$\textbf{Solutions of } X^2 - DY^2 = 1 \qquad (5.3)$$

(P5.4). Let $N(\epsilon_D) = 1$. If t, u are odd, then

$$S_1 = \left\{ \left(\frac{x_n}{2}, \frac{y_n}{2} \right) \Big| 3 \text{ divides } n \geq 1 \right\}.$$

If t, u are even, then

$$S_1 = \left\{ \left(\frac{x_n}{2}, \frac{y_n}{2} \right) \Big| n \geq 1 \right\}.$$

Let $N(\epsilon_D) = -1$. If t, u are odd, then

$$S_1 = \left\{ \left(\frac{x_n}{2}, \frac{y_n}{2} \right) \Big| 6 \text{ divides } n \geq 1 \right\}.$$

If t, u are even, then

$$S_1 = \left\{ \left(\frac{x_n}{2}, \frac{y_n}{2} \right) \Big| 2 \text{ divides } n \right\}.$$

$$\textbf{Solutions of } X^2 - DY^2 = -1 \qquad (5.4)$$

(P5.5). $N(\epsilon_D) = 1$. Then $S_{-1} = \emptyset$.

Let $N(\epsilon_D) = -1$. If t, u are odd, then

$$S_{-1} = \left\{ \left(\frac{x_n}{2}, \frac{y_n}{2} \right) \Big| 3 \text{ divides } n \geq 1, \ n \text{ odd} \right\}.$$

The Equations $EX^2 - DY^2 = \pm C$

If t, u are even, then

$$S_{-1} = \left\{ \left(\frac{x_n}{2}, \frac{y_n}{2}\right) \Big| n \geq 1 \text{ odd} \right\}.$$

Solutions of $X^2 - DY^2 = 4$ \hfill (5.5)

(P5.6). *If $N(\epsilon_D) = 1$, then $S_4 = \left\{ (x_n, y_n) | n \geq 0 \right\}$.*
If $N(\epsilon_D) = -1$, then $S_4 = \left\{ (x_n, y_n) | n \geq 1 \text{ is even} \right\}$.

Solutions of $X^2 - DY^2 = -4$ \hfill (5.6)

(P5.7). *If $N(\epsilon_D) = 1$, then $S_{-4} = \emptyset$.*
If $N(\epsilon_D) = -1$, then $S_{-4} = \left\{ (x_n, y_n) | n \geq 1 \text{ is odd} \right\}$.

Solutions of $EX^2 - DY^2 = \pm 1$ \hfill (5.7)

(where $E, D > 1$, E and D square-free).

These equations have been the object of considerable attention, especially by Stolt (1952, 1954) and Nagell (1953, 1955). An expository expository paper by Walker (1967) is a good reference. For an overview of these and related results, see Walsh (1988).

(P5.8). *If Equation (5.7) has a solution in integers, let (x_1, y_1) be the solution in positive integers for which y_1 is minimal (called the **fundamental solution**). Then, the solutions in positive integers are (x_n, y_n), where the numbers x_n, y_n are defined by the relation*

$$x_n \sqrt{E} + y_n \sqrt{D} = (x_1 \sqrt{E} + y_1 \sqrt{D})^n, \tag{5.8}$$

for all odd $n \geq 1$.

Note that (5.7) need not have a solution, as indicated in the next result:

(P5.9). *If the equation $X^2 - EDY^2 = -1$ has a solution in integers, then both equations (5.7) do not have a solution in integers.*

This happens when $N(\epsilon_{ED}) = -1$.

Solutions of $X^2 - DY^2 = C$ \hfill (5.9)

This equation is similar to (5.7) and was studied by Schepel (1935).

(P5.10). *Assume that $D > 1$ is square-free and that C is a non-zero square-free integer such that C divides $2D$ (the case $C = \pm 1$ has already been considered).*

*If Equation (5.9) has a solution in integers, let (x_1, y_1) be the solution in positive integers for which y_1 is minimal (called the **fundamental solution**). Then the solutions in positive integers are (x_n, y_n) for all $n \geq 1$ when $C > 0$, or for all odd $n \geq 1$ when $C < 0$. The numbers x_n, y_n are defined by the following relation:*

$$|C|^{\frac{n-1}{2}}(x_n + y_n\sqrt{D}) = (x_1 + y_1\sqrt{D})^n. \quad (5.10)$$

Part A

SPECIAL CASES

In this chapter, I shall study special cases of Catalan's problem, namely the sequences of mth or n^{th} powers, where m or n is at most 3. I shall also consider the sequence of powers of 2 or 3. And, I will take every opportunity to digress and study other interesting diophantine equations.

1. Preliminary Lemmas

I begin with the following obvious remark.

If p is a prime dividing m, $m = pm'$, and if $a \geq 1$, then $a^m = (a^{m'})^p$. So the sequence of mth powers is a subsequence of the sequence of pth powers. Thus, the sequence of all powers is the same as the sequence of all powers with prime exponents.

Similarly, if p, q are given primes, if $m = pm'$, $n = qn'$, and if $x, y \geq 1$ are such that x^m, y^n are consecutive, then $(x^{m'})^p$, $(y^{n'})^q$ are consecutive p^{th}, q^{th} powers. This leads to the study of consecutive powers with prime exponents.

The following lemmas, which are basic in this study, were given by Euler.

(A1.1). Lemma. *Let p, q be primes, $x, y \geq 2$ and $x^p - y^q = 1$.*

i) *If p is odd, then either*

$$\begin{cases} x-1 = a^q & \text{with } y = aa', \ p \nmid aa', \\ \frac{x^p-1}{x-1} = a'^q & \gcd(a,a') = 1, \end{cases}$$

or

$$\begin{cases} x-1 = p^{q-1}a^q & \text{with } y = paa', \ p \nmid a', \\ \frac{x^p-1}{x-1} = pa'^q & \gcd(a,a') = 1. \end{cases}$$

ii) *If q is odd, then either*

$$\begin{cases} y+1 = b^p & \text{with } x = bb', \ q \nmid bb', \\ \frac{y^q+1}{y+1} = b'^p & \gcd(b,b') = 1, \end{cases}$$

or

$$\begin{cases} y+1 = q^{p-1}b^p & \text{with } x = qbb', \ q \nmid b', \\ \frac{y^q+1}{y+1} = qb'^p & \gcd(b,b') = 1. \end{cases}$$

Proof. i) First, note that $y^q = x^p - 1 = (x-1)\frac{x^p-1}{x-1}$. By (P1.2), $\gcd\left(x-1, \frac{x^p-1}{x-1}\right) = 1$ or p. The first case leads to the first alternative. In the second case, $p|y$, by (P1.2), $p^2 \nmid \frac{x^p-1}{x-1}$, and this leads to the second alternative.

ii) The proof is similar. ∎

(A1.2). Lemma *Let q be an odd prime, $x, y \geq 1$ and $x^2 - y^q = 1$. Then, either*

$$\begin{cases} x - 1 = 2a^q \\ x + 1 = 2^{q-1}a'^q \end{cases}$$

or

$$\begin{cases} x + 1 = 2a^q \\ x - 1 = 2^{q-1}a'^q \end{cases}$$

with a odd, $\gcd(a,a') = 1$, and $y = 2aa'$.

Proof: Note that $y^q = x^2 - 1 = (x+1)(x-1)$. If x is even, then $\gcd(x+1, x-1) = 1$, so $x+1 = c^q$, $x-1 = d^q$ (for some positive integers c, d). Then subtracting, $2 = c^q - d^q$, which is impossible. Thus, x is odd and $\gcd(x+1, x-1) = \gcd(x+1, 2) = 2$. So

$$\begin{cases} x - 1 = 2^e c^q \\ x + 1 = 2^f d^q \end{cases}$$

with c, d odd, $\gcd(c, d) = 1$, $e + f = rq$ $(r \geq 1)$ and $\min\{e, f\} = 1$.

If $e = 1$, then $f = rq - 1 = q - 1 + (r-1)q$, and this leads to the first alternative, with $a = c$, $a' = 2^{r-1}d$.

If $f = 1$, then $e - rq - 1 = q - 1 + (r-1)q$, and this leads to the second alternative, with $a = d$, $a' = 2^{r-1}c$. ∎

2. The Sequence of Squares or Cubes

In 1738, Euler showed that 8, 9 are the only consecutive integers in the sequence of squares or cubes. More precisely, he showed that $x = 3, y = 2$ are the only positive rational number solutions of $X^2 - Y^3 = \pm 1$.

For this purpose, Euler established the following lemma:

(A2.1). Lemma *Let $b, c \geq 1$ such that $\gcd(b, c) = 1$. If $bc(c^2 \pm 3bc + 3b^2)$ is a square, then $b = 1$, $c = 1$ or $b = 1, c = 3$, and the expression must be equal to $bc(c^2 - 3bc + 3b^2) = 1$ or 9.*

Proof: First note that the discriminant of $X^2 \pm 3bX + 3b^2$ is $-3b^2 < 0$, so $c^2 \pm 3bc + 3b^2 > 0$ for all values of b, c.

Assume that the statement is false, so there exists a smallest non-zero square of the form $bc(c^2 \pm 3bc + 3b^2)$ with $(b, c) \neq (1, 1), (1, 3)$.

To begin, I show that $3 \nmid c$. Indeed, if $c = 3d$, then $3^2 bd(b^2 \pm 3bd + 3d^2)$ is a square, so $bd(b^2 \pm 3bd + 3d^2)$ is a square. But

$$0 < bd(b^2 \pm 3bd + 3d^2) = \frac{1}{9}bc(c^2 \pm 3bc + 3b^2)$$
$$< bc(c^2 \pm 3bc + 3b^2).$$

By the assumption, this implies that $(d,b) = (1,1)$ or $(1,3)$ hence $(b,c) = (1,3)$ or $(3,3)$, the latter possibility being excluded, because $gcd(b,c) = 1$. This shows that $3 \nmid c$.

Next, I observe that $gcd(b, c^2 \pm 3bc + 3b^2) = 1$ and $gcd(c, c^2 \pm 3bc + 3b^2) = 1$, since $3 \nmid c$.

So, from the unique factorization theorem, each factor $b, c, c^2 \pm 3bc + 3b^2$ is a square.

Thus, I write

$$c^2 \pm 3bc + 3b^2 = e^2,$$

and observe that $e \neq c$, because $b \neq c$, $gcd(b,c) = 1$ and $(b,c) \neq (1,1)$.

Let $\frac{c-e}{b} = \frac{m}{n}$ with $n \geq 1$, $m \neq 0$, $gcd(m,n) = 1$. Hence

$$e = c - \frac{m}{n}b$$

and

$$c^2 \pm 3bc + 3b^2 = \left(c - \frac{m}{n}b\right)^2.$$

Therefore

$$c\left(\frac{3m}{n} \pm 3\right) = b\left(\frac{m^2}{n^2} - 3\right),$$

hence

$$\frac{b}{c} = \frac{2mn \pm 3n^2}{m^2 - 3n^2} \quad \text{(note that } m^2 \neq 3n^2\text{)}.$$

Now I show that

$$gcd(2mn \pm 3n^2, m^2 - 3n^2) = 1 \text{ or } 3.$$

Indeed, if $2|2mn \pm 3n^2$ and $2|m^2 - 3n^2$ then $2|m$, $2|n$, which is a contradiction. If $p \neq 2$, $r \geq 1$ and $p^r|2mn \pm 3n^2$, $p^r|m^2 - 3n^2$, then $p^r|m(m \pm 2n)$. If $p|m \pm 2n$, then $p|2mn \pm 4n^2$ so $p|n$, and $p|m$, which is impossible. Thus $p^r|m, p \nmid n$ and $p^r|3n^2$, hence $p^r = 3$.

The Sequence of Squares or Cubes

Two cases are possible.

<u>Case 1.</u> $gcd(2mn \pm 3n^2, m^2 - 3n^2) = 1$.

Now $3 \nmid m$ and $c = \pm(m^2 - 3n^2)$. If $c = -(m^2 - 3n^2)$, then $c \equiv -m^2 \pmod{3}$, and since c is a square, then -1 mod 3 is a square, which is absurd. This shows that $c = m^2 - 3n^2$, hence $b = 2mn \pm 3n^2 = n(2m \pm 3n)$. Let $m^2 - 3n^2 = c = f^2$, with $f \geq 0$; so $f > 0$, $m > f$, and I write $\frac{m-f}{n} = \frac{u}{v}$ with $u \geq 1$, $v \geq 1$, $gcd(u,v) = 1$.

Hence $f = m - \frac{u}{v}n$, $m^2 - 3n^2 = \left(m - \frac{un}{v}\right)^2$ and $-3n^2 = -\frac{2mnu}{v} + \frac{u^2 n^2}{v^2}$, from which it follows that

$$\frac{2um}{vn} = \frac{u^2}{v^2} + 3, \quad \frac{m}{n} = \frac{u^2 + 3v^2}{2uv},$$

hence

$$\frac{b}{n^2} = \frac{2m}{n} \pm 3 = \frac{u^2 \pm 3uv + 3v^2}{uv}, \quad n^2 uv(u^2 \pm 3uv + 3v^2) = bu^2 v^2$$

is a square, so $uv(u^2 \pm 3uv + 3v^2)$ is a square.

It is not possible to have $(v, u) = (1, 1)$ or $(1,3)$. Indeed, with these values of u, v, $uv(u^2 + 3uv + 3v^2)$ is not a square. Secondly, $uv(u^2 - 3uv + 3v^2) = 1$ or 9 hence $n^2 u^2 v^2 = 1$, so $n = 1, b = 1, m = 2, c = m^2 - 3n^2 = 1$, which is contrary to the hypothesis.

Therefore, by the hypothesis,

$$uv(u^2 \pm 3uv + 3v^2) \geq bc(b^2 \pm 3bc + 3b^2).$$

I still need to make some observations.

First, $gcd(2uv, u^2 + 3v^2)$ divides 6: indeed, if $4 | 2uv$ and $4 | u^2 + 3v^2$, then $2 | uv$, hence

$$\begin{cases} 2|u, 2 \nmid v \text{ so } 2 \nmid u^2 + 3v^2, & \text{or} \\ 2|v, 2 \nmid u \text{ so } 2 \nmid u^2 + 3v^2, & \text{a contradiction.} \end{cases}$$

Also, if p is an odd prime, $r \geq 1$ and $p^r | 2uv$, $p^r | u^2 + 3v^2$, then $p^r | uv$. Hence from $gcd(u,v) = 1$,

$$\begin{cases} p^r | u, p \nmid v, & \text{so } p^r | 3v^2 \text{ hence } p^r = 3, \text{ or} \\ p^r | v, p \nmid u, & \text{so } p^r | u, \text{ an absurdity.} \end{cases}$$

From $2uvm = n(u^2 + 3v^2)$, it follows that $uv | 3n$. Indeed, $gcd(2uv, u^2 + 3v^2) = 2^r 3^s$, with $r, s = 0$ or 1. Then $\frac{2uv}{2^r 3^s}$ divides n, and therefore uv divides $3n$.

Now I show that $uv(u^2 \pm 3uv + 3v^2)$ divides $9b$.

Indeed, $n^2(u^2 \pm 3uv + 3v^2) = uvb$ divides $3nb = 3n^2(2m \pm 3n)$, so $uv(u^2 \pm 3uv + 3v^2)$ divides $3uv(2m \pm 3n)$, which in turn divides $9n(2m \pm 3n) = 9b$. So $bc(c^2 \pm 3bc + 3c^2) \leq 9b$, hence $c(c^2 \pm 3bc + 3b^2) \leq 9$. On the other hand, $c(c^2 \pm 3bc + 3b^2) \geq \frac{c^3}{4}$, because

$$\frac{3c^2}{4} \pm 3bc + 3b^2 = \frac{3}{4}(c^2 \pm 4bc + 4b^2) = \frac{3}{4}(c \pm 2b)^2 \geq 0.$$

Hence $c^3 \leq 36$ so $c \leq 3$ and since c is a square, then $c = 1$.

From this, it follows that $1 \pm 3b + 3b^2 \leq 9$, so $3b(b \pm 1) \leq 8$, and since b is a square, $b = 1$, and this is a contradiction.

<u>Case 2.</u> $gcd(2mn \pm 3n^2, m^2 - 3n^2) = 3$.

Now $3|m$, hence $3 \nmid n$. Let $m = 3k$, so $gcd(k,n) = 1$ and $gcd(2kn \pm n^2, 3k^2 - n^2) = 1$. Also,

$$\frac{b}{c} = \frac{2kn \pm n^2}{3k^2 - n^2}.$$

Therefore $c = \pm(3k^2 - n^2)$. But $3k^2 - n^2$ is not a square, otherwise $-n^2$ is a square modulo 3, so -1 is a square modulo 3, which is not true. Therefore, since c is a square, $c = n^2 - 3k^2$, and $b = \pm n^2 - 2kn = n(\pm n - 2k)$.

I write $n^2 - 3k^2 = c = f^2$ with $f \geq 0$; so $f > 0$, $n > f$. Let $\frac{n-f}{k} = \frac{u}{v}$ with $v \geq 1$, $u \geq 1$, $gcd(u,v) = 1$, so $n^2 - 3k^2 = f^2 = \left(n - \frac{u}{v}k\right)^2$ and

$$-3k^2 = -\frac{2nku}{v} + \frac{u^2 k^2}{v^2},$$

from which it follows that

$$\frac{2un}{vk} = \frac{u^2}{v^2} + 3, \qquad \frac{n}{k} = \frac{u^2 + 3v^2}{2uv},$$

hence

$$\frac{b}{n^2} = \pm 1 - \frac{2k}{n} = \pm 1 - \frac{4uv}{u^2 + 3v^2}.$$

Since b, u, v are positive, then the lower sign has to be excluded. Therefore

$$\frac{b}{n^2} = \frac{u^2 - 4uv + 3v^2}{u^2 + 3v^2},$$

and

$$n^2(u^2 - 4uv + 3v^2)(u^2 + 3v^2) = b(u^2 + 3v^2)^2$$

is a square not equal to 0. Thus

$$(u^2 - 4uv + 3v^2)(u^2 + 3v^2) = (u-v)(u-3v)(u^2 + 3v^2)$$

is a square, hence $(u-v)(u-3v)$ is positive and $u \neq v$, $u \neq 3v$.

Let

$$\begin{cases} t = \frac{|u-v|}{2} \\ s = \frac{|u-3v|}{2} \end{cases} \quad \text{if } u, v \text{ are both odd}$$

or

$$\begin{cases} t = |u - v| \\ s = |u - 3v| \end{cases} \quad \text{otherwise.}$$

Thus $t > 0$, $s > 0$ and also $gcd(s,t) = 1$, because $gcd(u,t) = 1$.

Now I show that $ts(s^2 - 3ts + 3t^2)$ is a square. First assume that u and v are both odd. Noting that $(u-v)(u-3v) > 0$, then

$$\begin{aligned}
ts(s^2 &- 3ts + 3t^2) \\
&= \frac{|u-v|}{2} \times \frac{|u-3v|}{2} \times \frac{1}{4}[(u-3v)^2 \\
&\quad - 3|u-v|\,|u-3v| + 3(u-v)^2] \\
&= \frac{1}{16}(u^2 - 4uv + 3v^2)[(u^2 - 6uv + 9v^2) \\
&\quad - 3(u^2 - 4uv + 3v^2) + 3(u^2 - 2uv + v^2)] \\
&= \frac{1}{16}(u^2 - 4uv + 3v^2)(u^2 + 3v^2),
\end{aligned}$$

so $ts(s^2 - 3ts + 3t^2)$ is a square.

Now assume that u or v is even. Then

$$\begin{aligned}
ts(s^2 &- 3ts + 3t^2) \\
&= |u-v|\,|u-3v|[(u-3v)^2 - 3|u-v|\,|u-3v| + 3(u-v)^2] \\
&= (u^2 - 4uv + 3v^2)[(u^2 - 6uv + 9v^2) \\
&\quad - 3(u^2 - 4uv + 3v^2) + 3(u^2 - 2uv + v^2)] \\
&= (u^2 - 4uv + 3v^2)(u^2 + 3v^2),
\end{aligned}$$

so again $ts(s^2 - 3ts + 3t^2)$ is a square.

Next, I show that $(t,s) \neq (1,1)$ and (1.3), otherwise $ts(s^2 - 3ts + 3t^2) = 1$ or 9.

If u, v are both odd then

$$(u^2 + 3v^2)^2 > (u^2 - 4uv + 3v^2)(u^2 + 3v^2) = 16 \text{ or } 16 \times 9,$$

with $u^2 + 3v^2$ dividing 16, or 16×9; it is easy to see that both cases are impossible with relatively prime distinct odd integers u, v.

If u or v is even then

$$(u^2 - 4uv + 3v^2)(u^2 + 3v^2) = 1 \text{ or } 9,$$

which is clearly impossible. Thus $(t,s) \neq (1,1)$, (1.3).
From the hypothesis,

$$(u^2 - 4uv + 3v^2)(u^2 + 3v^2) \geq ts(t^2 - 3ts + 3t^2) \geq bc(c^2 - 3bc + 3b^2).$$

Now I observe that, as in the first case, $gcd(2uv, u^2 + 3v^2)$ divides 6. From $(u^2 + 3v^2)k = 2uvn$ it follows that $u^2 + 3v^2$ divides $6n$, therefore

$$(u^2 + 3v^2)(u^2 - 4uv + 3v^2) = b\left(\frac{u^2 + 3v^2}{n}\right)^2$$

divides $36b$, hence

$$c(c^2 - 3bc + 3b^2) \leq 36.$$

On the other hand, since $c(c^2 - 3bc + 3b^2) \geq \frac{c^3}{4}$ (as seen in the first case), then $c^3 \leq 4 \times 36$, so $c \leq 5$ and since c is a square, then $c = 1$ or 4. From this it follows that $1 - 3b + 3b^2 \leq 36$ or $16 - 12b + 3b^2 \leq 9$, and since b is a square, $b = 1$. The values $b = 1$, $c = 4$ do not give a square for $bc(c^2 - 3bc + 3b^2)$, so $b = 1$, $c = 1$, which is contrary to the hypothesis. ∎

Euler used this lemma to prove:

(A2.2). i) If x, u are integers, $x \neq 0, u \geq 1$, $gcd(x,u) = 1$, and if $\left(\frac{x}{u}\right)^3 \pm 1$ is the square of a non-zero rational number, then $\frac{x}{u} = 2$.

ii) If $x, y \geq 1$ are integers such that $x^3 - y^2 = \pm 1$, then $x = 2$ and $y = 3$.

Proof: i) Assume that $\left(\frac{x}{u}\right)^3 \pm 1$ is a non-zero square, then $x^3 u \pm u^4$ is a non-zero integer and also a square. Let $z = x \pm u$, so $gcd(z, u) = 1$ and $z \neq \pm u$. Then

$$0 < x^3 u \pm u^4 = u(x \pm u)(x^2 \mp xu + u^2) = uz(z^2 \mp 3uz + 3u^2),$$

so $z \neq 0$. Since $z^2 \mp 3uz + 3u^2 \geq 0$ (because its discriminant is $9u^2 - 12u^2 = -3u^2 < 0$), then $z > 0$.

By the lemma, it follows that $(u, z) = (1, 1)$ or $(1,3)$ and the expression must be $uz(z^2 - 3uz + 3u^2)$, which corresponds to $\left(\frac{x}{u}\right)^3 + 1$. So $z = x + u$, hence $x = 2$, $u = 1$.

ii) This is a trivial consequence of (i). ∎

The above proof by Euler involves the method of descent. In his book of 1919, Bachmann wrote a proof of statement (ii) of (A2.2) without the method of descent; however, as pointed out by Nagell in 1924, this proof is incorrect.

In 1921, Nagell gave a proof that if x, y are integers and $x^2 - y^3 = 1$ then $x = \pm 3, y = 2$. It is based on the result of Legendre about the equation $X^3 + Y^3 = 2Z^3$ (see comment after (P3.10)).

Nagell's proof:

Suppose $x^2 - y^3 = 1$, so

$$y^3 = x^2 - 1 = (x - 1)(x + 1).$$

If y is odd, then $gcd(x - 1, x + 1) = 1$, hence

$$\begin{cases} x + 1 = a^3 \\ x - 1 = b^3 \end{cases} \text{ for some non-zero integers } a, b.$$

Hence $2 = a^3 - b^3$, and this is impossible.

If y is even, then $\gcd(x-1, x+1) = 2$, hence

$$\begin{cases} x+1 = 2a^3 \\ x-1 = 4b^3 \end{cases} \text{ or } \begin{cases} x-1 = 2a^3 \\ x+1 = 4b^3, \end{cases}$$

for some non-zero integers a, b.

Therefore, $\pm 2 = 2a^3 - 4b^3$, and $\pm 1 = a^3 - 2b^3$.

By Legendre's result, $a = b = \pm 1$. It follows that $x = \pm 3$ and $y = 2$. ∎

It is also very easy to give a direct proof that if x, y are non-zero integers, then $x^3 - y^2 \neq 1$. The method was given by Lebesque and holds more generally (see §A3).

For the convenience of the reader, I now give the proof in the special case.

Let $K = \mathbb{Q}(i)$, $i = \sqrt{-1}$, be the field of all Gaussian numbers $r + si$ (with $r, s \in \mathbb{Q}$). The ring of Gaussian integers is $\mathbb{Z}[i]$, consisting of all $r + si$, with $r, s \in \mathbb{Z}$.

I shall use the following facts:

a) The units of A are $\pm 1, \pm i$.

b) Every $\alpha \in \mathbb{Z}[i]$ is in a unique way (up to units) the product of prime elements.

Lebesque's proof:

Assume that x, y are non-zero integers and $x^3 - y^2 = 1$. If y is odd, then $x^3 = y^2 + 1 \equiv 2 \pmod{4}$, which is impossible. So y is even.

The decomposition

$$x^3 = y^2 + 1 = (y+i)(y-i)$$

takes place in $\mathbb{Z}[i]$.

If a prime π of $\mathbb{Z}[i]$ divides both $y+i, y-i$, then $\pi | 2i$. But i is a unit, so $\pi | 2$; since $2 | y$, then $\pi | y$ and also $\pi | y + i$, hence $\pi | i$, which is absurd.

Therefore $\alpha = gcd(y+i, y-i)$ is a unit and necessarily there exists $a + bi \in A$ such that

$$y + i = \alpha(a+bi)^3 = \alpha[(a^3 - 3ab^2) + (3a^2b - b^3)i].$$

It follows that

$$y - i = \bar{\alpha}(a-bi)^3 = \bar{\alpha}[(a^3 - 3ab^2) - (3a^2b - b^3)i].$$

This leads to
$$2i = 2b(3a^2 - b^2)i, \quad \text{or}$$
$$2i = 2a(a^2 - 3b^2)i.$$

Hence $b = \pm 1$ and $a = 0$, or $a = \pm 1$ and $b = 0$. Thus, in all cases, $y = 0$, which is a contradiction. ∎

3. The Equation $X^m - Y^2 = 1$

Now I give Lebesgue's result of 1850 concerning the equation $X^m - Y^2 = 1$; see also Cassels (1953) and Tang (1974), who reproduced Lebesgue's proof.

(A3.1). *The equation $X^m - Y^2 = 1$ (with $m \geq 2$) has no solution in positive integers.*

Proof: The result is trivial if m is even, so I assume that m is odd. Let $x, y \geq 1$ be integers such that $x^m = y^2 + 1$. If y is odd then $x^m \equiv 2 \pmod 4$, which is impossible when $m \geq 2$. Therefore y is even and x is odd.

Let $i = \sqrt{-1}$, then $x^m = y^2 + 1 = (y+1)(y-1)$. I shall show that $gcd(y+i, y-i)$ is a unit of the Gaussian field $\mathbb{Q}(i)$. Indeed, if a prime π of $\mathbb{Q}(i)$ divides $y+i$ and $y-i$, then it divides $2i$. Since i is a unit, π divides 2, and so $\pi|y$. But $\pi|y+i$, hence $\pi|i$, a contradiction.

It follows that the factor $y + i$ of x^m is an mth power, up to a unit: $y + i = (u+iv)^m i^s$, $0 \leq s \leq 3$ with $u, v \in \mathbb{Z}$. So its conjugate is $y - i = (u - iv)^m (-i)^s$. Hence $x^m = y^2 + 1 =$

$(u^2+v^2)^m$, therefore $x = u^2+v^2$. But x is odd, hence u or v is even. By subtracting,

$$2i = [(u+iv)^m - (u-iv)^m(-1)^s]i^s.$$

If $s = 2r$, comparing the coefficients of i on both sides:

$$1 = (-1)^r \left[mu^{m-1}v - \binom{m}{3}u^{m-3}v^3 + \ldots \pm v^m \right];$$

thus v divides 1, hence $v = \pm 1$, so v is odd.
If $s = 2r+1$, then similarly

$$1 = (-1)^r \left[u^m - \binom{m}{2}u^{m-2}v^2 + \ldots \pm muv^{m-1} \right];$$

thus u divides 1, hence $u = \pm 1$, so u is odd.

Letting $w = u$ (in the first case) or $w = v$ (in the second case), by the above remark, w is even and in both cases I rewrite the above equalities as follows:

$$1 - \binom{m}{2}w^2 + \binom{m}{4}w^4 - \ldots \pm mw^{m-1} = \pm 1.$$

With the sign $-$, this is impossible, since it would imply that 2 is divisible by w^2, which is absurd, because w is even.

With the sign $+$, dividing by w^2, I obtain

$$\binom{m}{2} - \binom{m}{4}w^2 + \ldots \pm mw^{m-3} = 0.$$

I show that this is also impossible.

Since w is even, then $\binom{m}{2}$ must be even. Let its 2-adic value be $v_2\left(\binom{m}{2}\right) = t \geq 1$. But

$$\binom{m}{2k}w^{2k-2} = \binom{m}{2}\binom{m-2}{2k-2} \times \frac{2}{2k(2k-1)} w^{2k-2},$$

and for every $k \geq 2$, $2^{2k-2} > k$, so $v_2(w^{2k-2}) \geq 2k-2 > v_2(k)$; hence for each $k \geq 2$, the 2-adic value of each summand $\binom{m}{2k}w^{2k-2}$ is at least $t + v_2\left(\frac{w^{2k-2}}{k}\right) \geq t+1$. Therefore the relation indicated above is not possible, and this concludes the proof of the theorem. ∎

In particular, as shown in §A2, this also gives a new proof that $x^3 - y^2 = 1$ is impossible if x, y are non-zero integers.

Here is a consequence of Lebesgue's result.

For every $n \geq 0$, let $F_n = 2^{2^n} + 1$ be the nth Fermat number. Fermat noted that F_n is a prime, for $n = 0, 1, 2, 3, 4$. Even though he believed that F_n is a prime for every n, Euler found that 641 divides F_5. No other prime Fermat number is known, and it has been conjectured that there exist infinitely many square-free Fermat numbers.

Using Lebesgue's result, a weaker assertion follows at once:

(A3.2). *A Fermat number is not a proper power.*

Proof: If $F_n = a^m$, with $m \geq 2$, then $a^m - 2^{2^n} = 1$, so $n = 0$, by the preceding result, and this is clearly impossible. ∎

4. The Result of Størmer on Fermat's Equation

In 1898, Størmer proved a very interesting result about the divisibility properties of solutions of Fermat's equation $X^2 - DY^2 = \pm 1$, where $D > 1$ is a square-free integer. He gave a much simpler proof in 1908.

An extension of his result was indicated by Mahler (1935) for the equation $X^2 - DY^2 = C$, and by Walker (1967) for the equation $EX^2 - DY^2 = 1$, with appropriate assumptions on C, D, E.

Here, I shall present only the result about the equation $X^2 - DY^2 = C$ (see (P4.10)). It includes Størmer's original result as a special case.

Let $D > 1$ be an integer which is not a square. Let C be a square-free non-zero integer such that C divides $2D$. Assume

The Result of Størmer

that the equation $X^2 - DY^2 = C$ has solutions in integers. Let (x_1, y_1) be the fundamental solution. The solutions in positive integers are (x_n, y_n) for every $n \geq 1$ (if $C > 0$) or for every odd $n \geq 1$ (if $C < 0$). The integers x_n, y_n are defined by the relation

$$|C|^{\frac{n-1}{2}}(x_n + y_n\sqrt{D}) = (x_1 + y_1\sqrt{D})^n. \tag{4.1}$$

The integer $m \geq 1$ (or $m \geq 1$, m odd when $C < 0$) satisfies the *property* of *Størmer* (with respect to (C, D), as above) whenever if a prime p divides y_m, then p divides D.

It is convenient to denote by $S_{(C,D)}$ the set of integers $m \geq 1$ (odd integers $m \geq 1$, if $C < 0$) which satisfy Størmer's property.

(A4.1). *With the above notations:*

$$S_{(C,D)} \subseteq \{1, 3\}.$$

Proof: The proof will be divided into several parts.

1°) Assume that $|C| = D$ or $2D$. Since C is square-free, so is D, and if $|C| = 2D$, then D is odd.

If $x_m^2 - Dy_m^2 = \pm D$, or $\pm 2D$, then $D | x_m^2$, and since D is square-free, then $D | x_m$. Hence

$$y_m^2 - D\left(\frac{x_m}{D}\right)^2 = \pm 1, \text{ or } \pm 2.$$

If $y_m \neq 1$, let p be a prime dividing y_m; by hypothesis on m, $p | D$. This is absurd, when the right-hand side is ± 1. In the other case, $p = 2$, so D would be even, which is again absurd.

Thus, I shall henceforth assume that

$$|C| \neq D, 2D.$$

2°) It is convenient to make some remarks.

Note that $gcd(y_n, C) = 1$ for every $n \geq 1$. Indeed, if $p \neq 2$ and $p|C$, since $C|2D$ then $p|D$, so $p|x_n^2$, hence $p^2|x_n^2$. If $p|y_n$, then $p^2|Dy_n^2$, hence $p^2|C = x_n^2 - Dy_n^2$; but this is absurd because C is square-free. Now, if $2|C$, $2|y_n$, since C is square-free, then $4 \nmid C$; from $x_n^2 - Dy_n^2 = C$ it follows that $x_n^2 \equiv 2 \pmod{4}$, which is impossible. This proves that $gcd(y_n, C) = 1$.

From relation (4.1) it follows that

$$|C|^{n-1}x_n = x_1^n + \binom{n}{2}x_1^{n-2}y_1^2 D + \binom{n}{4}x_1^{n-4}y_1^4 D^2$$
$$+ \ldots + ny_1^{n-1} D^{\frac{n-1}{2}}$$
$$|C|^{\frac{n-1}{2}}y_n = nx_1^{n-1}y_1 + \binom{n}{3}x_1^{n-3}y_1^3 D + \binom{n}{5}x_1^{n-5}y_1^5 D^2$$
$$+ \ldots + y_1^n D^{\frac{n-1}{2}}.$$
(4.2)

Thus y_1 divides y_n, because $gcd(C, y_1) = 1$. Let $y_n = y_1 z_n$, in particular, $z_1 = 1$. The above relation may then be rewritten as follows:

$$|C|^{\frac{n-1}{2}}z_n = nx_1^{n-1} + \binom{n}{3}x_1^{n-3}y_1^2 D + \binom{n}{5}x_1^{n-5}y_1^4 D^2$$
$$+ \ldots + y_1^{n-1} D^{\frac{n-1}{2}}.$$
(4.3)

Let $A = x_1^2 - C = Dy_1^2 > 1$, so $\sqrt{A} = y_1\sqrt{D}$. Also, $x_1^2 = C + Dy_1^2 \geq C + D > 0$. The last inequality is trivial when $C > 0$; if $-C > 0$ and $C + D \leq 0$, then $D \leq -C \leq 2D$ (because $C|2D$), so $|C| = D$ or $2D$, which has already been excluded.

From (4.1), it follows that

$$|C|^{\frac{n-1}{2}}(x_n + z_n\sqrt{A}) = (x_1 + \sqrt{A})^n, \qquad (4.4)$$

and this gives the following relation, analogous to (4.3):

$$|C|^{\frac{n-1}{2}} z_n = nx_1^{n-1} + \binom{n}{3}x_1^{n-3}A + \binom{n}{5}x_1^{n-5}A^2 + \ldots + A^{\frac{n-1}{2}}. \tag{4.5}$$

Now, I show that if $q|n$ then $z_q|z_n$. Indeed, let $n = q\ell$, then

$$|C|^{\frac{n-1}{2}}(x_n + z_n\sqrt{A}) = (x_1 + \sqrt{A})^n = \left(|C|^{\frac{q-1}{2}}(x_q + z_q\sqrt{A})\right)^\ell.$$

Hence

$$|C|^{\frac{n-1}{2}} z_n$$
$$= |C|^{\frac{q-1}{2}\ell} z_q \left(\ell x_q^{\ell-1} + \binom{\ell}{3}x_q^{\ell-3}z_q^2 A + \ldots + z_q^{\ell-1} A^{\frac{\ell-1}{2}}\right).$$

Since $z_q|y_q$ and $\gcd(y_q, C) = 1$, then $z_q|z_n$.

3°) After these preliminaries, let $m \geq 1$ (m odd if $C < 0$) be such that if p is any prime dividing y_m, then $p|D$.

First it will be shown that if $q|m$, q prime, then $z_q = q^r$ ($r \geq 1$).

Indeed, since $q|m$ then $z_q|z_m$. If $p|z_q$, then $p|z_m$ so $p|y_m$, hence by hypothesis, $p|D$, so $p|Dy_1^2 = A$.

From (4.3) (with $n = q$), it follows that

$$|C|^{\frac{q-1}{2}} z_q = qx_1^{q-1} + \binom{q}{3}x_1^{q-3}A + \binom{q}{5}x_1^{q-5}A^2$$
$$+ \ldots + A^{\frac{q-1}{2}}. \tag{4.6}$$

Thus p divides qx_1^{q-1}. But $x_1^2 = A + C$, $p|A$ and $p \nmid C$ (since $p|z_q$, $z_q|y_q$, $\gcd(y_q, C) = 1$), so $p \nmid x_1$, hence $p = q$. This shows that $z_q = q^r$ (where $r \geq 0$) and also that $q \nmid x_1$. As a matter of fact, $r \neq 0$. Otherwise, let $z_q = 1$. From (4.6), if $C > 0$, then

$$|C|^{\frac{n-1}{2}} > x_1^{q-1} = (x_1^2)^{\frac{q-1}{2}} = (A+C)^{\frac{q-1}{2}},$$

so $A < 0$, which is absurd. If $C < 0$, then

$$|C|^{\frac{q-1}{2}} > A^{\frac{q-1}{2}} = \left(x_1^2 - C\right)^{\frac{q-1}{2}};$$

but $x_1^2 - C \geq D > 0$, thus $-C > x_1^2 - C > -C$, and this is again absurd.

4°) Now I show that m is a power of 3. Indeed, let q be a prime dividing m. If $q > 3$, and $z_q = q^r$ with $r \geq 2$, then q^2 divides each summand of (4.6), but $q^2 \nmid qx_1^{q-1}$, which is a contradiction. Thus, if $q > 3$, then $r = 1$.

By (4.6), if $C > 0$, then

$$C^{\frac{q-1}{2}}q > qx_1^{q-1} = q(A + C)^{\frac{q-1}{2}},$$

so $C > A + C$ and $A < 0$, which is not true.

If $C < 0$, then

$$(-C)^{\frac{q-1}{2}}q > qx_1^{q-1} \geq q(C + D)^{\frac{q-1}{2}},$$

where $C + D > 0$; hence $-C > C + D$, so $D < -2C$. By hypothesis, $-C$ divides $2D$. Then $-C \leq 2D < -4C$, so $2D = -C$ or $-2C$, or $-3C$. The first two possibilities have been excluded. If $2D = -3C$, let p be any prime dividing y_m, so by hypothesis $p|D$, hence $p|3C$; but $\gcd(C, y_m) = 1$, so $p = 3$. Thus y_m is a power of 3. Since z_m divides y_m, then z_m is a power of 3. On the other hand, from $q|m$ it follows that $z_q|z_m$, so z_q is a power of 3. But $z_q = q > 3$, and this is absurd.

Now assume that $2|m$ (which is possible only when $C > 0$). Then $z_2 = 2^r$ (with $r \geq 1$). But by (7.6), $\sqrt{C}z_2 = 2x_1$. This implies that $C = 1$, $2^{r-1} = x_1$. Since $z_2|z_m$ and $2|z_m$, by the hypothesis on m, $2|A = x_1^2 - 1$. Therefore $2 \nmid x_1$, so $r = 1$, $x_1 = 1$, and finally, $A = 0$, which is absurd.

It follows that q must be equal to 3, and I have shown that necessarily m is a power of 3: $m = 3^t$. It remains to prove that $t = 0$ or 1.

The Result of Størmer

In order to continue the proof, the following fact is needed:

5°). Assume that $3n$ has the property of Størmer. Then n also has Størmer's property.
Indeed,

$$|C|^{\frac{3n-1}{2}}(x_{3n} + y_{3n}\sqrt{D})$$
$$= (x_1 + y_1\sqrt{D})^{3n} = \left(|C|^{\frac{n-1}{2}}(x_n + y_n\sqrt{D})\right)^3,$$

hence
$$|C|(x_{3n} + y_{3n}\sqrt{D}) = (x_n + y_n\sqrt{D})^3.$$

Thus
$$\begin{cases} |C|x_{3n} = x_n^3 + 3x_n y_n^2 D \\ |C|y_{3n} = 3x_n^2 y_n + y_n^3 D = y_n(3x_n^2 + y_n^2 D). \end{cases} \quad (4.7)$$

Now, if $p|y_n$, then by the above relation $p|y_{3n}|C|$. But $\gcd(y_n, C) = 1$, so $p \nmid C$, hence $p|y_{3n}$. By the hypothesis, $p|D$.
It is convenient to treat separately the equation

$$X^2 - 3Y^2 = -2. \quad (4.8)$$

6°). For the above equation, m has Størmer's property if and only if $m = 1$ or 3.
If m has Størmer's property, it was seen that $m = 3^t$, with $t \geq 1$. If $t \geq 2$, then by (5°), $m = 3^2$ has Størmer's property.
By direct calculation using relations (4.7), $x_1 = 1$, $y_1 = 1$, $x_3 = 5$, $y_3 = 3$ and $x_9 = 530$, $y_9 = 153 = 9 \times 17$. It is seen that $m = 1, 3$ have Størmer's property, but $m = 9$ doesn't have it any more.
From now on, it will be assumed that $(D, C) \neq (3, -2)$.

7°) Assume that $3n$ has Størmer's property. Then:

i) $3x_n^2 + Dy_n^2 = 3^s|C|$, with $s \geq 2$.

ii) $3|D$ but $3^2 \nmid D$.
iii) $3 \nmid y_n$.

Indeed:

i) First note that C divides $3x_n^2 + y_n^2 D$: if p is any prime dividing C, then $p \nmid y_n$, so $p | 3x_n^2 + y_n^2 D$, as follows from (4.7). Since C is square-free, then C divides $3x_n^2 + y_n^2 D$.

Now, let p divide $\frac{3x_n^2 + Dy_n^2}{|C|}$, so $p|y_{3n}$, hence $p|D$, so $p|3x_n^2$, and also $p|Dy_n^2 = x_n^2 - C$. If $p|x_n$ then $p|C$, so p^2 divides $\frac{3x_n^2 + Dy_n^2}{|C|} \times |C| = 3x_n^2 + Dy_n^2$, hence $p^2|Dy_n^2$ and finally, $p^2|x_n^2 - Dy_n^2 = C$, which is absurd. So $p \nmid x_n$ and therefore $p = 3$.

If $3|C| = 3x_n^2 + Dy_n^2 = 3(Dy_n^2 + C) + Dy_n^2 = 4Dy_n^2 + 3C$, and if $C > 0$, then $4Dy_n^2 = 0$, which is impossible. If $C < 0$, then $4Dy_n^2 = -6C$, so $2Dy_n^2 = -3C$; but $C|2D$, hence $\left(\frac{2D}{-C}\right)y_n^2 = 3$. This is possible only if $y_n^2 = 1$, so $y_n = 1$, $n = 1$ and $2D = -3C$. Therefore $x_1^2 - D = C$, $2x_1^2 - 2D = 2C$ and $2x_1^2 = -C$. This implies that $x_1 = 1$, $C = -2$, $D = 3$, a case which has been excluded.

ii) By (i), $3|Dy_n^2$, so either $3|D$ or $3|y_n$, and by (5°), $3|D$.

If $3^2|D$, then by (i), 3^2 divides $3^s|C| - Dy_n^2 = 3x_n^2$, so $3|x_n$ and therefore $3^2|x_n^2 - Dy_n^2 = C$, which is an absurdity.

iii) If $3|y_n$, then $3^2|y_n^2$ and by (i), $3^2|3x_n^2$. So $3|x_n^2$, hence $3^2|x_n^2 - Dy_n^2 = C$, again an absurdity.

8°) Now I shall show that 3 has Størmer's property if and only if the following conditions are satisfied:

i) 1 has Størmer's property.
ii) $3|D$ and there exists $u \geq 1$ such that

$$\frac{D}{3}y_1^2 = \frac{(3^u \pm 1)|C|}{4}. \qquad (4.9)$$

Indeed, assume that 3 has Størmer's property. By (5°), 1 has Størmer's property and by (7°) part (ii), $3|D$. Also

$$|C|y_3 = y_1(3x_1^2 + Dy_1^2), \qquad (4.10)$$

so by (7°) part (i), there exists $s \geq 2$ such that
$$3^s|C| = 3x_1^2 + Dy_1^2 = 3(Dy_1^2 + C) + Dy_1^2 = 4Dy_1^2 + 3C.$$
By (7°) part (ii),
$$4\left(\frac{D}{3}\right)y_1^2 + C = 3^{s-1}|C|,$$
hence
$$\frac{D}{3}y_1^2 = \frac{(3^{s-1} \mp 1)|C|}{4},$$
and so (ii) holds with $u = s - 1$.

Conversely, from the above relation, it follows that
$$3^s|C| = 4Dy_1^2 + 3C = 3x_1^2 + Dy_1^2.$$
By (4.10),
$$|C|y_3 = y_1 3^s |C|,$$
hence
$$y_3 = 3^s y_1.$$

If $p|y_3$, then either $p = 3$, so $p|D$, or $p \neq 3$, so $p|y_1$, and by hypothesis $p|D$. Thus 3 has Størmer's property.

9°). Now the proof will be completed. Assume that $m = 3^t$, with $t \geq 2$, has Størmer's property. By (7°) part (iii), $3 \nmid y_3$. By (7°) part (ii), $3|D$. Then, by (4.10), $3|y_1(3x_1^2 + y_1^2 D) = |C|y_3$. Therefore, $3|C$. By (8°) and (4.9), 3 divides $\frac{D}{3}y_1^2$. Since $\gcd(y_1, C) = 1$, then $3 \nmid y_1$, hence $3^2|D$, and this contradicts (7°) part (ii).

This shows that $m = 1$ or 3, concluding the proof. ■

As a corollary, here is Størmer's original result, which concerns the equations
$$X^2 - DY^2 = \pm 1.$$

(A4.2). *If $\epsilon = \pm 1$, then $S_{(\epsilon,D)} \subseteq \{1\}$. In other words, if $m > 1$, there exists a prime p such that $p|y_m$, but $p \nmid D$.*

Proof: In virtue of (A4.1), it suffices to show that $m = 3$ does not have Størmer's property.

In the proof of (A4.1), part (8°), it was shown that if 3 satisfies Størmer's property, then $3|D$ and $\frac{D}{3}y_1^2 = \frac{(3^u \pm 1)}{4}$ (with $u \geq 1$). Hence $4(x_1^2 \pm 1) = 3(3^u \pm 1)$, so $3^{u+1} = 4x_1^2 \pm 1$.

With the lower sign, it follows that

$$\begin{cases} 2X^1 + 1 = 3^r \\ 2X^1 - 1 = 3^s, \end{cases}$$

where $0 \leq s \leq r$ and $r + s = u + 1 \geq 2$. Then $2 = 3^s(3^{r-s} - 1)$ and so $s = 0$, $r = 1$ and $r + s = 1 \geq 2$, which is absurd.

With the upper sign, $3|4x_1^2 + 1$; but $4x_1^2 + 1 \equiv 1$ or $2 \pmod{3}$, and this is a contradiction. ∎

The result (A4.2) of Størmer will be used in an essential way in the study of the equation $X^2 - Y^n = 1$ (see §A6). It will also have interesting applications in Part C. As for (A4.1), it will be required in §A8.

In 1944, Skolem extended Størmer's result; a further extension was given by Nagell in 1955, as I describe now. Let C, D be as above and $S'_{(C,D)} = \{m \geq 1 | m \text{ odd}\} \cap S_{(C,D)}$. Thus $S'_{(C,D)} = S_{(C,D)}$ when $C < 0$. Also, by (A4.1), $S'_{(C,D)} \subseteq \{1, 3\}$.

(A4.3). *With the above hypothesis:*

1) *If $\epsilon = +1$ or -1 and $D > 5$, then $S'_{(4\epsilon,D)} \subseteq \{1\}$.*

2) *If $D \equiv 5 \pmod{8}$, then $S'_{(4\epsilon,D)} \subseteq \{1\}$.*

The proof is omitted.

Mahler called *singular* any pair (C, D) (satisfying the conditions indicated) such that $m = 1$ and $m = 3$ have the property of Størmer. Otherwise, (C, D) is called a *regular* pair.

Thus, by Størmer's result (A4.2), the pairs $(1, D), (-1, D)$ are regular. Mahler also showed that $(2, D)$ is regular, while $(-2, D)$ is singular exactly when $D = 3, 6, 123$.

In the same paper, Mahler proved that for every squarefree integer $C \neq 0$, there exist only finitely many integers $D > 0$, which are not squares, such that C divides $2D$ and (C, D) is singular.

These results will not be needed in the sequel.

5. The Attempts to Solve $X^2 - Y^n = 1$

I have already indicated that it took about 120 years after Catalan's conjecture was formulated to solve the equation $X^2 - Y^n = 1$, with $n > 3$. There were many partial results, false routes, until Chao Ko, in two papers of 1960, 1964, settled the question. Later, in 1976, a quite simple proof of the nonexistence of solutions was given by Chein.

In 1657, Frénicle de Bessy solved a problem which had been proposed by Fermat. His manuscript was discovered by Hofmann in 1943. The same result is stated without proof by Catalan (1885).

(A5.1). *If p is an odd prime and $n \geq 2$, the equation $X^2 - 1 = p^n$ has no integer solution. If $n > 3$, the equation $X^2 - 1 = 2^n$ has no integer solution.*

Proof: If x is an integer such that $p^n = x^2 - 1 = (x+1)(x-1)$ and $p \neq 2$, then $gcd(x+1, x-1) = 1$ and $x+1, x-1$ are powers of p. Thus $x - 1 = 1$, $x + 1 = p^n$, hence $p^n = 3$. Since $n \geq 2$, this is impossible.

Similarly, if $2^n = x^2 - 1 = (x+1)(x-1)$, then $gcd(x+1, x-1) = 2$, so $x - 1 = 2$, $x + 1 = 2^{n-1}$. Hence $x = 3, n = 3$, against the hypothesis. ∎

In 1932, Selberg dealt with the equation $X^4 - Y^n = 1$, where $n \geq 2$. His proof required the following result, proved by Størmer in 1899:

> If n is odd, x is a positive integer and $2x^n - 1$ is a
> square, then $x = 1$.

I shall postpone the proof of this proposition until §A11.
Here is Selberg's result:

(A5.2). *The equation $X^4 - Y^n = 1$ (with $n \geq 2$) has no solution in positive integers.*

Proof: Assume that x, y are positive integers such that $x^4 - y^n = 1$. It is clear that n cannot be even.

    Case 1. If y is odd, since $y^n = x^4 - 1 = (x^2 - 1)(x^2 + 1)$, then x is even, $\gcd(x^2 - 1, x^2 + 1) = 1$ and so

$$\begin{cases} x^2 + 1 = a^n \\ x^2 - 1 = b^n \end{cases}$$

with a, b odd, $1 \leq b < a$. Then

$$2 = a^n - b^n = (a - b)(a^{n-1} + a^{n-2}b + \ldots + ab^{n-2} + b^{n-1}).$$

Since $a - b$ is even, then $a - b = 2$ and

$$a^{n-1} + a^{n-2}b + \ldots + ab^{n-2} + b^{n-1} = 1,$$

hence $n = 1$, against the hypothesis.

    Case 2. If y is even, since $y^n = x^4 - 1 = (x^2 + 1)(x^2 - 1)$, then x is odd, $\gcd(x^2 + 1, x^2 - 1) = 2$. Therefore, since $x^2 \equiv 1 \pmod 4$,

$$\begin{cases} x^2 + 1 = 2a^n \\ x^2 - 1 = 2^{n-1}b^n \end{cases}$$

with a odd, $\gcd(a, b) = 1$. Since n is odd, by the above quoted result of Størmer, the only possible positive integers satisfying $x^2 + 1 = 2a^n$ are $x = 1$, $a = 1$, so $y = 0$, which is a contradiction. ∎

Attempts to Solve $X^2 - Y^n = 1$

Besides the result of Selberg, showing that $y^n + 1$ is never a fourth power, I now list several partial results obtained in attempts to establish that $y^n + 1$ is never a square.

In 1921, Nagell showed that if q is an odd prime, and if x, y are positive integers satisfying $x^2 - y^q = 1$, then y is even and q divides x. I shall give the proof later (see (A6.1)).

In 1934, Nagell proved that if a solution exists, then $q \equiv 1 \pmod{8}$. He also noted that, according to a general theorem of Thue (1908), for each prime $q \geq 3$, this equation has at most finitely many solutions in integers.

Obláth (1940, 1941) used methods of Vandiver, Lubelski and Nagell to obtain various partial results about the equation $x^2 - y^q = 1$, where q is a prime, $q \geq 3$. For example, he showed that if the equation has solution in positive integers, then $2^{q-1} \equiv 1 \pmod{q^2}$ and $3^{q-1} \equiv 1 \pmod{q^2}$. Obláth also used Thue's theorem to show that the equation has at most one solution in positive integers.

The above congruences are very restrictive. For example, Lehmer used the congruence

$$\frac{2^{q-1} - 1}{q} \equiv 1 + \frac{1}{3} + \frac{1}{5} + \ldots + \frac{1}{q-2} \pmod{q}$$

to show that, with the exceptions of $q = 1093$ and 3511, no prime $q < 6 \times 10^9$ satisfies $2^{q-1} \equiv 1 \pmod{q^2}$. Next, he noted that 1093 and 3511 do not satisfy the congruence $3^{q-1} \equiv 1 \pmod{q^2}$.

Inkeri and Hyyrö showed in 1961 that if $x^2 - y^q = 1$, then q^2 divides x and q^3 divides $y + 1$. Moreover, in 1951 and 1954, they improved the estimates of Obláth for x, y, showing that

$$x > 2^{q(q-2)}, \quad y > 4^{q-2},$$

and since $q > 6 \times 10^9$, it follows that $x > 10^{9 \times 10^{18}}$, $y > 10^{3 \times 10^9}$.

However, all of these results were superseded when Chao Ko established that $X^2 - Y^q = 1$ has no solution in positive integers. This is an instance when, with sharper insight but still with elementary methods, better results were obtained than those previously deduced using rather sophisticated theories.

Chein simplified the proof of Chao Ko's theorem, using Nagell's result of 1921 quoted above. This, in turn, required the result of Størmer (A4.2) on divisibility properties of the solutions of Fermat's equation $X^2 - DY^2 = 1$.

6. The Equation $X^2 - Y^n = 1$, $n \geq 3$

I shall now indicate Chein's proof of Chao Ko's theorem. It is based on the following result of Nagell (proved in 1921 and again in 1934). It should be noted that a proof published by Obláth (1941) contains a gap, as pointed out by Inkeri and Hyyrö in 1961: namely, the first relations in page 129 cannot be justified.

(A6.1). *If x, y are positive integers, $q \geq 3$ a prime number, and $x^2 - y^q = 1$, then $2|y$ and $q|x$.*

Proof: First I show that $2|y$. Indeed, if y is odd then x is even and $gcd(x+1, x-1) = 1$. Since $y^q = (x+1)(x-1)$ then there exist integers a, b, with

$$\begin{cases} x + 1 = a^q \\ x - 1 = b^q \end{cases}$$

with $a > b > 0$, a and b odd. Then $2 = a^q - b^q = (a-b)\frac{a^q - b^q}{a-b}$, so $a - b = 1$ or 2; hence $a - b = 2$ (since a, b are odd) and

$$1 = \frac{a^q - b^q}{a - b} = a^{q-1} + a^{q-2}b + \ldots + b^{q-1},$$

which is impossible.

Now I show that $q|x$. Indeed, if $q \nmid x$ from

$$x^2 = y^q + 1 = (y+1)\frac{y^q + 1}{y + 1},$$

The Equation $X^2 - Y^n = 1$, $n \geq 3$

by (P1.2), $\gcd(y+1, \frac{y^q+1}{y+1}) = 1$, so there exist integers $c > 1$, $d > 0$ such that

$$\begin{cases} y + 1 = c^2 \\ \frac{y^q+1}{y+1} = d^2 \end{cases}$$

Hence $x^2 - (c^2-1)[(c^2-1)^{\frac{q-1}{2}}]^2 = 1$; that is, $(x, (c^2-1)^{\frac{q-1}{2}})$ is a solution of $X^2 - (c^2-1)Y^2 = 1$. The fundamental solution is $(c, 1)$, because $c^2 - (c^2-1) = 1$. By Størmer's result (A4.2), $(c^2-1)^{\frac{q-1}{2}} = 1$ which is impossible. ∎

I am now ready to give Chein's elegant proof (1976) of Chao Ko's theorem (see 1960, 1964).

(A6.2). *The equation $X^2 - Y^n = 1$ with $n > 3$ has no solution in positive integers.*

Proof: If n is even, there is no solution. Since the only solution of $X^2 - Y^3 = 1$ is $x = 3, y = 2$, then there is no solution when n is a power of 3, $n > 3$. Therefore, it suffices to prove that if $q > 3$ is a prime, the equation $X^2 - Y^q = 1$ has no solution in positive integers. Suppose that x, y are positive integers such that $x^2 = y^q + 1$. By Lemma (A2.1), either

(I) $\begin{cases} x + 1 = 2a^q \\ x - 1 = 2^{q-1}b^q \end{cases}$ or (II) $\begin{cases} x + 1 = 2^{q-1}b^q \\ x - 1 = 2a^q, \end{cases}$

with a odd, $a, b > 0$, $\gcd(a, b) = 1$ and $y = 2ab$. Note that in the first case $a > b$.

Subtracting and dividing by 2,

$$a^q - 2^{q-2}b^q = \pm 1$$

according to the case. Then

$$(a^2)^q \mp (2b)^q = (a^q \mp 2)^2 = \left(\frac{x \mp 3}{2}\right)^2.$$

By (A6.1) $q|x$; since $q \neq 3$, then q does not divide $\frac{x \mp 3}{2}$, hence by (P1.2),
$$\gcd\left(a^2 \mp 2b, \frac{(a^2)^q \mp (2b)^q}{a^2 \mp 2b}\right) = 1.$$
Hence $a^2 \mp 2b = h^2$, where h divides $\frac{x \mp 3}{2}$; note that h is odd, hence 4 divides
$$a^2 - h^2 = \pm 2b,$$
so b is even. Then
$$(ha)^2 + b^2 = (a^2 \mp b)^2.$$

The integers $ha, b, a^2 \mp b$ are positive and relatively prime. They constitute a primitive solution of the Pythagorean equation $X^2 + Y^2 = Z^2$; hence there exist integers c, d, with $c, d \geq 1$ such that
$$\begin{cases} ha = c^2 - d^2 \\ b = 2cd \\ a^2 \mp b = c^2 + d^2. \end{cases}$$
Then $(c \pm d)^2 = (a^2 \mp b) \pm b = a^2$. In the first case
$$b - a = 2cd - (c + d) = (c-1)(d-1) + (cd - 1) > 0,$$
that is, $a < b$. However $a^q = 2^{q-2}b^q + 1 > b^q$, so $a > b$, a contradiction. In the second case,
$$b - a = 2cd - (c - d) = c(2d - 1) + d > 0.$$

So $a < b$. However, $a^q = 2^{q-2}b^q - 1 > b^q$, hence $a > b$, which is impossible. ∎

The following result was indicated without proof by Catalan (1885); a proof was published by Moret-Blanc in 1876.

The Equation $X^2 - Y^n = 1$, $n \geq 3$

(A6.3). *If $x, y \geq 2$ and $x^y - y^x = 1$ then $x = 3, y = 2$.*

Proof: First note that x, y are not both even. Also, x, y are not both odd, otherwise $x^y - y^x$ would be even.

If x is even, $x = 2u$, then $x^y - (y^u)^2 = 1$, contrary to Lebesque's result (A3.1). If y is even, $y = 2v$, then $(x^v)^2 - y^x = 1$, so by Chao Ko's result (A6.2), $x = 3$ and by (A2.2), $3^v = 3$, $v = 1, y = 2$. ∎

This proposition may also be proved without appealing to any previous result, as it was communicated to me by G. Skandalis (1982); another more complicated proof was published by Rotkiewicz (1960).

Skandalis' proof:
It suffices to prove that $|x^y - y^x| \neq 1$ for $x > y$, $(x, y) \neq (3, 2)$. It is trivial when $(x, y) = (4, 3)$. Let $x \geq 5$, $y = 2$, then

$$2^x - x^2 = (1+1)^x - x^2 \geq 2\left[1 + \binom{x}{1} + \binom{x}{2}\right] - x^2 = 2 + x > 1.$$

Now, let $z = x - y \geq 1$ and $y \geq 3$. Then

$$\frac{y^x}{x^y} = \frac{y^z}{\left(1 + \frac{z}{y}\right)^y} > \left(\frac{y}{e}\right)^z \geq \frac{3}{e} > 1 + \frac{1}{x^y},$$

because

$$\frac{1}{x^y} \leq \frac{1}{4^3} = \frac{1}{64} \quad \text{and} \quad \frac{3}{e} > \frac{3}{2.8} = 1 + \frac{1}{14};$$

therefore $|x^y - y^x| > 1$ and there is no solution with $y \geq 3$. ∎

7. The Equations $X^3 - Y^n = 1$ and $X^m - Y^3 = 1$, with $m, n \geq 3$

The equations $X^3 - Y^n = 1$ ($n \geq 3$) and $X^m - Y^3 = 1$ ($m \geq 3$) were studied by Nagell. To begin, he considered the equations

$$X^2 + X + 1 = Y^m \tag{7.1}$$

and

$$X^2 + X + 1 = 3Y^m, \tag{7.2}$$

where $m \geq 2$, and proved the following result (1921); when $m = 3$ the proof was completed by Ljunggren (1942, 1943). See also the paper by Estes, Guralnick, Schacher and Straus (1985), with a somewhat different proof.

(A7.1). *The solutions in integers x, y of the equation*

$$X^2 + X + 1 = Y^m$$

are the following:

a) *If m is even:* $(0, \pm 1)$, $(-1, \pm 1)$.
b) *If m is odd, $m \neq 3$:* $(0, 1), (-1, 1)$.
c) *If $m = 3$:* $(0, 1)$, $(-1, 1)$, $(18, 7)$, $(-19, 7)$.

Proof: To begin, note that if q is a prime dividing m with $m = qm'$, and if (x, y) is a solution of the equation $X^2 + X + 1 = Y^m$, then $(x, y^{m'})$ is a solution of the equation $X^2 + X + 1 = Y^q$. Thus, it suffices to determine the solutions of the latter equation.

Next, observe that if x, y are such that $x^2 + x + 1 = y^q$, then $3 \nmid y$. Indeed, $x^2 + x + 1 \not\equiv 0 \pmod{9}$, for every integer x; so if $3 | y$ then 9 divides $y^q = x^2 + x + 1$, which is a contradiction. It follows that $x \not\equiv 1 \pmod{3}$.

Note also that if $x^2 + x + 1 = y^q$, then

$$(-x-1)^2 + (-x-1) + 1 = x^2 + x + 1 = y^q.$$

$X^3 - Y^n = 1$ and $X^m - Y^3 = 1$

So, solutions appear in pairs (x, y), $(-x - 1, y)$. Since one of the integers $x - x - 1$ is congruent to -1 modulo 3, it suffices to determine the solutions (x, y) with $x \equiv -1 \pmod{3}$.

Let

$$\omega = \frac{-1 + \sqrt{-3}}{2}, \quad \omega^2 = \frac{-1 - \sqrt{-3}}{2}$$

be the primitive cubic roots of 1. Let $K = \mathbb{Q}(\omega)$ and $A = \mathbb{Z}[\omega]$.

Some basic facts about the arithmetic of the field K were recalled in §P2, before Lemma P3.1. Here, the following specific facts will be needed. The units of K are $\pm 1, \pm \omega, \pm \omega^2$, hence of the form $(-\omega)^s$, with $0 \le s \le 5$. The prime element $\lambda = 1 - \omega = \sqrt{-3}\omega^2$ is associated to $\sqrt{-3}$.

If $q \equiv \pm 1 \pmod{6}$ and $\alpha = \epsilon \tau^q$, where ϵ is a unit and $\tau \in \mathbb{Z}[\omega]$, then there exists $\theta \in \mathbb{Z}[\omega]$ such that $\alpha = \theta^q$. Indeed, if $q \equiv 1 \pmod{6}$, then $\alpha = \epsilon \tau^q = \epsilon^9 \tau^9 = (\epsilon\tau)^9$. If $q \equiv -1 \pmod{6}$, then $\alpha = \epsilon \tau^9 = \epsilon^{-9}\tau^9 = (\epsilon^{-1}\tau)^9$.

Let $x^2 + x + 1 = (x - \omega)(x - \omega^2)$ and $\alpha = x - \omega$. The conjugate of α is $\alpha' = x - \omega^2$ and $\alpha' - \alpha = \omega - \omega^2 = \sqrt{-3}$.

Now, I show that $gcd(\alpha, \alpha')$ is a unit. Indeed, if $\gamma \in \mathbb{Z}[\omega]$ and $\gamma|\alpha, \gamma|\alpha'$, then γ divides the difference $\sqrt{-3}$. So γ is associated with $\sqrt{-3}$ or γ is a unit. In the first case, $-3 \sim \gamma^2$ divides $\alpha\alpha' = x^2 + x + 1 = y^q$, so $3|y$, a contradiction. Thus γ is a unit.

From $y^q = \alpha\alpha'$ and $gcd(\alpha, \alpha') = \gamma \sim 1$, it follows that $\alpha = \epsilon\tau^q$, where ϵ is a unit of $\mathbb{Q}(\omega)$ and $\tau = a + b\omega \in \mathbb{Z}[\omega]$, with $a, b \in \mathbb{Z}$. Then $\alpha' = \epsilon'\tau'^q$ where ϵ' is the conjugate of ϵ, so $\epsilon' = \epsilon^{-1}$, and $\tau' = a + b\omega^2$. Then $y^q = \alpha\alpha' = (\tau\tau')^q$ and hence $y = \tau\tau' = a^2 - ab + b^2$. Since $gcd(\alpha, \alpha') \sim 1$, then $gcd(a, b) = 1$.

It is also possible to assume, without loss of generality, that $a \equiv 1 \pmod{3}$ and $b \equiv 0 \pmod{3}$. For this purpose, it suffices to replace τ by one of the elements $\pm\tau, \pm\omega\tau, \pm\omega^2\tau$.

Indeed,
$$\tau = a + b\omega$$
$$-\tau = -a - b\omega$$
$$\omega\tau = -b + (a-b)\omega$$
$$-\omega\tau = b + (b-a)\omega$$
$$\omega^2\tau = (b-a) - a\omega$$
$$-\omega^2\tau = (a-b) + a\omega.$$

Noting that $3 \nmid y$, it is not possible to have
$$\begin{cases} a \equiv 0 \pmod{3} \\ b \equiv 0 \pmod{3} \end{cases}$$
$$\begin{cases} a \equiv 1 \pmod{3} \\ b \equiv -1 \pmod{3} \end{cases}$$
$$\begin{cases} a \equiv -1 \pmod{3} \\ b \equiv 1 \pmod{3}. \end{cases}$$

A simple inspection shows that one of the elements $\pm\tau$, $\pm\omega\tau, \pm\omega^2\tau$ satisfies the required condition.

Note now that if $a + b\omega$ and $a' + b'\omega$ are such that $a \equiv a' \equiv 1 \pmod{3}$, $b \equiv b' \equiv 0 \pmod{3}$, then the product
$$c + d\omega = (a + b\omega)(a' + b'\omega) = (aa' - bb') + (ab' + a'b - bb')\omega$$
is such that
$$c \equiv aa' - bb' \equiv 1 \pmod{3},$$
and
$$d \equiv ab' + a'b - bb' \equiv 0 \pmod{3}.$$

Writing $\tau^q = c + d\omega$, then $c \equiv 1 \pmod{3}$, $d \equiv 0 \pmod{3}$. Since the unit ϵ satisfies
$$x - \omega = \epsilon\tau^q,$$

where $x \equiv -1 \pmod 3$, then necessarily $\epsilon = \omega^2 = 1 - \omega$.

Summarizing, $x - \omega = \omega^2(a+b\omega)^q$ with $a \equiv 1 \pmod 3$ and $b \equiv 0 \pmod 3$. Taking the conjugate, $x - \omega^2 = \omega(a + b\omega^2)^q$. From these two relations, eliminating x,

$$1 - \omega = (a + b\omega^2)^q - \omega(a + b\omega)^q.$$

I examine the various cases where $q = 2, 3, q > 3$.

1°) Let $q = 2$. If $x^2 + x + 1 = y^2$, then

$$(2x+1)^2 - (2y)^2 = -3,$$

and so

$$(2x + 1 - 2y)(2x + 1 + 2y) = -3.$$

Therefore, either

$$\begin{cases} 2x + 1 - 2y = \pm 1 \\ 2x + 1 + 2y = \mp 3 \end{cases}$$

or

$$\begin{cases} 2x + 1 - 2y = \mp 3 \\ 2x + 1 + 2y = \pm 1. \end{cases}$$

These relations lead at once to the solutions

$$(x, y) = (0, \pm 1), \ (-1, \pm 1).$$

2°) Let $q = 3$. If $y^3 = x^2 + x + 1 = (x - \omega)(x - \omega^2)$, as seen above,

$$1 - \omega = (a + b\omega^2)^3 - \omega(a + b\omega)^3$$
$$= a^3 + 3ab\omega^2 + 3ab\omega + b^3 - a^3\omega - 3a^2b\omega^2 - 3ab^2 - b\omega;$$

therefore,

$$1 = a^3 - 3ab^2 + b^3.$$

Ljunggren showed in 1942, after a careful analysis, that the only solutions (a, b) of the above equation are $(1, 0)$, $(0, 1)$, $(-1, -1)$, $(2, -1)$, $(1, 3)$, $(-3, -2)$; see (§A15).

Replacing these values of a, b in the equation $x - \omega = \omega^2(a + b\omega)^3$ or $x - \omega = -\omega(b + a\omega)^3$ yields correspondingly the values $x = -1$, $x = -19$ and $x = 0$ (which is trivial), $x = 18$. Finally, since $(-(1 + x), y)$ is also a solution of the same equation, then each of the two couples gives the other.

3°) Let q be a prime, $q > 3$.

It was seen that $x - \omega = \omega^2 \tau^q$. As indicated in the beginning of the proof, there exists $\theta = \frac{c+d\sqrt{-3}}{2} \in \mathbb{Z}[\omega]$, with $c \equiv d \pmod{2}$, such that $x - \omega = \theta^q$. Let $\theta' = \frac{c-d\sqrt{-3}}{2}$ be the conjugate of θ. Hence $\theta' - \theta = -d\sqrt{-3}$ and $\theta' - \theta$ is not a unit.

On the other hand, $\theta'^q - \theta^q = \omega - \omega^2 = \sqrt{-3}$, so $\theta' - \theta$ divides $\sqrt{-3}$, hence $\theta' - \theta \sim \sqrt{-3}$; that is $\theta' - \theta = (-\omega)^s \sqrt{-3}$, with $0 \leq s \leq 5$. Therefore $d = \pm 1$, hence c is odd, so

$$\theta = \frac{c \pm \sqrt{-3}}{2} = \frac{c \pm 1}{2} \pm \frac{-1 \pm \sqrt{-3}}{2} = z \pm \omega$$

where $z = \frac{c \pm 1}{2}$ is an integer.

It follows that $x - \omega = (z - \omega)^q$ or $x - \omega = (z + \omega)^q$; in this case,

$$(-x) + \omega = ((-z) - \omega)^q.$$

To conclude the proof, I shall determine the solutions in integers of the equations

$$T - \omega = (U - \omega)^q \qquad (7.3)$$

and

$$T + \omega = (U - \omega)^q, \qquad (7.4)$$

where $q \equiv \pm 1 \pmod 6$.

I shall show that the only solutions of the first equation are $t = 0$, $u = 0$ with $q \equiv 1 \pmod 6$, and $t = -1$, $u = -1$

with $q \equiv -1 \pmod 6$. Also, the only solutions of the second equation are $t = 1$, $u = 0$ with $q \equiv 1 \pmod 6$, $t = 0$, $u = -1$ with $q \equiv -1 \pmod 6$. These solutions lead to the only solutions $(x, y) = (0, 1), (-1, 1)$ of $X^2 + X + 1 = Y^q$.

The following special cases will be considered in turn: $u = 0$, $|u| = 1$ and finally $|u| \geq 2$.

Let $u = 0$ and $t \pm \omega = -\omega^q$. If $q \equiv 1 \pmod 6$, then the lower sign holds and $t = 0$. If $q \equiv -1 \pmod 6$, then $t \pm \omega = -\omega^2 = 1 + \omega$, so the upper sign holds and $t = 1$.

Now let $u = 1$ and $\lambda = 1 - \omega = \sqrt{-3}\omega^2$. If $t \pm \omega = (1-\omega)^q$, then $t \pm 1 = \lambda^q \pm \lambda = \lambda(\lambda^{q-1} \pm 1)$. So λ divides $t \pm 1$, but λ^2 does not divide $t \pm 1$. However $\lambda^2 = -3\omega$, so 3 divides $(t \pm 1)^2$, hence λ^2 divides $t \pm 1$, and this is a contradiction.

If $u = -1$ then $t \pm \omega = (-1-\omega)^q = \omega^{2q}$. If $q \equiv 1 \pmod 6$, then $t \pm \omega = -1 - \omega$, hence the lower sign holds and $t = -1$. If $q \equiv -1 \pmod 6$ then $t \pm \omega = \omega$, so the upper sign holds and $t = 0$.

From now on, I assume that $|u| \geq 2$ and I shall derive a contradiction. Since

$$t \pm \omega = (u - \omega)^q = \sum_{j=0}^{q} \binom{q}{j} u^j \omega^{q-j} (-1)^{q-j},$$

taking the conjugate,

$$t \pm \omega^2 = (u - \omega^2)^q = \sum_{j=0}^{q} \binom{q}{j} u^j \omega^{2(q-j)} (-1)^{q-j},$$

and subtracting, then

$$\pm(\omega^2 - \omega) = \sum_{j=0}^{q} \binom{q}{j} u^j (\omega^{2(q-j)} - \omega^{q-j})(-1)^{q-j}.$$

But for every $j \geq 0$, $\omega^{2(q-j)} - \omega^{q-j}$ is either 0 (when $q \equiv j \pmod 3$), or equal to $\pm(\omega^2 - \omega)$. Hence, dividing by

$\omega^2 - \omega$, it follows that

$$\pm 1 - (\pm 1) = \sum_{j=1}^{q-1} \epsilon_j \binom{q}{j} u^j,$$

where $\epsilon_j = 0$ or ± 1. The left-hand side is equal to $\delta = 0$ or ± 2.

Case 1. $q \equiv 1 \pmod{6}$.
   In this case $\omega^{2(q-1)} - \omega^{q-1} = 0$, hence $\epsilon_1 = 0$. Similarly,

$$\omega^{2(q-2)} - \omega^{q-2} = -(\omega^2 - \omega),$$

so $\epsilon_2 = 1$, and

$$\omega^{2(q-3)} - \omega^{q-3} = \omega^2 - \omega,$$

so $\epsilon_3 = 1$.
   Next, I observe that $\delta \neq \pm 2$. Indeed, if

$$\pm 2 = \sum_{j=2}^{q-1} \epsilon_j \binom{q}{j} u^j,$$

then u^2 divides 2, which is impossible, since $|u| \geq 2$. Thus $\delta = 0$, and I shall show that there exists a prime ℓ such that the ℓ-adic value

$$v_\ell\left(\binom{q}{2} u^2\right) < v_\ell\left(\binom{q}{j} u^j\right) \quad \text{for all } j \geq 3.$$

Hence the equality $0 = \sum_{j=2}^{q-1} \epsilon_j \binom{q}{j} u^j$ will also be impossible.

Since $\binom{q}{j} = \binom{q-2}{j-2}\binom{q}{2}\frac{2}{j(j-1)}$ for $j \geq 3$, it suffices to find ℓ such that $v_\ell\left(\binom{q-2}{j-2}\frac{2u^{j-2}}{j(j-1)}\right) > 0$, and for this it is enough to prove that $v_\ell(2u^{j-2}) > v_\ell(j(j-1))$, for all $j \geq 3$.

First assume that $|u| > 3$.

Let ℓ be a prime factor of u; note that if $|u|$ is a power of 3, then $\ell = 3$ and ℓ^2 divides u.

Let $e = v_\ell(2u^{j-2})$ and $f = v_\ell(j(j-1))$. Since $\gcd(j, j-1) = 1$ and $\ell^f | j(j-1)$, then $\ell^f | j$ or $\ell^f | j-1$, and in both cases, $\ell^f \le j$.

On the other hand, if $\ell \ge 5$, then $e \ge j-2$, because $|u|^{j-2} \ge \ell^{j-2} \ge 5^{j-2}$, hence $e = v_\ell(2u^{j-2}) \ge j-2$. Similarly, if $\ell = 3$, then $|u|^{j-2} \ge 3^{2(j-2)}$, or $e = v_3(2u^{j-2}) \ge 2(j-2)$. Also, if $\ell = 2$, then $e = v_2(2u^{j-2}) \ge j-1$. But, if $j \ge 3$, then $5^{j-2} \ge j$, $3^{2(j-2)} \ge j$, $2^{j-1} \ge j$. Hence $\ell^e > j \ge \ell^f$, and therefore $e \ge f$.

Now let $|u| = 3$.

If $u = -3$, then

$$0 = \sum_{j=2}^{q-1} \epsilon_j \binom{q}{j}(-3)^j.$$

Noting that $\epsilon_2 = \epsilon_3 = 1$ and dividing by 3^2, then

$$-\binom{q}{2} + \binom{q}{3}3 = \sum_{j=4}^{q-1} \epsilon_j \binom{q}{j}(-3)^{j-2},$$

and I shall show that this relation is impossible.

Let $q = 3^e h + 1$ with $3 \nmid h$, $e \ge 1$. Then the 3-adic value of the left-hand side is

$$v_3\left[\binom{q}{2}(-1 + (q-2))\right] = v_3\left(\frac{q(q-1)(q-3)}{2}\right) = e.$$

On the other hand,

$$\binom{q}{j}3^{j-2} = \binom{q}{2}\binom{q-2}{j-2}\frac{2 \times 3^{j-2}}{j(j-1)},$$

$$v_3\left(\binom{q}{2}\right) = e,$$

and if $j \geq 4$ then $3^{j-2} > j$, hence

$$v_3\left(\binom{q}{j}3^{j-2}\right) \geq v_3\left(\binom{q}{2}\right) + j - 2 - v_3(j(j-1)) > e.$$

Therefore the relation indicated above is impossible.

If $u = 3$, first assume that t satisfies the equation $T - \omega = (3-\omega)^q$. From $t-\omega = (3-\omega)^q$ it follows that $t-\omega^2 = (3-\omega^2)^q$ and multiplying, $t^2+t+1 = 13^q$. But $t-\omega \equiv -\omega^q \equiv -\omega \pmod 3$, hence $t \equiv 0 \pmod 3$. Similarly, $t - \omega = (4+\omega^2)^q \equiv \omega^2 \equiv -1 - \omega \pmod 4$, hence $t \equiv -1 \pmod 4$.

Let $t_1 = -t - 1$. Then $t_1^2 + t_1 + 1 = t^2 + t + 1 = 13^q$ and $t_1 \equiv -1 \pmod 3$, $t_1 \equiv 0 \pmod 4$, so $t_1 = 12s + 8$. Hence $(12s+8)(12s+9) = t_1(t_1+1) = 13^q - 1$.

Dividing by 12:

$$(3s+2)(4s+3) = \frac{13^q - 1}{13 - 1}.$$

If p is any prime dividing $\frac{13^q-1}{13-1}$ then by (P1.2), $p \neq 2, 3$, and by (P1.4), $p \equiv 1 \pmod q$. It follows that

$$\begin{cases} 3s + 2 \equiv 1 \pmod q \\ 4s + 3 \equiv 1 \pmod q, \end{cases}$$

so by subtracting, $s \equiv -1 \pmod q$, so $-1 \equiv 1 \pmod q$ and $q = 2$, which is a contradiction.

Finally, if $t + \omega = (3 - \omega)^q$, then taking the conjugate $t + \omega^2 = (3 - \omega^2)^q$, that is, $(t - 1) - \omega = (4 + \omega)^q$, hence $-(t-1)+\omega = (-4-\omega)^q$. But it was seen that $T+\omega = (U-\omega)^q$ has no solution (t, u), with u even, concluding the proof of this case.

<u>Case 2.</u> $q \equiv -1 \pmod 6$.

If $\delta = \pm 2$, then u divides 2, so $|u| = 2$. If $u = 2$, then

$$t \pm \omega = (2 - \omega)^q = (3 + \omega^2)^q \equiv \omega + 3q\omega^2$$
$$\equiv -3q + (1 - 3q)\omega \pmod 9.$$

Then $\pm 1 \equiv 1 - 3q \pmod{9}$, so $3q \equiv 0$ or $2 \pmod{9}$, which is impossible. If $u = -2$, then $t \pm \omega = -(2+\omega)^q$. But

$$|t \pm \omega| = \left|t \mp \frac{1}{2} \pm i\frac{\sqrt{3}}{2}\right|$$
$$= \sqrt{\left(t \mp \frac{1}{2}\right)^2 + \frac{3}{4}} = \sqrt{t^2 \mp t + 1}$$

and

$$|2 + \omega| = \left|2 - \frac{1}{2} + i\frac{\sqrt{3}}{2}\right| = \sqrt{3}.$$

Hence $t^2 + t + 1 = 3^q$, so $t = \frac{-1 \pm \sqrt{1-4(1-3)^q}}{2}$. Since t is an integer, then $4 \times 3^q - 3 = 3(4 \times 3^{q-1} - 1)$ is a square, so 3 divides $4 \times 3^{q-1} - 1$, which is impossible.

If $\delta = 0$, then $\omega^{2(q-1)} - \omega^{q-1} = \omega^2 - \omega$ so $\epsilon_1 = 1$, also $\omega^{2(q-2)} - \omega^{q-2} = 0$. Hence

$$0 = qu + \sum_{j=3}^{q-1} \epsilon_j \binom{q}{j} u^j.$$

To show that this equality is impossible, it suffices to show that there exists a prime ℓ such that $v_\ell(qu) < v_\ell\left(\binom{q}{j}u^j\right)$, that is, $v_\ell\left(\binom{q-1}{j-1}\frac{u^{j-1}}{j}\right) \geq 1$ for $j \geq 3$.

Let ℓ be a prime dividing u, then $\ell^{j-1} | u^{j-1}$; since

$$\ell^{j-1} \geq 2^{j-1} > j \text{ for } j \geq 3,$$

then $v_\ell(u^{j-1}) \geq j - 1 > v_\ell(j)$, proving the assertion.
This concludes the proof of the proposition. ∎

Nagell proved in the same paper (1921):

(A7.2). *The solutions in integers of the equation* $X^2 + X + 1 = 3Y^m$ *are the following:*

a) *If $m = 2$ the solutions are (x_n, y_n), where*

$$x_n = \pm\frac{\sqrt{3}}{4}\left[(2+\sqrt{3})^{2n+1} - (2-\sqrt{3})^{2n+1}\right] - \frac{1}{2},$$

$$y_n = \pm\frac{1}{4}\left[(2+\sqrt{3})^{2n+1} + (2-\sqrt{3})^{2n+1}\right]$$

for $n = 0, 1, 2, \ldots$; in particular,

b) *If $m \neq 2$ the equation has only the solutions $(1, \pm 1)$, $(-2, \pm 1)$ when m is even and $(1, 1), (-2, 1)$ when m is odd.*

Proof: a) Let $m = 2$ and let x, y be integers such that $x^2 + x + 1 = 3y^2$; x, y are non-zero and I may assume $y \geq 1$.

First, suppose that $x \geq 1$. Since $x \equiv 1 \pmod{3}$, I may write $2x + 1 = 3z$, hence $z \geq 1$ is odd and $9z^2 + 3 = 12y^2$, so $(2y)^2 - 3z^2 = 1$. Thus $2y \pm z\sqrt{3}$ are units of the field $\mathbb{Q}(\sqrt{3})$. A fundamental unit of this field is $2 + \sqrt{3}$, and the only roots of unity are ± 1.

Hence $2y + z\sqrt{3} = (2+\sqrt{3})^h$, for some integer h, and since $2y + z\sqrt{3} \geq 2 + \sqrt{3} > 1$, $h \geq 1$. Its conjugate is $2y - z\sqrt{3} = (2-\sqrt{3})^h$. But $\sqrt{3} \equiv (\sqrt{3})^h \pmod 2$, so h is odd (note that $\{1, \sqrt{3}\}$ is an integral basis of the field $\mathbb{Q}(\sqrt{3})$). Hence $h = 2n+1$, $n \geq 0$, and

$$y = \frac{1}{4}\left[(2+\sqrt{3})^{2n+1} + (2-\sqrt{3})^{2n+1}\right];$$

therefore

$$x = \frac{3z-1}{2} = \frac{\sqrt{3}}{4}\left[(2+\sqrt{3})^{2n+1} - (2-\sqrt{3})^{2n+1}\right] - \frac{1}{2}.$$

If $x < 0$ then $x' = -x$ satisfies the equation $x'^2 - x' + 1 = 3y^2$ and proceeding in the same way

$$x = -\frac{\sqrt{3}}{4}\left[(2+\sqrt{3})^{2n+1} - (2-\sqrt{3})^{2n+1}\right] - \frac{1}{2}.$$

It is also clear that x, y, as above, satisfy the given equation. Also, if $n = 0$, the corresponding solutions are $(1, \pm 1)$, $(-2, \pm 1)$.

b) $m \neq 2$. Clearly, if m is even, $(1, \pm 1)$, $(-2, \pm 1)$ are solutions, and if m is odd, then $(1, 1)$, $(-2, 1)$ are solutions. I show that these are all of the solutions.

1°) If $4 | m$ and x, y satisfy $x^2 + x + 1 = 3y^4$, then $x \equiv 1 \pmod{3}$, so I may write $2x + 1 = 3z$; hence, as before, $4y^4 - 1 = 3z^2$. Thus, z is odd and from the unique factorization of integers into primes,

$$\begin{cases} 2y^2 + 1 = a^2 \\ 2y^2 - 1 = 3b^2 \end{cases} \quad \text{or} \quad \begin{cases} 2y^2 - 1 = a^2 \\ 2y^2 + 1 = 3b^2 \end{cases}$$

for some odd integers $a, b \geq 1$, with $gcd(a, b) = 1$, and 3 not dividing a.

The first case is impossible since

$$2y^2 - 1 \equiv \pm 1 \pmod{3}.$$

In the second case

$$4y^2 - a^2 = 3b^2.$$

Since $3 \nmid a$, with the same argument,

$$\begin{cases} 2y + a = 3c^2 \\ 2y - a = d^2 \end{cases} \quad \text{or} \quad \begin{cases} 2y - a = 3c^2 \\ 2y + a = d^2 \end{cases}$$

for some odd integers $c, d \geq 1$, with $gcd(c, d) = 1$ and 3 not dividing d. Hence

$$\begin{cases} 4y = 3c^2 + d^2 \\ \pm 2a = 3c^2 - d^2. \end{cases}$$

This yields
$$2\left(\frac{3c^2+d^2}{4}\right)^2 - \left(\frac{3c^2-d^2}{2}\right)^2 = 1,$$
hence
$$9c^4 - 18c^2d^2 + d^4 = -8,$$
so
$$3c^2 = 3d^2 \pm 2\sqrt{2(d^4-1)}.$$

Therefore $2(d^4-1)$ is a square, say $d^4 - 1 = 2f^2$, hence
$$(d^2+1)(d+1)(d-1) = 2f^2.$$

If $d = 1$ then $c = d = 1$, $y = 1$ and $x = 1$ or $x = -2$.

If $d \neq 1$, then $f \neq 0$; since d is odd and $\frac{d+1}{2}, \frac{d-1}{2}, \frac{d^2+1}{2}$ are non-zero pairwise relatively prime integers, then
$$\begin{cases} d+1 = 2u^2 \\ d-1 = 2v^2 \\ \frac{d^2+1}{2} = 2w^2 \end{cases}$$
for some non-zero integers u, v, w, with $gcd(u, v) = 1$. Hence
$$u^4 + v^4 = \left(\frac{d+1}{2}\right)^2 + \left(\frac{d-1}{2}\right)^2 = \frac{d^2+1}{2} = 2w^2.$$

As shown in (A4.2), $u^2 = v^2 = w^2 = 1$. This implies that $d = 1$ but also that $d = 3$, which is a contradiction.

2°) If m is a power of 3 and the given equation has a solution, then there exist integers x, y such that $x^2 + x + 1 = 3y^3$. So $\left(\frac{x+2}{3}\right)^3 - \left(\frac{x-1}{3}\right)^3 = y^3$. Again, as shown in (P3.8), one of the above integers is equal to 0, and since $y \neq 0$, $x = 1, y = 1$ or $x = -2$, $y = 1$.

3°) Finally, if $m \neq 2$, $4 \nmid m$ and m is not a power of 3, there exists a prime $q > 3$ dividing m. If the given equation has a solution, then there exist integers x, y such that $x^2 + x + 1 = 3y^q$, hence $(x - \omega)(x - \omega^2) = 3y^q$. Putting $\alpha = x - \omega$, its conjugate is $\alpha' = x - \omega^2$ and $gcd(\alpha, \alpha') = \sqrt{-3}$. Indeed, if $\gamma \in \mathbb{Z}[\omega]$ and γ divides α, α' then γ divides $\alpha - \alpha' = \sqrt{-3}$. On the other hand $\sqrt{-3}$ divides $3y^q$, hence it divides $x-\omega$ or $x-\omega^2$, and therefore it divides both factors, since $\omega - \omega^2 = \sqrt{-3}$.

Let $\beta = \frac{\alpha}{\sqrt{-3}}$, its conjugate is $\beta' = -\frac{\alpha'}{\sqrt{-3}}$, hence $\beta\beta' = y^q$ and $gcd(\beta, \beta') = 1$. Then $\beta = \epsilon\tau^q$, where $\tau \in \mathbb{Z}[\omega]$ and ϵ is a unit of $\mathbb{Q}(\omega)$. As indicated in the beginning of the proof of the preceding proposition, I may choose $\theta \in \mathbb{Z}[\omega]$, such that $\beta = \theta^q$. Let θ' be the conjugate of θ, so

$$\theta^q + \theta'^q = \beta + \beta' = \frac{\alpha - \alpha'}{\sqrt{-3}} = -1.$$

Thus $\theta + \theta'$ divides 1, and since $\theta + \theta' \in \mathbb{Z}$, $\theta + \theta' = \pm 1$. Hence $\theta = \frac{\pm 1 + b\sqrt{-3}}{2}$ with $b \in \mathbb{Z}$, b odd. Therefore

$$-1 = \theta^q + \theta'^q = \theta^q + (\pm 1 - \theta)^q.$$

In the case of the upper sign,

$$-2 = \sum_{j=1}^{q-1} \binom{q}{j}(-1)^j \theta^j,$$

so q divides 2, hence $q = 2$, which has been excluded.

In the case of the lower sign,

$$-1 = -1 - \sum_{j=1}^{q-1} \binom{q}{j}\theta^j;$$

dividing by $q\theta$:

$$-1 = \theta \sum_{j=2}^{q-1} c_j \theta^{j-2},$$

with coefficients $c_j \in \mathbb{Z}$. Hence θ is a unit.

Therefore β and also β' are units. Hence $y^q = \beta\beta' = \pm 1$. From $x^2 + x + 1 = \pm 3$, it follows at once that $x = -2$ or 1. ■

With the preceding propositions, Nagell showed:

(A7.3). *If $m > 2$, then the equations $X^3 - Y^m = \pm 1$ have no solution in non-zero integers.*

Proof: If there is a solution, I may assume without loss of generality that $m = q$ is a prime, $q > 3$. Assume that x, y are non-zero integers such that $x^3 \mp 1 = y^q$. Then $(x^2 \pm x + 1)(x \mp 1) = y^q$. By (P1.2), $\gcd\left(\frac{x^3 \mp 1}{x \mp 1}, x \mp 1\right) = 1$ or 3, so $x^2 \pm x + 1 = a^q$ or $3a^q$.

Replacing x by $-x$ (in the case of the sign $-$), then $x^2 + x + 1 = a^q$ or $3a^q$.

By the preceding propositions, this implies that $x = \pm 1$ or -2, but then $x^3 \mp 1$ is not a qth power (with $q > 3$). ■

8. The Equation $\frac{X^n-1}{X-1} = Y^m$

Nagell and Ljunggren have also considered the similar equation

$$\frac{X^n - 1}{X - 1} = Y^m \qquad (8.1)$$

where $m \geq 2$ and $n > 3$ (the case $n = 3$ has already been settled).

It is clearly sufficient to consider the equations with $m = q$, a prime number. First, I shall give the results for the equation

$$\frac{X^n - 1}{X - 1} = Y^2. \qquad (8.2)$$

The Equation $\frac{X^n-1}{X-1} = Y^m$

It is easy to see that this equation has the following solutions (x, y) with $|x| \leq 1$ and $y > 0$: $(0, 1), (-1, 0)$ if n is even, $(-1, 1)$ if n is odd and $(1, \sqrt{n})$ if n is a square.

It remains to describe the solutions (x, y) with $|x| > 1$.

(A8.1). *If x, y are integers such that $|x| > 1$, $y > 0$ and $\frac{x^n-1}{x-1} = y^2$, with $n > 3$, then either $n = 4$, $x = 7$, $y = 20$ or $n = 5$, $x = 3$, $y = 11$.*

Proof: I shall divide the proof into several parts. The proof of (1°) is due to Nagell (1921); the proof of (2°) is a simplification of Nagell's proof in the same paper. The proof of (3°) is due to Ljunggren (1943).

1°) If $4|n$ and $|x| > 1$, $y > 0$ are such that $\frac{x^n-1}{x-1} = y^2$, then $n = 4$, $x = 7$, $y = 20$.

Indeed, write $n = 2^a m$, with m odd, $a \geq 2$. Let $z = x^m$. By §P1, (1.3),

$$y^2 = \frac{x^n - 1}{x - 1} = \frac{z^{2^a} - 1}{z - 1} \cdot \frac{x^m - 1}{x - 1}$$

$$= \frac{x^m - 1}{x - 1} \cdot (z + 1)(z^2 + 1) \ldots (z^{2^{a-1}} + 1). \qquad (8.3)$$

By (P1.9), if $0 \leq j < i$, then

$$\gcd(z^{2^j} + 1, z^{2^i} + 1) = \begin{cases} 1 & \text{when } z \text{ is even} \\ 2 & \text{when } z \text{ is odd.} \end{cases}$$

By (P1.2) (vii), if $0 \leq i$, then

$$\gcd\left(\frac{x^m - 1}{x - 1}, x^{2^i m} + 1\right) = 1.$$

Thus, if z is even, each factor in (8.3) is a square, in particular $z^2 + 1$ is a square, and this is impossible. If z is odd and $a = 2$, then from (8.3)

$$\begin{cases} z + 1 = 2e^2 \\ z^2 + 1 = 2f^2, \end{cases} \text{ where } e, f \text{ are integers.}$$

By (P3.6), $x^m = z = \pm 1$ or 7, hence $m = 1$, $n = 4$ and therefore $x = \pm 1$ (which has been excluded) or $x = 7$.

Now let $a > 2$. By the same argument, there exist integers e, f, such that
$$\begin{cases} z^2 + 1 = 2e^2 \\ z^4 + 1 = 2f^2, \end{cases}$$
and as before $z^2 = 1$ or 7, so $z^2 = 1$, $m = 1$, $x = \pm 1$, which was excluded.

2°) If $n > 3$ is even and $|x| > 1$, $y > 0$ are integers such that $\frac{x^n-1}{x-1} = y^2$, then $n = 4$, $x = 7$, $y = 20$.

Indeed, in virtue of (1°), I may assume $n = 2m$, where $m \geq 2$ is odd. Now
$$y^2 = \frac{x^m - 1}{x - 1}(x^m + 1)$$

and by (P1.2) (vii) $gcd\left(\frac{x-m-1}{x-1}, x^m + 1\right) = 1$, so there exist integers t, u such that $gcd(t, u) = 1$ and
$$\begin{cases} \frac{x^m-1}{x-1} = t^2 \\ x^m + 1 = u^2. \end{cases}$$

The last relation is impossible, according to the result of Ko (see (A6.2)).

3°) If $n > 3$ is odd and $|x| > 1$, y are integers such that $\frac{x^n-1}{x-1} = y^2$, then $n = 5$, $x = 3$, $y = 11$.

First assume that $x > 1$. The relation may be rewritten as
$$(x^n - 1)(x - 1) = [(x-1)y]^2,$$
hence
$$[(x-1)y]^2 - x(x-1)(x^{\frac{n-1}{2}})^2 = -(x-1).$$

The Equation $\frac{X^n-1}{X-1} = Y^m$

Write $x - 1 = t^2 s$, where s is a square-free integer. So

$$(t^2 sy)^2 - xt^2 s(x^{\frac{n-1}{2}}) = -t^2 s,$$

and therefore

$$(tsy)^2 - xs(x^{\frac{n-1}{2}})^2 = -s.$$

At this point it is important to observe that xs is not a square; indeed, $\gcd(x, s) = 1$ and if x is a square, then $s \neq 1$, so xs is not a square.

Consider the equation $X^2 - DY^2 = C$, with $D = xs$, $C = -s$ (square-free integer), and note that C divides $2D$. This equation has the fundamental solution $(ts, 1)$, because

$$t^2[(ts)^2 - (xs)1^2] = (t^2 s)^2 - xt^2 s$$
$$= (x-1)^2 - x(x-1) = -(x-1) = -t^2 s.$$

According to (P5.9) all the solutions in positive integers are (x_m, y_m), where $m \geq 1$, m odd, and

$$s^{\frac{m-1}{2}}(x_m + y_m \sqrt{xs}) = (ts + \sqrt{xs})^m.$$

In particular, there exists $m \geq 1$, m odd, such that

$$tsy = x_m \quad \text{and} \quad x^{\frac{n-1}{2}} = y_m.$$

Note that if a prime q divides y_m, then q divides $D = xs$. By (A4.1), necessarily $m = 1$ or 3. If $m = 1$, then $x^{\frac{n-1}{2}} = 1$, which is impossible because $x > 1$. If $m = 3$, then

$$x^{\frac{n-1}{2}} = y_3 = \frac{3t^2 s^2 + xs}{s}$$
$$= 3t^2 s + x = 3(x-1) + x = 4x - 3.$$

For $n > 3$, $x.x^{\frac{n-3}{2}} = 4x - 3$, so $3 = x(4 - x^{\frac{n-3}{2}})$, and therefore $x = 1$ (excluded) or $x = 3$; so $n = 5$ and necessarily $y = 11$.

Now assume that $x < -1$ and let $z = -x > 1$; then
$$y^2 = \frac{x^n - 1}{x - 1} = \frac{z^n + 1}{z + 1}$$
and
$$[(z+1)y]^2 - z(z+1)(z^{\frac{n-1}{2}})^2 = z + 1.$$
Let $z + 1 = t^2 s$, where s is square-free. Then
$$(tsy)^2 - zs(z^{\frac{n-1}{2}})^2 = s.$$

As before, consider the equation $X^2 - DY^2 = C$, with $D = zs$, $C = s$ (square-free integer) and note that C divides $2D$. Also, zs is not a square, and this is seen as in the preceding case. The fundamental solution is $(ts, 1)$, as is easily verified.

So, by (P5.9), there exists $m \geq 1$, m odd, such that $tsy = x_m$ and $z^{\frac{n-1}{2}} = y_m$.

Since every prime dividing y_m must divide $D = zs$, then, by (A4.1), it follows that $m = 1$ or $m = 3$. If $m = 1$ then $z^{\frac{n-1}{2}} = 1$, which is impossible. If $m = 3$, then a similar calculation gives
$$z^{\frac{n-1}{2}} = y_3 = 3t^2 s + z = 3(z+1) + z = 4z + 3.$$
For $n > 3$,
$$z \cdot z^{\frac{n-3}{2}} = 4z + 3 \text{ implies } z(z^{\frac{n-3}{2}} - 4) = 3,$$
so $z = 1$ (excluded) or $z = 3$ and $z^{\frac{n-3}{2}} = 5$, an absurdity. ∎

Now I shall present the known results concerning the equation
$$\frac{X^n - 1}{X - 1} = y^q \tag{8.4}$$
where q is an odd prime and $n > 3$ (the case $n = 3$ has already been discussed).

The Equation $\frac{X^n-1}{X-1} = Y^m$

The first result establishes a relation between Equation (8.4) and the equations

$$\frac{X^p - 1}{X - 1} = Y^q \tag{8.5}$$

$$\frac{X^p - 1}{X - 1} = pY^q \tag{8.6}$$

where p is any odd prime factor of n.

(A8.2). Let $n > 3$ be an integer not divisible by 4, q an odd prime, and x, y integers satisfying (8.4) with $|x| > 1$ and $y \neq 0$. If ℓ is an odd prime dividing n, $n = 2^c \ell^a d$, $\ell \nmid d$, then there exist non-zero integers $h \geq 1$ and $t_0, t_1, \ldots, t_a$ such that

$$\begin{cases} \frac{x^h-1}{x-1} = t_0^q \\ \frac{(x^h)^\ell - 1}{x^h - 1} = t_1^q \\ \ldots \\ \frac{(x^{h\ell^{a-1}})^\ell - 1}{x^{h\ell^{a-1}} - 1} = t_a^q \end{cases} \quad \text{or} \quad \begin{cases} = t_0^q \\ = \ell t_1^q \\ \ldots \\ = \ell t_a^q. \end{cases}$$

Moreover, if ℓ is the smallest odd prime factor of n, then $h = d$.

Proof: If n is even, then $n = 2k$, with k odd, and

$$y^q = \frac{x^n - 1}{x - 1} = \frac{x^{2k} - 1}{x^k - 1} \cdot \frac{x^k - 1}{x - 1} = (x^k + 1)\frac{x^k - 1}{x - 1}.$$

By (P1.2) (vii), $gcd\left(x^k + 1, \frac{x^k-1}{x-1}\right) = 1$, hence $\frac{x^k-1}{x-1} = t^q$.

If n is odd, let $k = n$, $t = y$. Also let p be the smallest prime factor of k and assume that $k = p^a m$, with $a \geq 1$, $p \nmid m$.
By §P1 (1.3),

$$t^q = \frac{x^k - 1}{x - 1} = \frac{x^m - 1}{x - 1} \cdot \Phi_p(x^m) \cdot \Phi_{p^2}(x^m) \ldots \Phi_{p^a}(x^m). \tag{8.7}$$

Note that $p \nmid \frac{x^m-1}{x-1}$. Indeed, let e be the order of x modulo p. If $p|\frac{x^m-1}{x-1}$, then $p|x^m - 1$, so $e|m$. But, $e|p - 1$, and since p is the smallest prime factor of k, $e = 1$, so $x \equiv 1 \pmod{p}$. By (P1.2) (ii), p divides

$$\gcd\left(\frac{x^m-1}{x-1}, x-1\right) = \gcd(m, x-1),$$

so $p|m$, which is absurd.

By (P1.9)

$$\gcd(x^m - 1, \Phi_{p^i}(x^m)) = 1 \text{ or } p,$$

hence

$$\gcd\left(\frac{x^m-1}{x-1}, \Phi_{p^i}(x^m)\right) = 1.$$

Next, if $0 \leq i < j \leq a$, then by (P1.9),

$$\gcd\left(\Phi_{p^i}(x^m), \Phi_{p^j}(x^m)\right) = \begin{cases} p & \text{if } x^m \equiv 1 \pmod{p} \\ 1 & \text{if } x^m \not\equiv 1 \pmod{p}. \end{cases}$$

First let $x^m \not\equiv 1 \pmod{p}$, so $x \not\equiv 1 \pmod{p}$. Then, from (8.7) and the above considerations, there exist non-zero integers $u, v_1, \ldots, v_a$ such that

$$\begin{cases} \frac{x^m-1}{x-1} = u^q \\ \Phi_p(x^m) = \frac{(x^m)^p-1}{x^m-1} = v_1^q \\ \ldots \\ \Phi_{p^a}(x^m) = \frac{(x^{mp^{a-1}})^p-1}{x^{mp^{a-1}}-1} = v_a^q. \end{cases} \qquad (8.8)$$

If $x^m \equiv 1 \pmod{p}$, then $p|\Phi_{p^i}(x^m)$; but $p^2 \nmid \Phi_{p^i}(x^m) = \frac{x^{mp^i}-1}{x^{mp^{i-1}}-1}$, in virtue of (P1.2). Thus, if $x^m \equiv 1 \pmod{p}$, then

The Equation $\frac{X^n-1}{X-1} = Y^m$

there exist non-zero integers $u, v_1, \ldots, v_a$, such that

$$\begin{cases} \frac{x^m-1}{x-1} = u^q \\ \Phi_p(x^m) = \frac{(x^m)^p-1}{x^m-1} = pv_1^q \\ \ldots \\ \Phi_{p^a}(x^m) = \frac{(x^{mp^{a-1}})^p-1}{x^{mp^{a-1}}-1} = pv_a^q \end{cases} \quad (8.9)$$

(and here $q|a$, since the product of the right-hand sides is t^q).

If $m > 1$, the argument should be repeated with the relation $\frac{x^m-1}{x-1} = u^q$. After finitely many iterations, it may be concluded that for every odd prime factor ℓ of n, one of the relations in one of the sets in the statement must hold. ∎

The next result indicates known cases where Equation (8.4), with q odd prime, $n > 3$, has no solution. Nagell proved parts (i) and (ii) in 1920, while the remaining parts were established by Ljunggren in 1943.

(A8.3). *The equation* (8.4) *with* $n > 3$ *and* q *an odd prime has no solution in integers* x, y, *with* $|x| > 1$, $y > 0$, *in the following cases:*

i) 4 *divides* n.
ii) 3 *divides* n.
iii) $n \not\equiv -1 \pmod{6}$ *and* $q = 3$.

Proof: i) Assume that 4 divides n and that there exist integers x, y with $|x| > 1$, $y \neq 0$, such that $\frac{x^n-1}{x-1} = y^q$. Let $n = 2^a m$, with m odd, $a \geq 2$. According to the proof of (1°) of (A7.4), letting $z = x^m$,

$$y^a = \frac{x^m-1}{x-1}(z+1)(z^2+1)\ldots(z^{2^{a-1}}+1), \quad (8.10)$$

with $gcd\left(\frac{x^m-1}{x-1}, x^{2^i m}+1\right) = 1$ for $0 \le i$, and

$$gcd(z^{2^j}+1, z^{2^i}+1) = \begin{cases} 1 & \text{when } z \text{ is even} \\ 2 & \text{when } z \text{ is odd,} \end{cases}$$

for $0 \le j < i$.

If z is even, the factors in the right-hand side of (8.10) are pairwise relatively prime. Then, each factor is a qth power, so there exists an integer t such that

$$z^2 + 1 = t^q.$$

However, by Lebesgue's result (A3.1), this is impossible.

If z is odd, the same argument shows that there exists an integer t such that $z^2 + 1 = 2t^q$. However, by Størmer's results (A11.1) to be shown in §A11, this is again impossible.

ii) Let $n = 3^a m$, with $a \ge 1$, $3 \nmid m$. By (i), I may assume that $4 \nmid n$. If $|x| > 1$, $y \ne 0$ and $\frac{x^n-1}{x-1} = y^q$, by (A8.2) there exist integers $t_0, t_1, \ldots, t_a \ne 0$, such that

$$\begin{cases} \frac{x^m-1}{x-1} = t_0^q \\ \frac{(x^m)^3-1}{x^m-1} = t_1^q \\ \ldots \\ \frac{(x^{m3^{a-1}})^3-1}{x^{m3^{a-1}}-1} = t_a^q \end{cases} \quad \text{or} \quad \begin{cases} = t_0^q \\ = 3t_1^q \\ \ldots \\ = 3t_a^q. \end{cases}$$

In the first case, by (A7.1), necessarily $q = 3$, $x^m = 18$ or -19; hence $m = 1$, $x = 18$ or -19. So $n = 3^a > 3$, hence $a \ge 2$ and therefore $\frac{(x^3)^3-1}{x^3-1} = t_2^3$. Then again, $x^3 = 18$ or -19, which is absurd.

In the second case, by (A7.2), necessarily $x^m = -2$, hence $m = 1$, $x = -2$. Thus $n = 3^a > 3$, so $a \ge 2$. Again $\frac{(x^3)^3-1}{x^3-1} = 3t_2^q$, hence $x^3 = -2$, which is impossible.

The Equation $\frac{X^n-1}{X-1} = Y^m$

iii) First let $n = 2m$. Since $\frac{x^m-1}{x-1}$ is odd and $gcd(x^m + 1, x^m - 1) = 1$ or 2, then $gcd\left(\frac{x^m-1}{x-1}, x^m + 1\right) = 1$. If $|x| > 1$, $y \neq 0$ are such that $y^3 = \frac{x^{2m}-1}{x-1} = \frac{x^m-1}{x-1}(x^m + 1)$, it follows that there exists $u \neq 0$ such that $x^m + 1 = u^3$. This contradicts (A7.3).

If $n \not\equiv -1 \pmod{6}$ and if there exist x, y such that $|x| > 1$, $y \neq 0$ and $\frac{x^n-1}{x-1} = y^3$, then $3 \nmid n$, hence $n = 6t + 1$ and I may write
$$x(x^{2t})^3 - (x-1)y^3 = 1.$$

However, in 1935, Nagell showed that if $|a| > 1$ the equation
$$aU^3 - (a-1)V^3 = 1 \qquad (8.11)$$
has only the trivial solution $u = v = 1$ (see §A15).

In the present case, this implies that $|x| = 1$, which is contrary to the hypothesis. ∎

As a curiosity, I indicate an interpretation of (A7.1) and (A7.4).

Let $b > 1$ be a base of numeration and N a natural number with k digits all equal to 1, when written in base b:
$$N = 11\ldots 1_{(b)}.$$

Such numbers are called *repunits*.

According to (A7.1), $N = 111_{(b)}$ (with 3 digits) is an mth power ($m \geq 2$) only when $b = 18$, $m = 3$.

Also, it follows from (A7.4) that if $k > 1$ and $N = 11\ldots 1_{(b)}$ (k digits) is a square, then either $b = 7$, $k = 4$ or $b = 3$, $k = 5$.

In particular, a repunit (different from 1) in base 10 is not a square.

In 1972, Inkeri showed that a repunit (different from 1) in base 10 cannot be a cube.

Rotkiewicz gave (in 1987) the following direct proof:

Assume that $\frac{10^k-1}{10-1} = y^3$; then $k \neq 2$ and if $k \geq 3$, it follows from (A8.3) that $k = 6h + 5$, $h \geq 0$. Therefore $10^{6(h+1)} - 10 = 90y^3$, and considering the congruence modulo 7, then $1 - 3 \equiv -y^3 \pmod{7}$, so $y^3 \equiv 2 \pmod{7}$ and $y^6 \equiv 4 \pmod{7}$, which is impossible. ∎

Recently R. Bond has kindly communicated to me that a repunit (different from 1) in base 10 is not a fifth power. Indeed, if $\frac{10^k-1}{10-1} = x^5$, with $k \geq 2$, then $x^5 \equiv 11 \pmod{100}$, so x is odd. Let $x = 20q + r$, with $1 \leq r < 20$. Then $x^5 \equiv r^5 \pmod{100}$. However, a simple calculation shows that if $1 \leq r \leq 19$, r odd, then $r^5 \not\equiv 11 \pmod{100}$.

More generally, as noted by Inkeri (private communication):

(A8.4). *If $k > 1$, if p is a prime, if $x > 1$ and $v_p(x) = 1$, then $\frac{x^k-1}{x-1}$ is not a pth power.*

Proof: If $\frac{x^k-1}{x-1} = y^p$ (with $y > 1$), then $x^k - x = (y^p - 1)(x - 1)$. Since $v_p(x) = 1$, then $v_p(x^k) > 1$, so $v_p(x^k - x) = 1$. But $p \nmid x - 1$, hence $v_p(y^p - 1) = 1$. Thus $y \equiv y^p \equiv 1 \pmod{p}$, and therefore $y^p \equiv 1 \pmod{p^2}$, hence $v_p(y^p - 1) \geq 2$, which is a contradiction. ∎

In particular, taking the basis $x = 10$, a repunit ($\neq 1$) cannot be a square, nor a 5th power.

Similar questions have been considered in relation to numbers with all digits equal to a given digit, in a base b.

In base 10, Obláth showed in 1956:

If $N = aa \ldots a_{(10)}$ (n digits equal to a base 10, where $1 < a \leq 9$) is a proper power, then $n = 1$, $a = 1, 4, 8$ or 9. Note that this result does not refer to repunits.

In 1972, Inkeri generalized the above result:

If $2 \leq b \leq 10$ and $N = aa\ldots a_{(b)}$ (n digits equal to a in base b, where $1 < a \leq b$) is a proper power, then $b = 7$, $a = 4$, $n = 4$ and N is a square.

In the same vein, R. Goormaghtigh proposed the problem of determining the integers which have all digits equal to a, respectively to a', when written in different bases of numeration (see Shorey and Tijdeman, 1976; Balasubramanian and Shorey, 1980; Shorey, 1984). I shall return to this question in §C9.

I shall conclude this section by investigating the existence of solutions of equation (8.1) with $y = p$, a prime number, now allowing $m = 1$.

Explicitly, given a prime p, to find integers $x > 1$, $m \geq 1$, $n \geq 3$, such that

$$\frac{x^n - 1}{x - 1} = p^m. \qquad (8.12)$$

As shown in the papers of Guralnick (1983, 1983), Perlis (1978), and Estes et al. (1985), this study is connected with the existence of simple groups having a p-complement (i.e., a subgroup of order prime to p and index a power of p), and with the existence of non-isomorphic number fields having the same Dedekind zeta function.

In the next proposition, I group several results about Equation (8.12), which may be found in papers by Suryanarayana (1967, 1970), Edgar (1971, 1985), and Estes et al. (1985).

(A8.5). *Let p be a prime, let $x > 1$, $m \geq 1$, $n \geq 3$ be integers satisfying equation (8.12).*

1. *n is a prime, equal to the order of x modulo p and $p \equiv 1 \pmod{n}$, hence $p \neq 2$.*
2. *If $x = r^b$, $b \geq 1$, then $b = n^3$, $e \geq 0$, $p \equiv 1 \pmod{n^{e+1}}$, and p is not a Fermat prime.*
3. *With the assumption of (2), and if r is a prime, then n does not divide m.*

Proof: 1) Assume that n is not a prime and let $n = hk$, where $h, k > 1$. Then

$$p^m = \frac{x^n - 1}{x - 1} = \frac{x^{hk} - 1}{x^k - 1} \cdot \frac{x^k - 1}{x - 1}.$$

Since $\frac{x^k-1}{x-1} > 1$ then p divides $\frac{x^k-1}{x-1}$, or $x^k \equiv 1 \pmod{p}$. Hence

$$\frac{x^{hk} - 1}{x^k - 1} = 1 + x^k + \ldots + x^{k(h-1)} \equiv h \pmod{p}.$$

But p divides $\frac{x^{hk}-1}{x^k-1}$, hence $p|h$. Since this is true for every divisor h of n, then n is a power of p.

If $n = 2^e (e \geq 2)$, then

$$\frac{x^4 - 1}{x - 1} = (x^2 + 1)(x + 1) = 2^f,$$

with $f \geq 1$. Therefore $x^2 + 1$ and $x + 1$ are powers of 2. But $x^2 \not\equiv -1 \pmod 4$, so $x^2 + 1 = 2$, hence $x = 1$, which is absurd.

If $n = p^e (e \geq 2, p$ odd), then by the above argument (with $h = p$, $k = p^{e-1} > 1$), $\frac{x^p-1}{x-1} = p^f$ $(f \geq 1)$. By (P2) (v), $p^2 \nmid \frac{x^p-1}{x-1}$, hence

$$p = \frac{x^p - 1}{x - 1} = 1 + x + \ldots + x^{p-1} \geq p;$$

this implies that $x = 1$, which is against the hypothesis.

Thus n is a prime and clearly $x^n \equiv 1 \pmod{p}$. If also $x \equiv 1 \pmod{p}$, by (P1.2) (ii), p divides n, so $p = n$. Then $p^m = \frac{x^p-1}{x-1}$, so as indicated above $m = 1$, $x = 1$, an absurdity. In conclusion, since n is a prime, then n is the order of x modulo p and by (P1.4), $p \equiv 1 \pmod{n}$, hence $p \neq 2$.

The Equation $\frac{X^n-1}{X-1} = Y^m$

2) Assume that $x = r^b$, where $b \geq 1$. If $b > 1$ let q be a prime, $q \neq n$, such that $q|b$, and write $b = qs$. Then

$$p^m = \frac{x^n-1}{x-1} = \frac{r^{bn}-1}{r^b-1} = \frac{r^{qsn}-1}{r^{qs}-1}$$

$$= \frac{r^{sn}-1}{r^s-1} \times \frac{1 + r^{sn} + r^{2sn} + \ldots + r^{(q-1)sn}}{1 + r^s + r^{2s} + \ldots + r^{(q-1)s}}$$

$$= \frac{r^{sn}-1}{r^s-1} \times \frac{\Phi_q(r^{sn})}{\Phi_q(r^s)}$$

$$= \frac{r^{sn}-1}{r^s-1} \times \Phi_{qn}(r^s),$$

by §P1, (1.4). Since $\frac{r^{sn}-1}{r^s-1} > 1$, then $p|r^{sn}-1$. But by the binomial, $r^{qsn}-1 = x^n-1$ has the primitive factor p, because n is the order of x modulo p (by (1)). This implies that $sn = qsn$, which is impossible. Thus, b is necessarily a power of n, say $b = n^e$, with $e \geq 0$.

Now I show that $p \equiv 1 \pmod{n^{e+1}}$.

By (1), $x \not\equiv 1 \pmod{p}$, $x^n \equiv 1 \pmod{p^{e+1}}$. So $r^{n^e} = r^b = x \not\equiv 1 \pmod{p}$, while $r^{n^{e+1}} = x^n \equiv 1 \pmod{p}$. Since n is a prime, then n^{e+1} is the order of n modulo p. By (P1.4), $p \equiv 1 \pmod{n^{e+1}}$.

Finally, if p is a Fermat prime, say $p = 2^{2^h}+1$, then n^{e+1} divides $p-1 = 2^{2^h}$, so $n = 2$, which is absurd.

3) Assume that $m = nk$ and let $f = p^k$, so $f^n = \frac{x^n-1}{x-1} \equiv 1 \pmod{x}$. But $f \not\equiv 1 \pmod{x}$, otherwise $p^m = f^n > x^n > \frac{x^n-1}{x-1}$, which is a contradiction. Since n is a prime by (1), then the order of f modulo x is equal to n. Therefore n divides $\varphi(x) = \varphi(r^b) = r^{b-1}(r-1)$. Note that n does not divide $r-1$, otherwise $x = r^b \equiv 1 \pmod{n}$, and so

$$p^m = f^n = \frac{x^n-1}{x-1} = 1 + x + \ldots + x^{n-1} \equiv 0 \pmod{n},$$

implying that $n = p$, which is not true, by (1). This shows that n divides r, and since r is a prime, then $n = r$. Since

$x = r^b = n^{n^e}$ and $f^n \equiv 1 \pmod{n^b}$, then $f \equiv 1 \pmod{n^{b-1}}$ by (P1.1). Thus $f - 1 = tn^{b-1}$, with $t \geq 1$.

Therefore $(tn^{b-1}+1)^n = f^n = \frac{x^n-1}{x-1}$, and so $(x-1)(tn^{b-1})^n < (x-1)f^n < x^n = n^{bn}$, hence

$$n^{n^e} - 1 = x - 1 < \left(\frac{n}{t}\right)^n.$$

If $e = 0$, then $x = n$ and $f \equiv f^n \equiv 1 \pmod{n}$, which is a contradiction.

If $e \geq 1$, then $n^n \leq x = n^{n^e} < \left(\frac{n}{t}\right)^n + 1$, so $e = t = 1$ and $f - 1 = n^{n-1}$. But n is odd, so f is even, hence $p = 2$, which is not true. This concludes the proof. ∎

To end the section it should be mentioned that in virtue of a general theorem of Siegel, the equation

$$\frac{X^n - 1}{X - 1} = X^{n-1} + X^{n-2} + \ldots + X + 1 = Y^q$$

with $q \geq 3$, $n \geq 3$, has at most finitely many solutions. But as seen above, it may well happen that the equation has no solution at all. I shall return to this question in Part C.

9. The Sequence of Powers of 2 or 3

I begin with a proof of the result of the famous medieval astronomer Levi ben Gerson (see Goldstein, 1985). For another proof, see Franklin (1923).

(A9.1). *8 and 9 are the only consecutive integers in the sequence of powers of 2 or 3.*

Proof: If $2^n - 3^m = 1$ with $n \geq 2$, $m \geq 2$, then $2^n \equiv 1 \pmod{3}$, so n is even, and I write $n = 2n'$. Thus

$$3^m = 2^{2n'} - 1 = (2^{n'} - 1)(2^{n'} + 1),$$

hence
$$\begin{cases} 2^{n'} - 1 = 3^{m'} \\ 2^{n'} + 1 = 3^{m-m'} \end{cases}$$
with $0 \le m' < m - m'$. Subtracting, $2 = 3^{m'}(3^{m-2m'} - 1)$, hence $m' = 0$, $n' = 1$, and $n = 2$ $m = 1$, which is contrary to the hypothesis.

If $3^m = 2^n = 1$ with $m \ge 2$, $n \ge 2$, then $n \ne 2$, so $n \ge 3$, hence $3^m \equiv 1 \pmod 8$. Therefore m must be even, $m = 2m'$. Thus $2^n = 3^{2m'} - 1 = (3^{m'} - 1)(3^{m'} + 1)$, hence
$$\begin{cases} 3^{m'} - 1 = 2^{n'} \\ 3^{m'} + 1 = 2^{n-n'} \end{cases}$$
with $0 \le n' < n - n'$. Subtracting $2 = 2^{n'}(2^{n-2n'} - 1)$, hence $n' = 1$, $n = 2n' + 1 = 3$, and $m = 2$. ∎

In 1936, Herschfeld considered the difference $2^x - 3^y$. Using a result of Pillai (1931), which is not of an elementary nature (see (C6.4) and (C6.5)), he derived the following proposition:

(A9.2). *If $|d|$ is sufficiently large, there exists at most one pair of integers (x, y) such that $2^x - 3^y = d$.*

However, for small values of d this is not true. With elementary methods, Herschfeld showed:

(A9.3). *The only pairs of positive integers (x, y) such that $2^x - 3^y = d$, with $|d| \le 10$, are given in the table.*

(x, y)	d
(2,1)	1
(1,1), (3.2)	−1
(3,1), (5,3)	5
(2,2)	−5
(4,2)	7
(1,2)	−7

Proof: Clearly $2^x - 3^y \neq \pm 2, \pm 3, \pm 4, \pm 6, \pm 8, \pm 9$. First I note that if $2^x - 3^y = d$ with $|d| \leq 10$ and $x \geq 3$, then $d = -1$, 5 or 7.

Indeed, since $3^y \equiv -d \pmod{8}$ and $3^n \equiv 1$ or $3 \pmod{8}$, it follows that $d \equiv 7$ or $5 \pmod{8}$, hence $d = -1, 5$ or 7.

Thus, if $d = 1, -5$ or -7, then $x \leq 2$ and the following are all of the possible solutions:

$$2^2 - 3 = 1$$
$$2^2 - 3^2 = -5$$
$$2 - 3^2 = -7.$$

Now let $d = -1, 5$ or 7.

Assume that $2^x - 3^y = -1$, so $3^y \equiv 1 \pmod{2^x}$. If $x > 2$, then the order of 3 mod 2^x is equal to 2^{x-2}. Indeed, by (P1.1), since $v_2(3 - (-1)) = 2$, then $v_2(3^{2^{x-2}} - 1) = x$. This means that $3^{2^{x-2}} \equiv 1 \pmod{2^x}$, but $3^{2^{x-3}} \not\equiv 1 \pmod{2^x}$.

From $3^y \equiv 1 \pmod{2^x}$ it follows that $2^{x-2} \leq y$.

Now, if $x \geq 4$, then

$$3^y - 1 \geq 3^{2^{x-2}} - 1 \geq 3^x - 1 > 2^x,$$

which is contrary to the hypothesis. This shows that $x \leq 3$, and then the only possibilities are $2 - 3 = -1$, $2^3 - 3^2 = -1$.

Assume next that $2^x - 3^y = 5$. There are the obvious possibilities: $2^3 - 3 = 5$, $2^5 - 3^3 = 5$ (for $y \leq 3$), and I show that there are no others. Indeed, if $y > 3$ then $x > 6$ and $3^y \equiv -5 \pmod{2^6}$. But the order of 3 modulo 2^6 is equal to 16, and also $3^{11} \equiv -5 \pmod{2^6}$. Hence $y = 11 + 16k$ with $k \geq 0$. Considering congruences modulo 17, since $3^{16} \equiv 1 \pmod{17}$, then $3^y \equiv 3^{11} \equiv -10 \pmod{17}$, so $2^x \equiv 3^y + 5 \equiv -5 \pmod{17}$. However, a simple calculation shows that no power of 2 satisfies the above congruence.

Finally, assume that $2^x - 3^y = 7$. If $y \leq 3$, the only possibility is $2^4 - 3^2 = 7$. If $y > 3$ then $x > 6$. Taking

congruences modulo 3 and 4, it follows that x and y are even. Then
$$(2^{x/2} - 3^{y/2})(2^{x/2} + 3^{y/2}) = 7,$$
so necessarily $2^{x/2} - 3^{y/2} = 1$ and $2^{x/2} + 3^{y/2} = 7$, and this is impossible. ■

Thus, if $x > 5$ or $y > 3$ then $|2^x - 3^y| > 10$.

The same method was used by Herschfeld to show that if $x > 8$ or $y > 5$, then $|2^x - 3^y| > 100$.

In §C6 I shall return to the consideration of the sequence of powers of natural numbers a or b, and I shall indicate more general results.

10. Interlude

In the preceding sections, I have studied many special cases of Catalan's problem. With the exception of a few results, easy to state but much more difficult to establish, it was possible to prove all of the theorems stated, using relatively easy and well-known facts about the arithmetic of special number fields. I found it expedient not to interrupt the mainstream and opted to postpone to the end of Part A the discussion of the results still unproved. Basically, they concern the diophantine equations

$$X^3 - 3XY^2 + Y^3 = 1 \qquad (10.1)$$

and

$$2X^n - 1 = Z^2 \quad (n \text{ odd}). \qquad (10.2)$$

The first equation deals with a special instance of the problem of representation of integers by binary cubic forms. Even though this is outside the scope of this book, I shall not resist giving supplementary information, beyond the immediate needs on this problem.

With respect to equation (10.2), it will be handled in detail in §A11.

For the sake of completeness, this equation, with n even, will also be considered. It is easily reducible to the case where n is a power of 2. For $n = 2$, it becomes the equation

$$2X^2 - 1 = Z^2, \tag{10.3}$$

which is one of Fermat equations so it has already been dealt with in (P1.4), (e). There are infinitely many solutions in integers.

For $n = 4$, it is a most interesting equation:

$$2X^4 - 1 = Z^2. \tag{10.4}$$

It has only finitely many solutions in integers, and they were completely determined by Ljunggren in 1942.

As it turns out, this equation has a connection with a problem of Fermat, originating in an observation about Diophantus. I will discuss Fermat's problem, which led to the equation

$$2X^4 - Y^4 = Z^2. \tag{10.5}$$

The history of this equation is of interest. The complete solution was first given by Lagrange.

From an entirely distinct source, related to the speedy calculation of decimals of the number π, Gravé asked to determine all possible identities of the form

$$m \arctan \frac{1}{x} + n \arctan \frac{1}{y} = k \frac{\pi}{4} \tag{10.6}$$

where m, n, k, x, y are integers.

Størmer solved this problem, except that he erroneously attributed to Lagrange the complete determination of all of the solutions of (10.4). As I said, this was accomplished only by Ljunggren, and what Ljunggren proved supported Størmer's claim. I shall devote Sections 12 and 14 to Gravé's problem and Ljunggren's result. At the same time, I will venture into the theory of biquadratic equations, but the treatment is not complete and the methods and calculations are too laborious.

11. The Equation $2X^n - 1 = Z^2$

Now I give the proof of Størmer's result (1899), which was required in establishing (A5.2).

(A11.1). *If $n > 2$ is not a power of 2, then the only solution in positive integers of $2X^n - 1 = Z^2$ is $x = 1, z = 1$.*

Proof: By hypothesis, n has an odd factor; therefore there is no loss of generality to assume that n is odd. Assume that x and z are positive integers such that $2x^n - 1 = z^2$, and assume that $z \geq 2$, hence $x > 1$. Note that z is odd. Write $z = 2t + 1$, with $t \geq 1$. Then

$$x^n = t^2 + (t+1)^2 = [t + (t+1)i][t - (t+1)i].$$

In the ring $\mathbb{Z}[i]$ of Gaussian integers, $t+(t+1)i$, $t-(t+1)i$ are relatively prime. Indeed, if a prime π of $\mathbb{Z}[i]$ divides both elements, then π divides $2t$, $2(t+1)$ and also π divides $x^n = t^2 + (t+1)^2$. Since $t^2 + (t+1)^2$ is odd, $\pi \nmid 2$, so $\pi | t$ and $\pi | t+1$, which is impossible.

It follows that $t + (t+1)i$ is, up to a unit ϵ, an nth power:

$$t + (t+1)i = \epsilon(a + bi)^n, \qquad (11.1)$$

with $a, b \in \mathbb{Z}$, $\epsilon \in \{\pm 1, \pm i\}$. The conjugate is $t - (t+1)i = \bar{\epsilon}(a - bi)^n$, hence

$$x^n = t^2 + (t+1)^2 = (a^2 + b^2)^n,$$

so $x = a^2 + b^2$.

I shall multiply both sides of the equality (11.1) by $(1-i)^n$. Since

$$(a + bi)(1 - i) = (a + b) - (a - b)i,$$

the right-hand side becomes

$$\epsilon[(a+b)-(a-b)i]^n = \epsilon[(a+b)A+(a-b)Bi],$$

where A and B are integers. Since $1-i = -i(1+i)$, then

$$(1-i)^{\frac{n-1}{2}} = (-i)^{\frac{n-1}{2}}(1+i)^{\frac{n-1}{2}},$$

and so multiplying by $(1-i)^{\frac{n-1}{2}}$, it follows that $(1-i)^{n-1} = (-1)^{\frac{n-1}{2}} 2^{\frac{n-1}{2}}$ and $(1-i)^n = (1-i)(-2)^{\frac{n-1}{2}}$. Hence the left-hand side becomes

$$[t+(t+1)i](1-i)(-2)^{\frac{n-1}{2}} = [(2t+1)+i](-2)^{\frac{n-1}{2}}$$
$$= \left(2^{\frac{n-1}{2}}z + 2^{\frac{n-1}{2}}i\right)(-1)^{\frac{n-1}{2}},$$

and so

$$2^{\frac{n-1}{2}}z + 2^{\frac{n-1}{2}}i = \epsilon_1[(a+b)A+(a-b)Bi],$$

where ϵ_1 is a unit of $\mathbb{Z}[i]$.

According to the possible values of ϵ_1, $(a+b)A = \pm 2^{\frac{n-1}{2}}$. Since $x = a^2 + b^2$ is odd, then a, b have different parity. Hence $a \pm b$ is odd, and therefore the above relations give $a+b = \pm 1$ or $a - b = \pm 1$. Now I show that

$$a + bi = \epsilon_2[1 + 2r(1 \pm i)],$$

where r is an integer and ϵ_2 is a unit of $\mathbb{Z}[i]$.

Indeed, if a is even, $a = 2a_1$, then

$$a + bi = 2a_1 \pm (\pm 1 - 2a_1)i = \pm i[\pm 1 \pm 2a_1(1 \pm i)].$$

If b is even, $b = 2b_1$, then similarly

$$a + bi = \pm 1 \pm 2b_1 + 2b_1 i = \pm i[\pm 1 \pm 2b_1(1 \pm i)].$$

Substituting in a previous relation,
$$t + (t+1)i = \epsilon_3[1 + 2r(1 \pm i)]^n$$
where ϵ_3 is a unit.

We write $[1 + 2r(1 \pm i)]^n = u \pm vi$. Since
$$(1 \pm i)^4 = (\pm 2i)^2 = -4,$$
then
$$u = 1 + \binom{n}{1}2r - 2\binom{n}{3}(2r)^3 + 4\sum_{k=4}^{n} a_k \binom{n}{k}(2r)^k,$$
$$v = \binom{n}{1}2r + 2\binom{n}{2}(2r)^2 + 2\binom{n}{3}(2r)^3 + 4\sum_{k=4}^{n} b_k \binom{n}{k}(2r)^k,$$
and a_k, b_k are integers.

Since $t + (t+1)i = \epsilon_3(u \pm vi)$, then
$$u \pm v = \pm 1.$$

There are several cases to be considered.

If $u + v = 1$, then $1 + 4nr + 8hr = 1$ (with h integer). This implies that $n + 2h = 0$, so n would be even, a contradiction.

If $u \pm v = -1$, then $1 + 4h = -1$ (with h integer), which is impossible.

If $u - v = 1$, then
$$1 - 2\binom{n}{2}(2r)^2 - 4\binom{n}{3}(2r)^3 + 4\sum_{k=4}^{n}(a_k - b_k)\binom{n}{k}(2r)^k = 1,$$
hence
$$\binom{n}{2} = -2\binom{n}{3}2r + 2\sum_{k=4}^{n}(a_k - b_k)\binom{n}{k}(2r)^{k-2}.$$

I write $n - 1 = 2^s d$, with $s \geq 1$, d odd, and compute the 2-adic values of both sides.

To begin,

$$v_2\left(\frac{n(n-1)}{2}\right) = s - 1,$$

$$v_2\left(2\binom{n}{3}2r\right) = v_2\left(\frac{n(n-1)(n-2)2r}{3}\right) \geq s + 1.$$

As it is well known, for every k,

$$v_2(k!) = \frac{k - s_k}{2 - 1} \leq k - 1$$

(s_k denotes the sum of the digits of the 2-adic development of k). Hence

$$v_2\left(2(a_k - b_k)\binom{n}{k}(2r)^{k-2}\right) \geq 1 + s - v_2(k!) + k - 2$$
$$\geq k - 1 + s - (k - 1) = s.$$

Therefore the equality expressing $\binom{n}{2}$ would imply that $v_2\binom{n}{2} \geq s$, which is a contradiction, and concludes the proof. ∎

12. π and Gravé's Problem

Gravé's problem originated from attempts to calculate decimals of π. The history of calculation of π is fascinating but falls outside the scope of this book. For more information on this matter, the reader may consult the following sources: Wrench (1960), Borwein and Borwein (1986), Castellanos (1988), as well as the special issue dedicated to π, in 1980, of the "Le Petit Archimède."

I shall evoke only the points which relate directly to the topic of this section.

J. Gregory gave in 1671 the power series for the arc tangent function (for $|x| \leq 1$):

$$\arctan x = x - \frac{x^3}{3} + \frac{x^5}{5} - \frac{x^7}{7} + \ldots;$$

for $x = 1$, this gives the following value indicated by G.W. Leibniz in 1674:

$$\frac{\pi}{4} = 1 - \frac{1}{3} + \frac{1}{5} - \frac{1}{7} + \ldots .$$

This series converges too slowly: for three decimal figures it requires 2000 terms! However, the convergence may be speeded up by using the addition formula for the arc tangent.
Since

$$\tan(x+y) = \frac{\tan x + \tan y}{1 - \tan x \tan y},$$

taking $x = \arctan u$, $y = \arctan v$, then

$$\arctan u + \arctan v = \arctan \frac{u+v}{1-uv}.$$

For example, if $1 = \frac{u+v}{1-uv}$, letting $u = \frac{1}{2}$, then $v = \frac{1}{3}$, and this gives the formula of C. Hutton (1776):

$$\frac{\pi}{4} = \arctan \frac{1}{2} + \arctan \frac{1}{3}. \tag{12.1}$$

This series still converges too slowly. The speed of convergence may be improved by taking $u = \frac{1}{3}$ and $\frac{1}{2} = \frac{u+v}{1-uv}$, which implies that $v = \frac{1}{7}$; similarly, with $u = \frac{1}{5}$ and $\frac{1}{3} = \frac{u+v}{1-uv}$, then $v = \frac{1}{8}$. Hence

$$\arctan \frac{1}{2} = \arctan \frac{1}{3} + \arctan \frac{1}{7}$$

$$\arctan \frac{1}{3} = \arctan \frac{1}{5} + \arctan \frac{1}{8}.$$

This gives at once the following formulas of Hutton (1776) and Euler (1779):

$$\frac{\pi}{4} = 2\arctan\frac{1}{2} - \arctan\frac{1}{7} \tag{12.2}$$

$$\frac{\pi}{4} = 2\arctan\frac{1}{3} + \arctan\frac{1}{7}. \tag{12.3}$$

This last formula is easily rewritten in the following form, attributed to G. von Vega, 1794:

$$\frac{\pi}{4} = 2\arctan\frac{1}{5} + \arctan\frac{1}{7} + 2\arctan\frac{1}{8}. \tag{12.4}$$

Beginning with $u = \frac{120}{119}$, from $1 - \frac{u+v}{1-uv}$, it follows that $v = -\frac{1}{239}$; hence

$$\frac{\pi}{4} = \arctan\frac{120}{119} - \arctan\frac{1}{239}.$$

This yields the formula of J. Machin (1706):

$$\frac{\pi}{4} = 4\arctan\frac{1}{5} - \arctan\frac{1}{239}. \tag{12.5}$$

This series in the above formula converges much more rapidly. Thus, Machin (1706) computed 100 decimals of π, F. de Lagny (1719) found 112 correct decimals, while von Vega obtained 126 correct decimals in 1789.

The same method was used by the German prodigy calculator Z. Dase. In 1844, at age 20, he calculated (in his head) correctly 200 decimals of π, all this in about 2 months! Incidentally, Dase also computed an extended table of primes, from 6,000,000 to 9,000,000 in 1861.

The astronomer T. Clausen went to 248 decimals in 1847. Further efforts in the last century were, among others, by W. Lehman in 1853, up to 261 decimals, and W. Shanks in 1873, with 707 decimals (but, of these, only 527 were correct, as found out later).

The calculations used formulas like the ones indicated above, or some appropriate combinations involving arc tangents of several values. For example:

$$\frac{\pi}{4} = 12 \arctan \frac{1}{18} + 8 \arctan \frac{1}{57} - 5 \arctan \frac{1}{239}, \quad (12.6)$$

due to Gauss;

$$\frac{\pi}{4} = 8 \arctan \frac{1}{10} - \arctan \frac{1}{239} - 4 \arctan \frac{1}{515}, \quad (12.7)$$

due to S. Klingenstierna (1730); and

$$\frac{\pi}{4} = 6 \arctan \frac{1}{8} + 2 \arctan \frac{1}{57} + \arctan \frac{1}{239}, \quad (12.8)$$

due to C. Størmer (1896).

In 1962, D. Shanks and J.W. Wrench, Jr. used Størmer's formula to calculate 100,000 decimal digits of π. Gauss' formula (12.6) was used by J. Guilloud and M. Bouyer to reach for the first time 1 million digits.

The formulas indicated above, and those containing more than three arc tangents, were obtained by combining the formulas with only two arc tangents. This prompted Gravé to formulate his problem, which I will spell out below.

Before leaving the topic of calculation of decimals of π, I wish to mention the latest advances.

E. Salamin devised in 1976 a rapidly converging algorithm to calculate decimals of π. It is based on the arithmetic-geometric mean of Gauss. This method was further refined by the brothers Borwein.

In 1983, Y. Tamura and Y. Kanada computed $2^{23} = 8,388,608$ decimals of π with Salamin's algorithm. In 1986, D.H. Bailey used Borwein's algorithm to calculate 29,360,000 decimal digits. In 1989, D. Chudnovski and G. Chudnovski computed π to 1,011,196,691 digits; this was reported in *Focus*, Vol. 9, No. 5, 1989. For their calculation, the brothers Chudnovski used a formula similar to one given by Ramanujan, namely

$$\frac{1}{\pi} = \frac{6541681608}{640320^{3/2}} \sum_{k=0}^{\infty} \left(\frac{13591409}{545140134} + k\right) \frac{(6k)!}{(3k)!(k!)^3} \cdot \frac{(-1)^k}{(640320)^{3k}}$$

I received a communication from D. Sato (November 1989) that Y. Kanada spent 74 hours and 30 minutes to compute 1,073,740,000 digits of π—which is more than did the Chudnovski brothers—but may no longer be a record at the moment I am writing.

After this introduction, I turn to Gravé's problem, which was solved by Størmer in 1885. He gave a simpler proof in 1899.

Gravé's problem:

G: *To find all integers m, n, k, x, y such that $x \neq 0$, $y \neq 0$ and*

$$m \arctan \frac{1}{x} + n \arctan \frac{1}{y} = k\frac{\pi}{4}.$$

First, I shall consider the analogous simpler problem:

G': *To find all integers m, k, a, b such that $a \neq 0$ and*

$$m \arctan \frac{b}{a} = k\frac{\pi}{4}.$$

To exclude trivial cases, I assume that k, m, b are non-zero. Moreover, it suffices to find the solutions with $k > 0, m > 0$ and $gcd(a, b) = 1$, $gcd(k, m) = 1$.

(A12.1). *The only positive integers k, m, a, b such that $gcd(k, m) = 1$, $gcd(a, b) = 1$ and*

$$m \arctan \frac{b}{a} = k\frac{\pi}{4},$$

are $k = m = a = b = 1$.

Proof: Since
$$a + ib = \sqrt{a^2 + b^2} e^{i \arctan \frac{b}{a}}$$
and
$$1 - i = \sqrt{2} e^{-i\frac{\pi}{4}},$$
then
$$m \arctan \frac{b}{a} - k \frac{\pi}{4} = 0$$
exactly when $\lambda = (a + ib)^m (1 - i)^k$ is a real number.

Note that if $\alpha \in \mathbb{Z}[i]$ is a prime Gaussian integer and α divides λ, then $\alpha \sim 1 + i$.

Indeed, if α is not associated with $1 + i$ then $\alpha \nmid 2$; but $1 - i = -i(1+i)$, so $\alpha \nmid 1 - i$, hence $\alpha | (a+ib)^m$, and so $\alpha | a + ib$. But $\lambda = \bar{\lambda}$, hence $\alpha | a - ib$, so $\alpha | 2a$, $\alpha | 2ib$, therefore $\alpha | a$ and $\alpha | b$. Since $\gcd(a, b) = 1$ there exist $\ell, k \in \mathbb{Z}$ such that $\ell a + kb = 1$, hence $\alpha | 1$, which is absurd.

If $a + bi$ is a unit, then $a + bi = \pm 1$ or $\pm i$, so $a = \pm 1, b = \pm 1$.

If $a + bi$ is not a unit, then the only prime divisors of $a + bi$ are associated to $1 + i$. Now note that $(1 + i)^2 = 2i$ does not divide $a + bi$, otherwise 2 divides a and also b, which is absurd. Therefore $a + bi = \epsilon(1 + i)$, where ϵ is a unit. And I conclude again that $a = \pm 1$, $b = \pm 1$. It follows that $k = m = 1$. ∎

Now I am ready to consider Gravé's problem.

To exclude trivial cases and the problem (G') already treated, it suffices to find the solutions with $x \neq 0, y \neq 0$, $k > 0, m > 0, n > 0, x \neq \pm 1, y \neq \pm 1$ and $x \neq y$.

A solution is primitive if $\gcd(m, n) = 1$. Every solution is easily determined, once all primitive solutions are found. Indeed, assume that
$$m \arctan \frac{1}{x} + n \arctan \frac{1}{y} = k \frac{\pi}{4}$$

and let $d = \gcd(m, n) \geq 1$. Write $m = dm_1, n - dn_1$, so that

$$d\left(m_1 \arctan \frac{1}{x} + n_1 \arctan \frac{1}{y}\right) = k\frac{\pi}{4}.$$

By the addition formula for the arctangent function, there exist integers a, b, such that

$$d \arctan \frac{b}{a} = k\frac{\pi}{4}.$$

If $e = \gcd(d, k) \geq 1$, $d = ed'$, $k = ek'$, then

$$d' \arctan \frac{b}{a} = k'\frac{\pi}{4}.$$

If $a = 0$ or $b = 0$, then $\arctan \frac{b}{a}$ is a multiple of $\frac{\pi}{2} = 2\frac{\pi}{4}$. If $a \neq 0$ and $b \neq 0$, I may assume $\gcd(a, b) = 1$ and by (G'), $d' = k' = a = b = \pm 1$, so $\arctan \frac{b}{a}$ is equal to an odd multiple of $\frac{\pi}{4}$. Thus, in both cases

$$m_1 \arctan \frac{1}{x} + n_1 \arctan \frac{1}{y} = k_1 \frac{\pi}{4}$$

with $\gcd(m_1, n_1) = 1$ and k_1 an integer.

Now I shall determine all primitive solutions of Gravé's problem:

(A12.2). *Gravé's problem has four primitive solutions, which correspond to the relations*:

$$\arctan \frac{1}{2} + \arctan \frac{1}{3} = \frac{\pi}{4}$$
$$2 \arctan \frac{1}{2} - \arctan \frac{1}{7} = \frac{\pi}{4}$$
$$2 \arctan \frac{1}{3} + \arctan \frac{1}{7} = \frac{\pi}{4}$$
$$4 \arctan \frac{1}{5} - \arctan \frac{1}{239} = \frac{\pi}{4}.$$

π and Gravé's Problem

Proof: Since

$$x + i = \sqrt{x^2 + 1}\, e^{i \arctan \frac{1}{x}}$$
$$y + i = \sqrt{y^2 + 1}\, e^{i \arctan \frac{1}{y}}$$

and

$$1 - i = \sqrt{2}\, e^{-i\frac{\pi}{4}},$$

then

$$m \arctan \frac{1}{x} + n \arctan \frac{1}{y} = k\frac{\pi}{4}$$

if and only if $\lambda = (x+i)^m (y+i)^n (1-i)^k$ is a real number. Then $\lambda = \bar{\lambda} = (x-i)^m (y-i)^n (1+i)^k$.

Let $a, b, c, d, n \geq 0$, $s \geq 0$ be integers such that $1+i \nmid a+ib$, $1+i \nmid c+id$ and

$$\begin{cases} x + i = (1+i)^r (a + ib) \\ y + i = (1+i)^s (c + id). \end{cases}$$

Then

$$\begin{cases} x^2 + 1 = 2^r (a^2 + b^2) \\ y^2 + 1 = 2^s (c^2 + d^2). \end{cases}$$

Since $x^2 \equiv 0$ or $1 \pmod 4$, then $r = 0$ or 1, and similarly $s = 0$ or 1. Also $a + ib$ is not a unit, otherwise $a^2 + b^2 = 1$ and $x^2 + 1 = 2^r$ ($r = 0$ or 1), so $x = 0$ or ± 1, which was excluded. Similarly, $c + id$ is not a unit.

Note that $\gcd(a + ib, a - ib) = 1$ because if α is a prime dividing $a + ib$ and $a - ib$, then α is not associated to $1 + i$, hence $\alpha \nmid 2$; but $\alpha | 2a$, $\alpha | 2ib$, so $\alpha | a$, $\alpha | b$, which is impossible because $\gcd(a, b) = 1$. Similarly, $\gcd(c + id, c - id) = 1$.

Since $1 - i = -i(1 + i)$, then

$$(1+i)^{rm}(a+ib)^m(1+i)^{sn}(c+id)^n(1-i)^k$$
$$= (1-i)^{rm}(a-ib)^m(1-i)^{sn}(c-id)^n(1+i)^k$$
$$= (-i)^{rm}(1+i)^{rm}(a-ib)^m(-i)^{sn}(1+i)^{sn}(c-id)^n i^k (1-i)^k,$$

then
$$(a+ib)^m(c+id)^n = i^\ell(a-ib)^m(c-id)^n$$
for some integer ℓ. It follows that
$$(a+ib)^m = \epsilon(c-id)^n$$
where ϵ is a unit.

Since the ring of Gaussian integers is a unique factorization domain, and $gcd(m,n) = 1$, there exist integers u, v such that
$$\begin{cases} a+ib = \epsilon_1(u+iv)^n \\ c-id = \epsilon_2(u+iv)^m, \end{cases}$$
and therefore
$$\begin{cases} x+i = \epsilon_1(1+i)^r(u+iv)^n \\ y+i = \bar\epsilon_2(1+i)^s(u-iv)^m. \end{cases}$$

Let $\min\{m,n\} = p$, then
$$\begin{cases} x^2+1 = 2^r(u^2+v^2)^n \\ y^2+1 = 2^s(u^2+v^2)^m \end{cases}$$
and also
$$(xy-1) + (x+y)i$$
$$= \epsilon_1\bar\epsilon_2(1+i)^{r+s}(u^2+v^2)^p(u+iv)^{n-p}(u-iv)^{m-p}.$$

Hence $x+y \equiv 0 \pmod{u^2+v^2}$.

For convenience, let $w = u^2+v^2$. First note that if $w = 1$, then $x^2+1 = 1$ or 2 and $y^2+1 = 1$ or 2. Since $x \neq 0, y \neq 0$ and $x \neq y$, this is excluded. Thus $w \neq 1$.

If $n \geq 2$ then $r \neq 0$, by Lebesgue's result (A3.1), so $r = 1$ and $x^2+1 = 2w^n$. If n is not a power of 2, this is impossible with $x \neq \pm 1$, by (A11.1). If n is a power of 2, it is necessary to appeal to the following result of Ljunggren (1942):

The only solutions in positive integers of

$$2X^4 - 1 = Z^2$$

are $(x, y) = (1, 1)$ and $(13, 239)$. In particular, $2X^8 - 1 = Z^2$ has only the solution $(x, y) = (1, 1)$.

This result was attributed erroneously by Størmer to Lagrange, who studied extensively the equation $2X^4 - Y^4 = Z^2$; Lagrange never made such an assertion, nor can it be derived from his work. The proof of Ljunggren's result is delicate and I shall discuss this matter in §A16.

In the present situation, from $x^2 + 1 = 2w^n$, n a power of 2 and $w \neq 1$, it follows that n is not a multiple of 8; thus $n = 2$ or 4.

In the same way, if $m \geq 2$, then $m = 2$ or 4.

By changing the roles of x and y, and noting that $x \neq y$ and $\gcd(m, n) = 1$, the only possibilities are

(I) $\begin{cases} x^2 + 1 = w \\ y^2 + 1 = 2w \end{cases}$ so $r = 0$, $s = 1, n = m = 1$

(II) $\begin{cases} x^2 + 1 = w \\ y^2 + 1 = 2w^2 \end{cases}$ so $r = 0$, $s = 1$, $n = 1, m = 2$

(III) $\begin{cases} x^2 + 1 = w \\ y^2 + 1 = 2w^4 \end{cases}$ so $r = 0$, $s = 1$, $n = 1$, $m = 4$

(IV) $\begin{cases} x^2 + 1 = 2w \\ y^2 + 1 = 2w^2 \end{cases}$ so $r = s = 1$, $n = 1$, $m = 2$

(V) $\begin{cases} x^2 + 1 = 2w \\ y^2 + 1 = 2w^4 \end{cases}$ so $r = s = 1$, $n = 1$, $m = 4$

with $w|x+y$.

Case (I).
$$\begin{cases} x+i &= \epsilon_1(u+iv) \\ y+i &= \epsilon_2(1+i)(u-iv) = \epsilon_3(1+i)(x-i) \\ &= \epsilon_3[(x+1)+(x-1)i] \end{cases}$$

If $\epsilon_3 = \pm 1$, then $1 = \pm(x-1)$, $y = \pm(x+1)$ (with corresponding signs), so $x = 2, y = 3$. If $\epsilon_3 = \pm i$, then $1 = \pm(x+1)$ and $y = \mp(x-1)$, hence $x = -2, y = 3$. Then $w = x^2 + 1 = 5$, and since $w = 5$ divides $x+y$, necessarily $x = 2$ (thus $\epsilon_3 \neq \pm i$), $y = 3$.

Case (II).
$$\begin{cases} x+i &= \epsilon_1(u+iv), \\ y+i &= \bar{\epsilon}_2(1+i)(u-iv)^2 = \epsilon_3(1+i)(x-i)^2 \\ &= \epsilon_3[(x^2+2x-1)+(x^2-2x-1)i], \end{cases}$$

hence $x^2 \pm 2x - 1 = \pm 1$; this implies that $x = \pm 2$, $w = x^2 + 1 = 5$, $y^2 = 2w^2 - 1 = 49$, $y = \pm 7$, 5 divides $x+y$, so $y = \mp 7$.

Case (III). By Ljunggren's result, $y = \pm 239$, $w = 13$ and $x^2 + 1 = 13$, which is impossible.

Case (IV).
$$\begin{cases} x+i &= \epsilon_1(1+i)(u+iv), \\ x+i &= \bar{\epsilon}_2(1+i)(u-iv)^2 = \bar{\epsilon}_2(1+i)\frac{(x-i)^2}{\epsilon_1^2(1-i)^2} \\ &= \frac{\epsilon_3}{4}(1+i)^3(x-i)^2 = \frac{\epsilon_3}{2}(1+i)(x-1-2xi) \end{cases}$$

hence

$$2y + 2i = \epsilon_3[(x^2+2x-1)+(x^2-2x-1)i].$$

As in case (III), $x^2 \pm 2x - 1 = \pm 2$; this implies that $x = \pm 3$ or ± 1. Since $2w = x^2 + 1$ and $w \neq 1$, then $x = \pm 3$, $w = 5$, $y^2 = 49$, and since 5 divides $x + y$, $y = \pm 7$.

<u>Case (V)</u>. As in case (III), $y = \pm 239$, $w = 13$, $x^2 + 1 = 26$, so $x^2 = 25$, and since 13 divides $x + y$, $x = \mp 5$.

k is determined from the fact that $\lambda = (x+i)^m(y+i)^n(1-i)^k$ is a real number.

If $x = 2, y = 3, m = n = 1$, it follows that $k = 1$.

If $x = 2, y = -7, m = 2, n = 1$, then $k = 1$. However, if $x = -2, y = 7, m = 2, n = 1$, then λ is not real for $k \neq 0$.

If $x = 3, y = 7, m = 2, n = 1$, then $k = 1$; however, if $x = -3, y = -7, = 2, n = 1$, then λ is not real for $k \neq 0$.

Similarly, if $x = 5, y = -239, m = 4, n = 1$, then $k = 1$ while $x = -5, y = 239$ has to be rejected.

So the only primitive solutions of Gravé's problem are the ones indicated in the statement. ∎

In conclusion, the well-known formulas by Hutton, von Vega and Machin are essentially the only possible ones!

In his long memoir of 1896, Størmer considered the more general problem of determination of integer solutions $m_1, \ldots, m_r, a_1, \ldots, a_r, b_1, \ldots, b_r, k$ such that $b_1, \ldots, b_r > 0$, $gcd(a_1, b_1) = \ldots = gcd(a_r, b_r) = 1$ and

$$m_1 \arctan \frac{a_1}{b_1} + \ldots + m_r \arctan \frac{a_r}{b_r} = k\frac{\pi}{4}.$$

This is a much more complicated question. For the case where $r = 3$, $a_1 = a_2 = a_3 = 1, k \neq 0$, Størmer indicated 102 particular solutions, of which 6 were proper solutions, in the sense that

$$b_1, b_2, b_3 \notin \{\pm 2, \pm 3, \pm 5, \pm 7, \pm 239\}.$$

These are

$$3\arctan\frac{1}{4} + 3\arctan\frac{1}{13} - \arctan\frac{1}{38} = \frac{\pi}{4}$$

$$3\arctan\frac{1}{4} + \arctan\frac{1}{20} + \arctan\frac{1}{1985} = \frac{\pi}{4}$$

$$5\arctan\frac{1}{6} - \arctan\frac{1}{43} - 2\arctan\frac{1}{117} = \frac{\pi}{4}$$

$$5\arctan\frac{1}{6} - 3\arctan\frac{1}{43} + 2\arctan\frac{1}{68} = \frac{\pi}{4}$$

$$5\arctan\frac{1}{6} - 3\arctan\frac{1}{117} - \arctan\frac{1}{68} = \frac{\pi}{4}$$

$$5\arctan\frac{1}{8} + 2\arctan\frac{1}{18} + 3\arctan\frac{1}{57} = \frac{\pi}{4}.$$

The problem is intimately related to the study of factorization of numbers of the form $1 + x^2$ and, as such, it involves the Gaussian primes. But I shall not say any more here and conclude by stating

Størmer's conjecture:

> There exist only finitely many solutions in integers $x_1, x_2, x_3 > 0$, $m_1, m_2, m_3, k \neq 0$, of the equation
>
> $$m_1 \arctan\frac{1}{x_1} + m_2 \arctan\frac{1}{x_2} + m_3 \arctan\frac{1}{x_3} = k\frac{\pi}{4}.$$

13. A Problem of Fermat on Pythagorean Triangles and the Equation $2X^4 - Y^4 = Z^2$

In his *Observations on Diophantus*, and in a letter of May 31, 1643 to Brûlard de St. Martin, Fermat proposed the following problem, related to Bachet's comments on problem 24

Fermat's Problem on Pythagorean Triangles

of Book VI of Diophantus (see Fermat's *Oeuvres*, volume III, page 269 and volume II, page 259):

> To find a Pythagorean triangle (that is, a right angled triangle, with sizes a, b, c measured in integers) such that the hypotenuse c is the square of an integer and the sum $a+b$ is also the square of an integer.

Clearly, if such a triangle exists, then $a \neq b$. Also, if d is any integer, $d > 1$, then the triangle with sides ad^2, bd^2, cd^2 also has the properties indicated, and vice versa. Thus, it is enough to find the primitive solutions, namely the triples of integers (a, b, c), such that $1 \leq a < b < c$ and no square $d^2 > 1$ divides a, b, c simultaneously.

Denote by S_F the set of primitive solutions of Fermat's problem.

Fermat asserted that the problem has infinitely many primitive solutions, of which the smallest is

$$\begin{cases} a = & 4565486027761 \\ b = & 1061652293520 \\ c = & 4687298610289. \end{cases} \quad (13.1)$$

However, Fermat left no written proof of his assertion.

It is very easy to reduce the solution of Fermat's problem to the solution of the diophantine equation

$$2X^4 - Y^4 = Z^2. \quad (I)$$

(A13.1). Lemma. *There is a bijection between the set S_F and the set of triples of positive integers (x, y, z) such that $z < y^2$, $2x^4 - y^4 = z^2$ and $\gcd(x, y) = 1$. Moreover, if (a, b, c), $(a', b', c') \in S_F$ correspond to (x, y, z), (x', y', z'), respectively, then $a + b < a' + b'$ if and only if $y < y'$.*

Proof: Let $(a, b, c) \in S_F$ and define the positive integers x, y, z by the relations
$$\begin{cases} a + b = y^2 \\ c = x^2 \\ b - a = z. \end{cases} \quad (13.2)$$

Then
$$2x^4 - y^4 = 2(a^2 + b^2) - (a+b)^2 = (b-a)^2 = z^2,$$

and clearly $z < y^2$. Moreover, $gcd(x, y) = 1$. Indeed, first observe that if a, b are even, then so are $a + b = y^2$ and $c = x^2$, hence $16|c^2 = a^2 + b^2$, $16|(a+b)^2$ and therefore $4|b-a$. If $a \equiv b \equiv 2 \pmod 4$ then $a^2 \equiv b^2 \equiv 4 \pmod 6$, hence $c^2 = a^2 + b^2 \equiv 8 \pmod{16}$, but this is impossible. If $a \equiv b \equiv 0 \pmod 4$, then $4|c$ and (a, b, c) would not be a primitive solution of Fermat's problem.

I have therefore shown that a, b cannot both be even. Now, if P is a prime which divides x and y, then p^2 divides $z = b - a$, hence $p^2|2a, p^2|2b$, and so $p \neq 2$ and in fact, $p^2|a, p^2|b$; hence $p^4|c^2 = a^2 + b^2$, so $p^2|c$ and $(a, b, c,) \notin S_F$.

Since
$$\begin{cases} a = \frac{1}{2}(y^2 - z) \\ b = \frac{1}{2}(y^2 + z) \\ c = x^2 \end{cases} \quad (13.3)$$

then the mapping $(a, b, c,) \mapsto (x, y, z)$ is injective.

Conversely, let (x, y, z) be given and satisfying the conditions indicated. Then $y \equiv z \pmod 2$, as follows from $2x^4 - y^2 = z^2$. Let a, b, c be defined by the above relations. So a, b are integers. Also $z < y^2$ implies that $0 < a < b$. Since $a^2 + b^2 = \frac{1}{2}(y^4 + z^2) = x^4 = c^2$, then $b < c$. It is also clear from (13.2) that $(a, b, c) \in S_F$ and its image by the mapping is (x, y, z).

Fermat's Problem on Pythagorean Triangles

The last assertion is now immediate. ∎

Following Fermat, Euler considered the problem of finding rational numbers x such that

$$ax^4 + bx^3 + cx^2 + dx + e$$

(with a, b, c, d, e integers) is the square of a rational number. He indicated a general method which, from a known x, led to the determination of another rational number x' such that $ax'^4 + bx'^3 + cx'^2 + dx' + e$ is also a rational square. This method is presented in his "Algebra," part II, Chapter IX. In particular, he considered the special case $ax^4 + b$ and also applied the method to obtain the solution (13.1) of Fermat's problem (see loc. cit., Chapter XIV, art. 240). However, as pointed out by Lagrange, Euler's method did not yield all the solutions, nor could Euler establish that (13.1) was the smallest solution of Fermat's problem.

Here I shall not go into Euler's general method, but just present the particular case of the expression $ax^4 + b$ and Euler's specific determination of the solution (13.1) of Fermat's problem.

Method to find rational numbers x such that $ax^4 + b$ is the square of a rational number

Let a, b be non-zero integers, and let h, k be positive rational numbers such that $ah^4 + b = k^2$.

The purpose is to determine another rational number x, such that $ax^4 + b$ is also the square of a rational number. Let $y = \frac{x-h}{x+h}$, hence $x = \frac{h(1+y)}{1-y}$ and therefore

$$ax^4 + b = \frac{ah^4(1+y)^4 + b(1-y)^4}{(1-y)^4}$$
$$= \frac{k^2 + 4(k^2 - 2b)y + 6k^2 y^2 + 4(k^2 - 2b)y + k^2 y^4}{(1-y)^4}.$$

This expression should be the square of a number of the form
$$\frac{k + ty - ky^2}{(1-y)^2},$$
where t is a rational number to be determined.

Since
$$(k + ty - ky^2)^2 = k^2 + 2kty + t^2y^2 - 2k^2y^2 - 2kty^3 + k^2y^4,$$
then
$$4(k^2 - 2b)y + 6k^2y^2 + 4(k^2 - 2b)y^3$$
should be equal to
$$2kty + t^2y^2 - 2k^2y^2 - 2kty^3.$$

Put
$$4(k^2 - 2b) = 2kt,$$
so
$$t = \frac{2k^2 - 4b}{k}.$$

Hence
$$6k^2 + 4(k^2 - 2b)y = t^2 - 2k^2 - 2kty,$$
thus
$$y(4k^2 - 8b + 2kt) = t^2 - 8k^2,$$
hence
$$y(8k^2 - 16b) = \frac{-4k^2 - 16bk^2 + 16b^2}{k^2}$$
and
$$y = \frac{-k^4 - 4bk + 4b^2}{k^2(2k^2 - 4b)}.$$

Then
$$1 + y = \frac{k^4 - 8bk^2 + 4b^2}{k^2(2k^2 - 4b)},$$
$$1 - y = \frac{3k^4 - 4b^2}{k^2(2k^2 - 4b)},$$

so
$$x = \frac{h(1+y)}{1-y} = \frac{h(k^4 - 8bk^2 + 4b^2)}{3k^4 - 4b^2}.$$

Then $ax^4 + b = z^2$, where
$$z = \frac{k + ty - ky^2}{(1-y)^2}, \qquad t = \frac{2k^2 - 4b}{k}.$$

I consider the special case where $a = 2, b = -1$, which will be relevant in the sequel.

If $h = 1$, then $2h^4 - 1 = k^2$, where $k = 1$. Using Euler's method, let
$$x = \frac{1+8+4}{-1} = -13 \qquad t = \frac{2+4}{1} = 6$$
$$y = \frac{-1+4+4}{2+4} = \frac{7}{6} \qquad z = \frac{1+7-\frac{49}{36}}{\frac{1}{36}} = 239.$$

The same method may be applied, beginning now with $h = 13$, and it leads to the value
$$x = \frac{42422452969}{9788425919},$$
which is not an integer.

Euler's partial solution of Fermat's problem

Euler proceeded as follows to determine two relatively prime positive integers a, b such that $a + b$ is a square and $a^2 + b^2$ is a biquadrate. Since $a^2 + b^2$ is a square, then $a \not\equiv b \pmod 2$. Say a is odd and b is even.

As is well known, there exist relatively prime positive integers p, q, such that
$$\begin{cases} a = p^2 - q^2 \\ b = 2pq. \end{cases}$$

Then $a^2+b^2 = (p^2+q^2)^2$, $p \not\equiv q \pmod 2$, hence $b \equiv 0 \pmod 4$; since $a+b$ is a square, then $p^2 - q^2 + 2pq = a+b \equiv 1 \pmod 4$, so q is even and p is odd. Since $a^2 + b^2$ is a biquadrate, then $p^2 + q^2$ is a square, hence there exist relatively prime positive integers r, s, such that

$$\begin{cases} p = r^2 - s^2 \\ q = 2rs. \end{cases}$$

Then $p^2 + q^2 = (r^2 + s^2)^2$ and $a^2 + b^2 = (r^2 + s^2)^4$. So

$$\begin{cases} a = (r^2 - s^2)^2 - 4r^2s^2 = r^4 - 6r^2s^2 + s^4 \\ b = 4r^3s - 4rs^3, \end{cases}$$

hence

$$a + b = r^4 + 4r^3s - 6r^2s^2 - 4rs^3 + s^4,$$

and this number should be the square of another integer m. Letting $m = r^2 + 2rs + s^2$, then

$$m^2 = r^4 + 4r^2s^2 + s^4 + 4r^3s + 4rs^3 + 2r^2s^2,$$

and this implies

$$-6r^2s^2 - 4rs^s = 6r^2s^2 + 4rs^3,$$

hence $-3r = 2s$, which cannot be satisfied by positive integers. If $a+b$ is the square of $n = r^2 - 2rs + s^2$, then

$$n^2 = r^4 + 4r^2s^2 + s^4 - 4r^3s - 4rs^3 + 2r^2s^2,$$

and this implies that

$$-6r^2s^2 + 4r^3s = 6r^2s^2 - 4r^3s,$$

Fermat's Problem on Pythagorean Triangles

hence $2r = 3s$ and $r = \frac{3}{2}s$. Taking $r = 3, s = 2$, this leads to $a = -119$, which is not acceptable.

Euler then considered $r = \frac{3}{2}s + t$, where t is to be determined so that $a + b$ is a square. Now

$$r^2 = \frac{9}{4}s^2 + 3st + t^2$$
$$r^3 = \frac{27}{8}s^3 + \frac{27}{4}s^2t + \frac{9}{2}st^2 + t^3$$
$$r^4 = \frac{81}{16}s^4 + \frac{27}{2}s^3t + \frac{27}{2}s^2t^2 + 6st^3 + t^4.$$

Hence

$$a + b = \left(\frac{81}{16}s^4 + \frac{27}{2}s^3t + \frac{27}{2}s^2t^2 + 6st^3 + t^4\right)$$
$$+ \left(\frac{27}{2}s^4 + 27s^3t + 18s^2t^2 + 4st^3\right)$$
$$- \left(\frac{27}{2}s^4 + 18s^3t + 6s^2t^2\right)$$
$$- \left(6s^4 + 4s^3t\right) + s^4$$
$$= \frac{1}{16}s^4 + \frac{37}{2}s^3t + \frac{51}{2}s^2t^2 + 10st^3 + t^4.$$

Thus

$$16(a + b) = s^4 + 296s^3t + 408s^2t^2 + 160st^3 + 16t^4$$

should be the square of $s^2 + 148st - 4t^2$, which is

$$s^4 + 296s^3t + 21896s^2t^2 - 1184st^3 + 16t^4.$$

This forces $408s + 160t = 21896s - 1184t$, hence $21488s = 1344t$, that is, $1343s = 84t$.

So, taking $s = 84$, $t = 1344$, then

$$r = \frac{3}{2}s + t = 1469$$

and

$$a = r^4 - 6r^2s^2 + s^4 = 4565486027761$$
$$b = 4r^3s - 4rs^3 = 1061652293520.$$

This is Fermat's solution.

Euler's method, of course, does not allow us to infer that the above are the smallest possible values for a, b.

In 1777, Lagrange used the method of infinite descent to determine all the integer solutions (x, y, z) of the equation $2X^4 - Y^4 = Z^2$. His proof, which was very ingenious, related the solutions of the above equation and those of the similar equations $X^4 - 2Y^4 = Z^2$ and $X^4 + 8Y^4 = Z^2$. There were three new features in Lagrange's work: first, the use of the method of infinite descent to show that solutions exist; second, the use of this method to describe all solutions; third, the use of the descent method involving solutions of more than one equation.

Subsequently, in 1853, Lebesgue gave a simplified proof of Lagrange's result, with the method of descent, but involving only the solutions of the equation being studied, $2X^4 - Y^4 = Z^2$.

I shall present Lebesgue's proof, made somewhat more precise.

I now turn to the determination of all of the solutions of $2X^4 - Y^4 = Z^2$. A triple of integers (x, y, z) is a solution of this equation if $2x^4 - y^4 = z^2$ and x, y, z are non-zero. A solution is primitive when $gcd(x, y) = 1$. Every solution of the equation is of the form (dx, dy, d^2z) where d is any non-zero integer and (x, y, z) is a primitive solution. The trivial solution is $(1, 1, 1)$.

For convenience, it suffices to consider the set S of all non-trivial primitive solutions (x, y, z), with x, y positive and $z \equiv 1 \pmod 4$.

To begin, an easy remark:

(**A13.2**). **Lemma.** *Let (x, y, z) be a primitive solution of* (I). *Then:*

i) x, y, z *are odd;*

Fermat's Problem on Pythagorean Triangles

ii) *The following conditions are equivalent:*

$$y^2 = |z|$$
$$x = 1$$
$$x = y = |z| = 1$$
$$xy = |z|.$$

Proof: i) If y is even then z is even, thus 4 divides $2x^4$, hence x is also even, contrary to the assumption that $gcd(x,y) = 1$. So y is odd and z is also odd. Therefore

$$2x^4 \equiv y^4 + z^2 \equiv 1 + 1 \equiv 2 \pmod{4},$$

so $x^4 \equiv 1 \pmod 2$ and x is odd.

ii) If $y^2 = |z|$ then $2x^4 = 2y^4$, so $x = y$, and since $gcd(x,y) = 1$, then $x = y = 1$.

If $x = 1$, then $2 = y^4 + z^2$, so $y = |z| = 1$. Clearly, if $x = y = |z| = 1$, then $xy = |z|$.

Finally, if $xy = |z|$, then $2x^4 - y^4 = z^2 = x^2y^2$, hence $2x^4 = y^2(x^2 + y^2)$ and since y is odd, $gcd(x,y) = 1$, then $x = y = 1$, so $y^2 = |z|$. ∎

Now, I come to the main result, establishing maps between solutions:

(A13.3). Lemma.

i) There exists a mapping

$$\Psi : S \to S \cup \{(1,1,1)\}$$

such that if $\Psi(x,y,x)) = (r,s,t)$ then $r < x$; moreover, $(13, 1, -239)$ is the only primitive solution in S such that $\Psi((13, 1, -239)) = (1, 1, 1)$.

ii) There exist mappings $\Phi : S \to S$, $\Phi' : S \to S$ such that if $\Phi((r,s,t)) = (x,y,z)$ then $r < x$, if $\Phi'((r,s,t,)) = $

(x', y', z') then $r < x'$. Moreover, $\Phi((r, s, t)) \neq \Phi'(r, s, t)$ for every $(r, s, t) \in S$.

iii) $\Psi \circ \Phi$ and $\Psi \circ \Phi'$ are equal to the identity map on S.

iv) If $(x, y, z) \in S$ and $(x, y, z) \neq (13, 1, -239)$, then $(x, y, z) = \Phi\Psi((x, y, z))$ or $(x, y, z) = \Phi'\Psi((x, y, z))$.

Proof: i) Let $(x, y, z) \in S$, so $x > 1, y > 1, z \equiv 1 \pmod{4}$. Then $\frac{y^2+z}{2}$ is odd, $\frac{y^2-z}{2}$ is even and $gcd\left(\frac{y^2+z}{2}, \frac{y^2-z}{2}\right) = 1$, as easily seen because $gcd(x, y) = 1$.

From $2x^4 - y^4 = z^2$, it follows that

$$\left(\frac{y^2+z}{2}\right)^2 + \left(\frac{y^2-z}{2}\right)^2 = x^4,$$

hence there exist relatively prime integers a, b such that

$$\begin{cases} \frac{y^2+z}{2} = a^2 - b^2 \\ \frac{y^2-z}{2} = 2ab \\ x^2 = a^2 + b^2. \end{cases}$$

It follows that a, b are not both odd, otherwise $x^2 \equiv 1+1 = 2 \pmod{4}$, which is not possible.

Since $y^2 = a^2 - b^2 + 2ab$, and y is odd, then a is odd and b is even. Even though only $|a|, |b|$ are determined by the above relations, the sign of ab is determined. But I may choose in a unique way a, b such that $a + b > 0$.

Again, there exist relatively prime integers h, k, such that

$$\begin{cases} a = h^2 - k^2 \\ b = 2hk \\ x = h^2 + k^2; \end{cases}$$

Fermat's Problem on Pythagorean Triangles

note that $|h|, |k|$ are uniquely defined and so is the sign of hk; moreover $h \not\equiv k \pmod 2$, hence $4|b$.

From $y^2 + 2b^2 = (a+b)^2$, by (P3.5), there exist relatively prime integers e, f such that

$$\begin{cases} y = |2e^2 - f^2| \\ a + b = 2e^2 + f^2 \\ b = 2ef; \end{cases}$$

note that f is odd, and since $4|b$, then e is even; therefore $|e|, |f|$ are uniquely defined and so is the sign of ef.

Then

$$h^2 - k^2 + 2hk = a + b = 2e^2 + f^2,$$

therefore h is odd, k is even.

Writing $h = e\frac{f}{k} = e\frac{n}{m}$, $k = f\frac{e}{h} = f\frac{n'}{m'}$, where $\gcd(m, n) = \gcd(m', n') = 1$, then $hk = ef$ implies $nn' = mm'$, hence $n' = m$, $m' = n$. Thus $k = f\frac{m}{n}$. It follows that $m|e$, $n|f$ and I write $e = mr$, $f = ns$, where r, s are relatively prime integers; hence $h = nr$, $k = ms$. Note that n, s are odd, since f is odd. Choose the signs, by taking $r > 0, s > 0$; then k, e, m have the same sign, and also h, f, n have the same sign. Finally, choose $n > 0$, and since the sign of hk is determined, then n, h, k, e, f are uniquely determined.

From $h^2 - k^2 + 2hk = 2e^2 + f^2$, it follows that

$$(h^2 - f^2 + 2hk - (k^2 + 2e^2) = 0,$$

i.e.,

$$n^2(r^2 - s^2) + 2rsmn - m^2(s^2 + 2r^2) = 0.$$

The discriminant of this expression is equal to

$$r^2 s^2 + (r^2 - s^2)(s^2 + 2r^2) = 2r^4 - s^4;$$

it is odd and since $\frac{n}{m}$ is rational, it is a square, so there exists a unique integer t such that $t^2 = 2r^4 - s^4$ and $t \equiv 1 \pmod 4$. Thus $(r, s, t) \in S$.

Next, I note that

$$r \le r^2 < n^2 r^2 + m^2 s^2 = h^2 + k^2 = x.$$

It also follows that

$$\frac{n}{m} = -\frac{rs - t}{r^2 - s^2} \quad \text{or} \quad -\frac{rs + t}{r^2 - s^2}.$$

Define the mapping

$$\Psi : S \to S \cup \{(1, 1, 1)\}$$

by letting $\Psi((x, y, z)) = (r, s, t)$, where r, s, t have been defined by the above procedure.

Now, I determine the solutions $(x, y, z) \in S$ such that $\Psi((x, y, z)) = (1, 1, 1)$.

As Euler indicated, $(13, 1, -239) \in S$, and I calculate its image by Ψ. According to the proof, $a = 5, b = 12, h = 3, k = 2, e = 2, f = 3, m = 2, n = 3, r = 1, s = 1$. Thus, the image of $(13, 1, -239)$ is equal to $(1,1,1)$.

Conversely, assume that $(x, y, z) \in S$ has image equal to $(1, 1, 1)$. Then $h = f = n > 0$, $k = e = m$ and $h^2 - k^2 + 2hk = 2e^2 + f^2 = 2k^2 + h^2$, so $2hk = 3k^2$; hence $h = 3, k = 2$ and $x = 13$, $y = |2e^2 - f^2| = 1$, so $z = -239$.

ii) For convenience, let S^* be the set of all triples of integers (r, s, t), such that r, s are positive and relatively prime and $2r^4 - s^4 = t^2$. Thus, $S \subseteq S^*$. Then, $r^2 - s^2 \ne 0$ and $rs \ne t$, by Lemma A12.1. Let

$$d = \gcd(rs - t, r^2 - s^2),$$
$$\begin{cases} m = \frac{r^2 - s^2}{d} \\ n = -\frac{rs - t}{d}. \end{cases}$$

Fermat's Problem on Pythagorean Triangles

The sign of d is chosen so that $n > 0$; then m is also determined. Clearly $m, n \neq 0$ and $\gcd(m, n) = 1$. Let
$$\begin{cases} h = nr \\ k = ms, \end{cases}$$
hence $h, k \neq 0$ and $\gcd(h, k) = 1$.

Indeed, if a prime p divides m and h, then it divides r, hence it would divide s, which is absurd. Similarly, if a prime p divides s and h, then it divides n, so p divides t, and therefore p divides $2r^4$; since s is odd, then $p \neq 2$, so $p|r$, which is absurd.

Also, $h \neq k$, otherwise $h = k = 1$, so $r = s = 1$, which is against the hypothesis. Let
$$\begin{cases} e = mr \\ f = ns, \end{cases}$$
hence $e, f \neq 0$ and $\gcd(e, f) = 1$ (the verification is similar). Then
$$\begin{cases} hk = ef \\ h^2 + 2hk - k^2 = 2e^2 + f^2. \end{cases}$$
Indeed,
$$d^2(h^2 + 2hk - k^2)$$
$$= r^2(rs - t)^2 - 2rs(rs - t)(r^2 - s^2) - s^2(r^2 - s^2)^2$$
$$= r^4 s^2 - 2r^3 st + 2r^6 - r^2 s^4 - 2r^4 s^2 + 2r^2 s^4 + 2r^3 st$$
$$\quad - 2rs^3 t - r^4 s^2 + 2r^2 s^4 - s^6$$
$$= 2r^6 - 2r^4 s^2 + 3r^2 s^4 - 2rs^3 t - s^6,$$
while
$$d^2(2e^2 + f^2)$$
$$= 2r^2(r^2 - s^2)^2 + s^2(rs - t)^2$$
$$= 2r^6 - 4r^4 s^2 + 2r^2 s^4 + r^2 s^4 - 2rs^3 t + 2r^4 s^2 - s^6$$
$$= 2r^6 - 2r^4 s^2 + 3r^2 s^4 - 2rs^3 t - s^6.$$

From $hk = ef$ it follows that h, k are not both odd, otherwise e, f are both odd and $h^2 + 2hk - k^2 \equiv 2 \pmod 4$, $2e^2 + f^2 \equiv 3 \pmod 4$, which is absurd.

Let
$$\begin{cases} a = h^2 - k^2 \\ b = 2hk. \end{cases}$$

Then b is even, $b \neq 0$, and a is odd, because $h \not\equiv k \pmod 2$. It follows that $gcd(a, b) = 1$, because if a prime p divides a and b, then $p \neq 2$ and if $p|h$ necessarily $p|k$, and if $p|k$ then $p|h$, both cases being impossible.

It follows that
$$a + b = h^2 - k^2 + 2hk = 2e^2 + f^2,$$

so $a + b > 0$.

Let
$$y = |2e^2 - f^2|,$$

then
$$y^2 + 2b^2 = (2e^2 - f^2)^2 + 8e^2 f^2 = (2e^2 + f^2)^2 = (a+b)^2.$$

Let $x = h^2 + k^2$, then
$$x^2 = a^2 + b^2 \quad \text{and} \quad y^2 = a^2 - b^2 + 2ab.$$

Finally, let $z = a^2 - b^2 - 2ab$, hence $z \equiv 1 \pmod 4$. Therefore
$$\begin{aligned} 2x^4 - y^4 &= 2(a^4 + 2a^2 b^2 + b^4) \\ &\quad - (a^4 + b^4 + 4a^2 b^2 - 2a^2 b^2 + 4a^3 b - 4ab^3) \\ &= a^4 + b^4 + 2a^2 b^2 - 4a^3 b + 4ab^3 \\ &= (a^2 - b^2 - 2ab)^2 = z^2. \end{aligned}$$

Fermat's Problem on Pythagorean Triangles   159

This shows that $(x,y,z) \in S$. Moreover,
$$r \le r^2 < n^2 r^2 + m^2 s^2 = h^2 + k^2 = x.$$

Now define the mapping $\Phi^* : S^* \to S$, by putting $\Phi^*((r,s,t)) = (x,y,z)$. Next define the mappings Φ, $\Phi' : S \to S$, as follows:
$$\Phi((r,s,t)) = \Phi^*((r,s,t))$$
and
$$\Phi'((r,s,t,)) = \Phi^*((r,s,-t)).$$

Now, I show that if $(r,s,t) \in S$ then $\Phi((r,s,t)) \ne \Phi'((r,s,t))$, that is,
$$\Phi^*((r,s,t)) \ne \Phi^*((r,s,-t)).$$

Assume the contrary.

Starting with (r,s,t), and according to the procedure in the definition of $\Phi^*((r,s,t)) = (x,y,z)$,
$$\begin{cases} x^2 = a^2 + b^2 \\ y^2 = a^2 + 2ab - b^2 \\ z = a^2 - 2ab - b^2 \end{cases}$$
and
$$x = h^2 + k^2$$
$$a = h^2 - k^2$$
$$b = 2hk$$
$$h = mr$$
$$k = ms$$
$$m = \frac{r^2 - s^2}{d}$$
$$n = -\frac{rs - t}{d} > 0$$
$$d = \gcd(rs - t, r^2 - s^2).$$

The condition that $n > 0$ completely determines d, m, h, k, a and b.

Similarly, if
$$\Phi^*((r, s, -t)) = (x, y, z),$$
then
$$\begin{cases} x^2 = a'^2 + b'^2 \\ y^2 = a'^2 + 2a'b' - b'^2 \\ z = a'^2 - 2a'b' - b'^2 \end{cases}$$
and
$$x = h'^2 + k'^2$$
$$a' = h'^2 - k'^2$$
$$b' = 2h'k'$$
$$h' = m'r$$
$$k' = m's$$
$$m' = \frac{r^2 - s^2}{d'}$$
$$n' = -\frac{rs + t}{d'} > 0$$
$$d' = gcd(rs + t, r^2 - s^2).$$

Similarly, $n' > 0$ completely determines d', h', k', a' and b'.

From
$$\begin{cases} x = a^2 + b^2 = a'^2 + b'^2 \\ y^2 = a^2 - b^2 + 2a = a'^2 - b'^2 + 2a'b' \\ z = a^2 - b^2 - 2ab = a'^2 - b'^2 - 2a'b', \end{cases}$$
it follows that $a^2 = a'^2, b^2 = b'^2$ and $ab = a'b'$. Since $a + b > 0$, $a' + b' > 0$, $|a| = |a'|$, $|b| = |b'|$, then necessarily $a = a', b =$

… Fermat's Problem on Pythagorean Triangles … 161

b'. From the uniqueness of the solution of the Pythagorean equation, $|h| = |h'|$, $|k| = |k'|$, $hk = h'k'$; from $h = nr > 0$, $h' = n'r > 0$, then $h = h'$, $k = k'$, therefore $n = n'$, $m = m'$, so $d = d'$ and therefore $rs + t = rs - t$, hence $t = 0$, which is impossible.

iii) Let $(r, s, t) \in S^*$; I shall show that

$$\Psi \circ \Phi^*((r, s, t)) = (r, s, (-1)^{\frac{t-1}{2}} t).$$

Assuming this has been established, if $(r, s, t) \in S$, then

$$\Psi \circ \Phi((r, s, t)) = (r, s, t)$$

and

$$\Psi \circ \Phi'((r, s, t)) = \Psi \circ \Phi^*((r, s, -t))$$
$$= (r, s, (-1)^{\frac{-t-1}{2}}(-t)) = (r, s, t).$$

Thus, let $(r, s, t) \in S^*$. According to the definition of $(x, y, z) = \Phi^*((r, s, t))$, consider successively

$$d = \gcd(r^2 - s^2, rs - t)$$
$$m = \frac{r^2 - s^2}{d}$$
$$n = -\frac{rs - t}{d} > 0$$
$$h = nr > 0$$
$$k = ms$$
$$e = mr$$
$$f = ns > 0$$
$$a = h^2 - k^2$$
$$b = 2hk = 2ef$$
$$x = h^2 + k^2$$
$$x^2 = a^2 + b^2$$
$$y^2 = a^2 + 2ab - b^2$$
$$z = a^2 - 2a - b^2;$$

in particular $z \equiv 1 \pmod 4$ and also

$$\frac{y^2 + z}{2} = a^2 - b^2, \qquad \frac{y^2 - z}{2} = 2ab.$$

According to the definition of $\Psi((x,y,z)) = (r', s', t')$, I consider a', b' satisfying

$$\frac{y^2 + z}{2} = a'^2 - b'^2, \qquad \frac{y^2 - z}{2} = 2a'b',$$

$$x^2 = a'^2 + b'^2 \quad \text{and} \quad a' + b' > 0;$$

moreover $y^2 = a'^2 - b'^2 + 2a'b'$; also $a' = h'^2 - k'^2$, $b' = 2h'k'$, $x = h'^2 + k'^2$ and $y = |2e'^2 - f'^2|$, $a' + b' = 2e'^2 + f'^2$, $b' = 2e'f'$, and finally $h' = n'r'$, $k' = m's'$, $e' = m'r'$, $f' = n's'$, where $r' > 0$, $s' > 0$ and also $n' > 0$, $f' > 0$.

Since $|a'| = |a|$, $|b'| = |b|$, as solutions of the Pythagorean equation, and $a + b > 0$, $a' + b' > 0$, $ab = a'b'$, then $a = a'$, $b = b'$. Similarly, $|h| = |h'|$, $|k| = |k'|$, $hk = h'k'$. But $h > 0$, $h' > 0$, hence $h = h'$ and so $k = k'$.

Similarly, $|e| = |e'|, |f| = |f'|$, $ef = e'f'$. But $f > 0$, $f' > 0$, so $f = f'$, hence $e = e'$. This implies $nr = n'r'$, $ms = m's'$, $ns = n's'$, $mr = m'r'$, hence $\frac{n}{m} = \frac{n'}{m'}$, and since $\gcd(m, n) = \gcd(m', n') = 1$ and $n > 0$, $n' > 0$, then $m = m'$, $n = n'$, and hence $r = r'$, $s = s'$. Finally, $|t'| = |t|$ and since $r' \equiv 1 \pmod 4$, then $t' = (-1)^{\frac{t-1}{2}} t$.

iv) Let $(x, y, z) \in S$, $(x, y, z) \neq (13, 1, -239)$, and let $(r, s, t) = \Psi((x, y, z))$. According to the definition of the mapping Ψ, the numbers a, b, h, k, e, f, m, n are introduced. As was observed, either

$$\frac{n}{m} = -\frac{rs - t}{r^s - s^2} \quad \text{or} \quad -\frac{rs + t}{r^2 - s^2}.$$

In the first case, the computation of $\Phi((r, s, t))$ yields (x, y, z), while in the second case $\Phi'((r, s, t))$ is equal to (x, y, z).

Fermat's Problem on Pythagorean Triangles

The proof is similar to the one in (iii) and the details may therefore be omitted. ■

From this lemma, I deduce the following proposition, which describes all of the solutions of equation (I):

(A13.4). *If $(x, y, z) \in S$, there exists a unique sequence $(\Phi_1, \ldots, \Phi_k)$ (with $k \geq 0$) of mappings from S to S, where $\Phi_i \in \{\Phi, \Phi'\}$, such that*

$$(x, y, z) = \Phi_k \circ \ldots \circ \Phi_1((13, 1, -239)).$$

Proof: Assume that there exists $(x, y, z) \in S$ which is not of the form indicated, and consider one such solution for which x is minimal. Then $(x, y, z) \neq (13, 1, -239)$. Let $\Psi((x, y, z)) = (r, s, t)$, so $(r, s, t) \neq (1, 1, 1)$, hence $(r, s, t) \in S$ and moreover, $r < x$. By minimality of x,

$$(r, s, t) = \Phi_k \circ \ldots \circ \Phi_1((13, 1, -239))$$

where $k \geq 0$ and each $\Phi_i \in \{\Phi, \Phi'\}$. But $(x, y, z) = \Phi((r, s, t))$ or $(x, y, z) = \Phi'((r, s, t))$, so $(x, y, z) = \Phi_{k+1}((r, s, t))$, where $\Phi_{k+1} = \Phi$ or Φ'. Then $(x, y, z) = \Phi_{k+1} \circ \Phi_k \circ \ldots \circ) \Phi_1((13, 1, -239))$, which is a contradiction.

Now, I prove the uniqueness. Let

$$(x, y, z) = \Phi_k \circ \ldots \circ \Phi_1((13, 1, -239))$$
$$= \Phi'_h \circ \ldots \circ \Phi'_1((13, 1, -239)),$$

where $k \geq 0, h \geq 0$, and each $\Phi'_i, \Phi'_j \in \{\Phi, \Phi'\}$. Asssume that $k \geq h$, and proceed by induction on h. If $h = 0$ and $k \geq 1$, then $x > 13$, which is absurd. More generally, let $(r, s, t) = \Psi((x, y, z))$, so

$$(r, s, t) = \Phi_{k-1} \circ \ldots \circ \Phi_1((13, 1, -239))$$
$$= \Phi'_{h-1} \circ \ldots \circ \Phi'_1((13, 1, -239)),$$

hence $k - 1 = h - 1$ and $\Phi_i = \Phi'_i$ for $i = 1, \ldots, h - 1$.

If $(x, y, z) = \Phi((r, s, t))$, then $\Phi_h = \Phi'_h = \Phi$. If $(x, y, z) = \Phi'((r, s, t))$, then $\Phi_h = \Phi'_h = \Phi'$, and this proves the uniqueness. ∎

I illustrate with some numerical calculations. First, I shall determine $\Phi((13, 1, -239))$ and $\Phi'((13, 1, -239))$, following the various steps in the definition of Φ, Φ';

$\Phi((13, 1, -239)) = (1525,\ 1343,\ 2750257)$

$\Phi'((13, 1, -239)) = (2165017, 2372159, 3503833734241).$

According to Lemma A13.1, since $1343^2 < 2750257$ the solution $\Phi((13, 1, -239))$ does not correspond to a solution of Fermat's problem. On the other hand, $\Phi'((13, 1, -239))$ corresponds to the solution (a, b, c) of Fermat's problem, with smallest sum $a + b$, and this is none other than (13.1).

For a discussion of historical aspects and attempts to solve Fermat's problem, see the paper by Hofmann (1969).

It is also worth noting that once the integral solutions of (I) have been determined, all solutions in rational numbers are immediately known. For example, if $(a, b, c) \in S$, then for any factorization $c = df$, let $x = \frac{a}{d}, y = \frac{b}{d}, z = \frac{f}{d}$, so that $2x^4 - y^4 = z^2$. In this way, all the rational solutions of (I) may be obtained. Indeed, if x, y, z are rational numbers such that $2x^4 - y^4 = z^2$, write them with the least common denominator, $x = \frac{m}{d}, y = \frac{n}{d}, z = \frac{p}{d}$, with $d \geq 1$, $\gcd(m, n, p, d) = 1$. Let $e = \gcd(m, n)$, and define $a = \frac{m}{e}, b = \frac{n}{e}$, then $e^2 | pd$. Therefore, if $c = \frac{pd}{e^2}$, I conclude that $2a^4 - b^4 = c^2$ and that $(a, b, \pm c) \in S$.

14. The Equations $X^4 \pm 2^m Y^4 = \pm Z^2$ and $X^4 \pm Y^4 = 2^m Z^2$

As was just seen, Fermat's problem led to equation (I):

$$2X^4 - Y^4 = Z^2. \qquad (I)$$

Several similar quartic equations having been studied by the classical authors, it appears worthwhile to describe systematically their results.

The equations in question fall into five types:

$$X^4 + 2^m Y^4 = Z^2 \quad (A)$$
$$X^4 - 2^m Y^4 = Z^2 \quad (B)$$
$$2^m X^4 - Y^4 = Z^2 \quad (C)$$
$$X^4 + Y^4 = 2^m Z^2 \quad (D)$$
$$X^4 - Y^4 = 2^m Z^2 \quad (E)$$

where $m \geq 0$ is an integer. Clearly, for types (A), (B), (C) it may be assumed, without loss of generality, that $m = 0, 1, 2, 3$, while for the other types, $m = 0, 1$. Table 14.1 gives, for each equation, the set of solutions in integers; a solution (x, y, z) is trivial when $xyz = 0$.

Equations of Type (A)

(A14.1). *If x, y, z are non-zero integers, $\gcd(x, y, z) = 1$ and $x^4 + 2^m y^4 = z^2$, then $m \equiv 3 \pmod{4}$, and the solution (x, y, z) may be obtained from a solution (u, v, w), with u, v, w odd, of the equation $2U^4 - V^4 = W^2$, by the process indicated below.*

Proof: First note that, without loss of generality, it may be assumed that y is odd. Indeed, if $y = 2^{m'}y'$ with $m' \geq 1$, y' odd, then (x, y', z) is a solution of $X^4 + 2^{m+4m'}Y^4 = Z^2$. This being the case, it follows that $\gcd(x, z)$ is a power of 2. Note also that $x \equiv z \pmod 2$.

<u>Case 1</u>. Let x and z be odd. Since

$$z^2 - x^4 = (z + x^2)(z - x^2) = 2^m y^4$$

and

$$\gcd(z + x^2,\ z - x^2) = 2,$$

Table 14.1

TYPE	EQUATION	SOLUTIONS	AUTHORS
A_0	$X^4 + Y^4 = Z^2$	trivial	Fermat
A_1	$X^4 + 2Y^4 = Z^2$	trivial	Euler
A_2	$X^4 + 4Y^4 = Z^2$	trivial	Euler
A_3	$X^4 + 8Y^4 = Z^2$	see below	Euler, Lagrange, Lebesgue
B_0	$X^4 - Y^4 = Z^2$	trivial	Fermat
B_1	$X^4 - 2Y^4 = Z^2$	see below	Euler, Lagrange, Lebesgue
B_2	$X^4 - 4Y^4 = Z^2$	trivial	Euler
B_3	$X^4 - 8Y^4 = Z^2$	trivial	Lebesgue
$C_0 = B_0$	$X^4 - Y^4 = Z^2$	—	—
$C_1 = I$	$2X^4 - Y^4 = Z^2$ equation (I)	see study of	Euler, Lagrange, Lebesgue
C_2	$4X^4 - Y^4 = Z^2$	trivial	Euler
C_3	$8X^4 - Y^4 = Z^2$	trivial	Lebesgue
$D_0 = A_0$	$X^4 + Y^4 = Z^2$	—	—
D_1	$X^4 + Y^4 = 2Z^2$	trivial	Euler
$E_0 = B_0$	$X^4 - Y^4 = Z^2$	—	—
E_1	$X^4 - Y^4 = 2Z^2$	trivial	Euler

then $2 \le m$ and there exist integers p, q such that

(a) $\begin{cases} z + x^2 = 2p^4 \\ z - x^2 = 2^{m-1}q^4 \\ y = pq \end{cases}$ or (b) $\begin{cases} z - x^2 = 2p^4 \\ z + x^2 = 2^{m-1}q^4 \\ y = pq. \end{cases}$

So p and q are odd, $gcd(p, q) = 1$ and

$$\pm x^2 = p^4 - 2^{m-2}q^4.$$

Consider first case (a).

By (P3.2), $m \ne 2$. Considering the above relation modulo 8, since x, p, q are odd, then $m \ne 3, 4$. So $m \ge 5$. From $p^4 - x^2 = 2^{m-2}q^4$ and $gcd(p^2 - x, p^2 + x) = 2$, there exist integers r, s such that

$$\begin{cases} p^2 \pm x = 2r^4 \\ p^2 \mp x = 2^{m-3}s^4 \\ q = rs. \end{cases}$$

So r, s are odd, $gcd(r, s) = 1$, and

$$p^2 = r^4 + 2^{m-4}s^4.$$

Thus, (r, s, p) is a solution of an equation of the same type, with $m - 4$, instead of m, and note also that $gcd(r, s, p) = 1$ and s is odd.

Now consider the case (b). Then

$$p^4 + x^2 = 2^{m-2}q^4.$$

The left-hand side is congruent to 2 modulo 4, thus $m = 3$.

So, by the above considerations, either case (b) will take place or the case (a) takes place repeatedly, leading to a relation of the type
$$p^2 = r^4 + 2^n s^4$$
where $n = 0, 1, 2$ or 3, and r, s, p are odd, $gcd(r, s, p) = 1$.

By (P3.1), $n \neq 0$. Also since $p^2 \equiv r^4 \equiv 1 \pmod{8}$, then $n = 3$. Thus $p^2 = r^4 + 8s^4$. It follows that there exist integers h, k such that

$$(a') \begin{cases} p + r^2 = 2h^4 \\ p - r^2 = 4k^4 \\ s = hk \end{cases} \quad \text{or} \quad (b') \begin{cases} p = r^2 = 2h^4 \\ p + r^2 = 4k^4 \\ s = hk. \end{cases}$$

In case (a'),
$$r^2 = h^4 - 2k^4.$$

Note that h, k, r are odd, hence $r^2 \equiv h^4 \equiv k^4 \equiv 1 \pmod 4$, so the above relation is impossible.

In case (b'), it follows that (h, k, r) is a solution of $U^4 - 2V^4 = W^2$.

<u>Case 2</u>. Let x and z be even. Let 2^n (with $n \geq 1$) be the highest power of 2 which divides x and z, so $x = 2^n x_1$, $z = 2^n z_1$, and either x_1 or z_1 is odd.

From $2^{2n} z_1^2 = 2^{4n} x_1^4 + 2^m y^4$, it follows that $m \geq 2n$ and
$$z_1^2 = 2^{2n} x_1^4 + 2^{m-2n} y^4.$$

If z_1 is odd, then $m = 2n$ and y is odd, so $y^4 + 2^{2n} x_1^4 = z_1^2$, with x_1, y, z_1 odd, $gcd(x_1, y, z_1) = 1$. By the first case, necessarily $2n \equiv 3 \pmod 4$, which is impossible.

If z_1 is even, then x_1 is odd. There are three possibilities.

If $2n < m - 2n$, then $z_1 = 2^n z_2$ and $z_2^2 = x_1^4 + 2^{m-4n} y^4$, hence z_2 is odd. By the first case, $m - 4n \equiv 3 \pmod 4$ and the solution (x_1, y, z_2) may be obtained from a solution (u, v, w), with odd integers u, v, w, of the equation $U^4 - 2V^4 = W^2$.

If $2n = m - 2n$, then $z_2^2 = x_1^4 + y^4$, which is impossible.

If $2n > m - 2n$ and $m - 2n$ is odd, then 2^{m-2n} is the exact power of 2 dividing z_1^2, an absurdity. So $m - 2n$ is even, hence m is even,

$$z_1 = 2^{\frac{m-2n}{2}} z_2$$

and

$$z_2^2 = 2^{4n-m} x_1^4 + y^4,$$

with y, x_1, z_2 odd. By the first case, necessarily $4n - m \equiv 3 \pmod{4}$, which is impossible with m even. ■

Equations of Type (B)

(A14.2). *If x, y, z are non-zero integers, $\gcd(x, y, z) = 1$ and $x^4 - 2^m y^4 = z^2$, then $m \equiv 1 \pmod 4$ and the solution (x, y, z) may be obtained from a solution (u, v, w), with u, v, w odd, of the equation $2U^4 - V^4 = W^2$, by the process indicated below.*

Proof: As in the preceding proof, it may be assumed without loss of generality that y is odd, and so $\gcd(x, z)$ is a power of 2. Also, $x \equiv z \pmod 2$.

Case 1. Let x and z be odd.

Since $x^4 \equiv z^2 \equiv 1 \pmod 8$, then $m \geq 3$. Also $\gcd(x^2 + z, x^2 - z) = 2$ and

$$x^4 - z^2 = (x^2 + z)(x^2 - z) = 2^m y^4,$$

hence there exist integers p, q such that

$$\begin{cases} x^2 \pm z = 2p^4 \\ x^2 \mp z = 2^{m-1} q^4 \\ y = pq. \end{cases}$$

Then $x^2 = p^4 + 2^{m-2}q^4$, with x, p, q odd. By (A14.1), $m - 2 \equiv 3 \pmod{4}$, hence $m \equiv 1 \pmod{4}$. By the preceding result, the solution (p, q, x) may be obtained from a solution (u, v, w), with u, v, w odd, of the equation $2U^4 - V^4 = W^2$, by the process indicated.

<u>Case 2</u>. Let x and z be even. Note that $\gcd(x^2 + z, x^2 - z)$ is a power of 2, say equal to 2^k, with $k \geq 1$. From

$$x^4 - z^2 = (x^2 + z)(x^2 - z) = 2^m y^4$$

it follows that $2k \leq m$. Also, there exist integers p, q such that

$$\begin{cases} x^2 \pm z = 2^k p^4 \\ x^2 \mp z = 2^{m-k} q^4 \\ y = pq. \end{cases}$$

Then $x^2 = 2^{k-1}p^4 + 2^{m-k-1}q^4$. Note that $k - 1 \leq m - k - 1$, so 2^{k-1} divides x^2.

If $k - 1$ is odd, then $k = 2h$; if $k - 1 < m - k - 1$, then 2^{k-1} would be the exact power of 2 dividing x^2, which is impossible. If $k - 1 = m - k - 1$, then $m = 2k = 4h$ and $x^4 - (2^h y)^4 = z^2$, which is impossible by (P3.2).

If $k - 1$ is even, then $x = 2^{\frac{k-1}{2}} x_1$ and $x_1^2 = p^4 + 2^{m-2k} q^4$. By (A14.1), $m - 2k \equiv 3 \pmod 4$, with k odd, so $m \equiv 1 \pmod 4$ and the solution (p, q, x_1) may be obtained from a solution (u, v, w) with u, v, w odd, of the equation $2U^4 - V^4 = W^2$. ∎

Equations of Type (C)

(A14.3). *If x, y, z are non-zero integers, $\gcd(x, y, z) = 1$ and $2^m x^4 - y^4 = z^2$, then $m \equiv 1 \pmod 4$ and the solution (x, y, z) may be obtained from a solution (u, v, w), with u, v, w odd, of the equation $2U^4 - V^4 = W^2$, by the process indicated below.*

Proof: Arguing as before, it may be assumed that x is odd and noting that $y \equiv z \pmod 2$, there are two cases.

<u>Case 1.</u> Let y and z be odd. First, $m = 0$ is impossible by (P3.2). If $m \geq 2$, considering the relation modulo 4, $-1 \equiv 1 \pmod 4$, which is impossible. Thus $m = 1$ and the equation becomes $2X^4 - Y^4 = Z^2$.

<u>Case 2.</u> Let y and z be even. Denote by $n \geq 1$ the largest integer such that 2^n divides both y and z; let $y = 2^n y_1$, $z = 2^n z_1$, so y_1 or z_1 is odd. Then $2n \leq m$ and the relation becomes

$$2^{m-2n} x^4 - 2^{2n} y_1^4 = z_1^2.$$

If z_1 is odd then $m = 2n$ and $x^4 - 2^{2n} y_1^4 = z_1^2$. By (A14.1), this is impossible since $2n \not\equiv 1 \pmod 4$.

If z_1 is even, then y_1 is odd, and there are three possibilities.

If $2n < m = 2n$, then $z_1 = 2^n z_2$ and z_2 is odd, so $2^{m-4n} x^4 - y_1^4 = z_2^2$, with y_1, z_2 odd. By the first case, this is impossible.

If $2n = m - 2n$, then $z_1 = 2^n z_2$ and $x^4 - y_1^4 = z_2^2$, which is again impossible.

If $2n > m - 2n$ and $m - 2n$ is odd, then 2^{m-2n} would be the exact power of 2 dividing z_1^2, an absurdity. So $m - 2n$ is even, hence m is even. Also, $z_1 = 2^{\frac{m-2n}{2}} z_2$ and $x^4 - 2^{4n-m} y_1^4 = z_2^2$. By (A35), this implies that $4n - m \equiv 1 \pmod 4$, which is an absurdity. ∎

I shall return later in this section to the equation $2U^4 - V^4 = W^2$.

Equations of Type (D)

(A14.4). *If $m \geq 0$ and $x^4 + y^4 = 2^m z^2$ then $xyz = 0$.*

Proof: It may be assumed that $m = 0$ or 1. If $m = 0$, this was shown in (P3.3). Let $m = 1$.

Assume that x, y, z are non-zero integers such that $x^4 + y^4 = 2z^2$, then

$$(x^2 + y^2)^2 = 2(z^2 + x^2y^2)$$
$$(x^2 - y^2)^2 = 2(z^2 - x^2y^2).$$

Hence, multiplying

$$\left(\frac{x^4 - y^4}{2}\right)^2 = z^4 - x^4y^4,$$

which is impossible. ■

Equations of Type (E)

(A14.5). *If $m \geq 0$ and $x^4 - y^4 = 2^m z^2$, then $xyz = 0$.*

Proof: It may be assumed that $m = 0$ or 1. By (A3), it is impossible if $m = 0$. Assume that $m = 1$ and that there exist non-zero integers x, y, z such that $x^4 - y^4 = 2z^2$, i.e.,

$$y^4 - 2z^2 = x^4.$$

By (P3.5), there exist relatively prime positive integers u, v such that

$$\begin{cases} y^2 = |2u^2 - v^2| \\ z = 2uv \\ x^2 = 2u^2 + v^2. \end{cases}$$

Hence $\pm x^2 y^2 = 4u^4 - v^4$. According to (A14.2) and (A14.3), this is impossible. ■

The preceding study indicates that the most interesting equation encountered was

$$2X^4 - Y^4 = Z^2. \tag{I}$$

Also, the solutions of the equations

$$X^4 - 2Y^4 = Z^2 \qquad \text{(II)}$$

and

$$S^4 + 8T^4 = U^2 \qquad \text{(III)}$$

are obtainable from those of (I).

In his paper of 1777, Lagrange described very explicitly this relationship; this I shall do now, without entering into details of the proofs.

To describe the results of Lagrange, it suffices to consider primitive solutions; since the coordinates of solutions may be taken positive or negative, it is enough to restrict the attention to the following sets of primitive solutions:

S_I: set of solutions (x, y, z) of (I), such that $\gcd(x, y) = 1$ and $z \equiv 1 \pmod{4}$,

S_{II}: set of solutions (x, y, z) of (II), such that $\gcd(x, y) = 1$ and $x \geq 1$, $z \equiv 1 \pmod{4}$,

S_{III}: set of solutions (x, y, z) of (III), such that $\gcd(x, y) = 1$ and $z \geq 1$, $x \equiv 1 \pmod{4}$.

The *size* of a triple of integers (a, b, c) is defined to be equal to $\max\{|a|, |b|\}$.

The solutions of size 1 are:

$$(\pm 1, \pm 1, 1) \in S_I,$$
$$(1, 0, 1) \in S_{II},$$
$$(1, 0, 1),\ (1, \pm 1, 3) \in S_{III}.$$

Lagrange constructed maps between the various sets S_I, S_{II}, S_{III}, with the property of increasing the size.

(A14.6). *There is a surjective map $\theta_{III,I} : S_{III} \to S_I$, such that if $(s, t, u) \neq (1, 0, 1), (1, 1, 3)$, then the size of $\theta_{III,I}(s, t, u)$ is bigger than the size of (s, t, u).*

Sketch of the proof:

Let $d = |gcd(u - 3st, s^2 - 8t^2)|$ and define

$$\begin{cases} m = \frac{u-3st}{d} \\ n = \frac{s^2-8t^2}{d}, \end{cases}$$

$$\begin{cases} p = ns \\ q = ns - 4mt, \end{cases}$$

and finally

$$\begin{cases} x = ms + nt \\ y = ms - nt \\ z = (ms - nt)^2 - 2pq. \end{cases}$$

Then $(x, y, z) \in S_I$. Define $\theta_{III,I}(s, t, u) = (x, y, z)$ and note that the size of (x, y, z) is bigger than the size of (s, t, u).

Conversely, given $(x, y, z) \in S_I$, $(x, y, z) \neq (\pm 1, \pm 1, 1)$, by the proof of Lemma (A13.3), there exist p, q such that

$$\begin{cases} 2x^2 + z - y^2 = 2p^2 \\ 2x^2 - z + y^2 = 2q^2. \end{cases}$$

Next, there exist relatively prime integers $m, n \neq 0$, such that

$$\frac{x+y}{p} = \frac{p-q}{x-y} = \frac{2m}{n}.$$

Again, there exist integers s, t such that

$$\begin{cases} x + y = 2ms \\ p = ns \\ p - q = 4mt \\ x - y = 2nt. \end{cases}$$

Let $u = \frac{m(s^2 - 8t^2)}{x} + 3st$. Then $(s, t, u) \in S_{III}$ and $\theta_{III,I}(s, t, u) = (x, y, z)$.

The verification of the details of the proof is omitted. ■

(A14.7). *There is a surjective map $\theta_{III,II} : S_{III} \to S_{II}$ such that if $t \neq 0$, then the size of $\theta_{III,II}(s, t, u)$ is bigger than the size of (s, t, u).*

Sketch of the proof:
Let
$$\begin{cases} x = y \\ y = 2st \\ z = s^4 - 8t^4. \end{cases}$$

Then $(x, y, z) \in S_{II}$. Define
$$\theta_{III,II}(s, t, u) = (x, y, z)$$

and note that the size of (x, y, z) is bigger than the size of (s, t, u).

Conversely, given $(x, y, z) \in S_{II}$, there exist p, q such that $p \equiv 1 \pmod 4$ and
$$\begin{cases} x^2 + z = 2p^4 \\ x^2 - z = 2q^4 \\ y = 2pq. \end{cases}$$

Let s, t, u be defined by
$$\begin{cases} s = p \\ t = q \\ u = x. \end{cases}$$

Then
$$(s,t,u) \in S_{III}$$
and
$$\theta_{III,II}(s,t,u) = (x,y,z).$$

Once again, the details are omitted. ∎

(A14.8). *There exist maps $\theta_{I,III} : S_I \to S_{III}$ and $\theta_{II,III} : S_{II} \to S_{III}$ such that*
  i) *if $(x,y,z) \in S_I$ with $|x| \geq 2$, then the size of $\theta_{I,III}(x,y,z)$ is bigger than the size of (x,y,z); if $(x',y',z') \in S_{II}$ and $y' \neq 0$, then the size of $\theta_{II,III}(x',y',z')$ is bigger than the size of (x',y',z').*
  ii) $\theta_{I,III}(S_I) \cup \theta_{II,III}(S_{II}) = S_{II}$.

Sketch of the proof:
Let $(x,y,z) \in S_I$ and let s,t,u be
$$\begin{cases} s = z \\ t = xy \\ u = 2x^4 + y^4. \end{cases}$$

Then $(s,t,u) \in S_{III}$. Define $\theta_{I,III}(x,y,z) = (s,t,u)$.
Let $(x',y',z') \in S_{II}$ and let s,t,u be
$$\begin{cases} s = z' \\ t = x'y' \\ u = x'^4 + 2y'^4. \end{cases}$$

Then $(s,t,u) \in S_{III}$. Define
$$\theta_{II,III}(x',y',z') = (s,t,u).$$

The assertion (i) is proved with no difficulty.

Conversely, given $(s, t, u) \in S_{III}$, there exist p, q such that

$$\text{(I)} \begin{cases} u + s^2 = 4p^4 \\ u - s^2 = 2q^4 \\ t = pq \end{cases} \quad \text{or} \quad \text{(II)} \begin{cases} u + s^2 = 2p^4 \\ u - s^2 = 4q^4 \\ t = pq. \end{cases}$$

In case (I), $(p, q, s) \in S_I$ and

$$\theta_{I,III}(p, q, s) = (s, t, u).$$

In case (II), $(p, q, s) \in S_{II}$ and

$$\theta_{II,III}(p, q, s) = (s, t, u). \qquad \blacksquare$$

Combining the various results above, Lagrange concluded:

(A14.9). *Each one of the equations* (I), (II), (III) *has infinitely many primitive solutions in integers.*

They may all be obtained from the solutions of size 1 by applying the maps $\theta_{III,I}$, $\theta_{III,I}$, $\theta_{I,III}$, $\theta_{II,III}$.

15. Representation of Integers by Binary Cubic Forms

The equation

$$X^3 - 3XY^2 - Y^3 = 1, \qquad (15.1)$$

treated by Ljunggren, is an example of an equation appearing in the study of representation of integers by binary cubic forms.

Let a, b, c, d be integers, and let

$$F(X, Y) = aX^3 + bX^2Y + cXY^2 + dY^3$$

be the binary cubic form, i.e., cubic homogeneous polynomial in the indeterminates X, Y, with coefficients a, b, c, d. For simplicity, it is denoted $F = \langle a, b, c, d \rangle$.

If k is any integer, consider the diophantine equation

$$F(X, Y) = k. \tag{15.2}$$

The questions to be settled are the following:

1) How many solutions has the equation? Infinitely many, finitely many or none?

2) If there are only finitely many solutions, to find the number of solutions, or at least to find a good upper estimate for the number of solutions. Better still, to find an effective bound for the absolute values of x, y, for any solution in integers (x, y).

3) If possible, to describe an algorithm to compute all of the solutions, and hopefully, an efficient algorithm.

For cubic equations of the form (15.2), these questions are not yet completely answered, despite a number of deep and ingenious papers on this topic.

It will be instructive to summarize the most important results obtained thus far on these equations, even though this has only an incidental bearing on the main topic treated in this book. Due to the intricate nature of the methods, it is out of the question to enter here into any specific details apart from a sketch of Ljunggren's treatment of equation (15.1), which intervened in Section §A7.

For each binary from $F(X, Y)$ with integer coefficients, and for each integer k, let

$$S_{F,k} = \{(x, y) | x, y \text{ are integers and } F(x, y) = k\}$$

denote the set of solutions in integers of $F(X, Y) = k$.

If $F(X,Y)$ is a reducible binary cubic form, i.e., $F = GH$, where
$$G(X,Y) = mX + nY$$
and
$$H(X,Y) = pX^2 + qXY + rY^2$$
(where m, n, p, q, r are integers), then it is clear that
$$S_{F,k} = \bigcup_{de=k} (S_{G,d} \cap S_{H,e})$$
(union for all pairs (d, e) of integers such that $de = k$).

The determination of the set of solutions $S_{G,d}$ of $mX + nY = d$ is easy; the set is non-empty if and only if $gcd(m, n)$ divides d, and in this case, $S_{G,d}$ is infinite. On the other hand, the determination of the set of solutions $S_{H,e}$ of $pX^2 + qXY + rY^2 = e$ is the object of the theory of representation of binary quadratic forms. There may be no solution, only finitely many solutions (in the case of definite forms), or infinitely many solutions (in the case of indefinite forms). For an up-to-date presentation of this topic, see my own paper (1990).

In view of the above discussion, it will be assumed from now on that the form $F(X,Y)$ is irreducible.

The binary cubic forms $F(X,Y)$ and $G(X,Y)$ are said to be *equivalent* whenever there exist integers m, n, p, q such that $mq - np = 1$ and
$$G(X,Y) = F(mX + nY, pX + qY).$$
In this situation, F is irreducible if and only if G is irreducible. Moreover, for every integer k there is a bijection between the set of solutions in integers of $G(X,Y) = k$ and the set of solutions in integers of $F(X,Y) = k$. This remark is in general used to replace a given form by a simpler one, and the representations of k by one form and by the other form correspond to each other bijectively.

Just as irreducible binary quadratic forms are divided into definite and indefinite forms which behave quite differently, here also the behavior of binary cubic forms is dictated by the sign of the discriminant.

In general, if $P(X)$ is any polynomial of degree $n \geq 1$, with leading coefficient $a \neq 0$, and if $P'(X)$ denotes its derivative, the discriminant of $P(X)$ is

$$\text{Discr}(P) = (-1)^{\frac{n(n-1)}{2}} \frac{1}{a} \text{Res}(P, P'),$$

(where Res denotes the resultant). Explicitly, in the case where

$$P(X) = aX^3 + bX^2 + cX + d,$$

then

$$P'(X) = 3aX^2 + 2aX + c,$$

and

$$\text{Discr}(P) = \det \begin{pmatrix} a & b & c & d & 0 \\ 0 & a & b & c & d \\ 3a & 2b & c & 0 & 0 \\ 0 & 3a & 2b & c & 0 \\ 0 & 0 & 3a & 2b & c \end{pmatrix}.$$

An easy calculation gives

$$\text{Discr}(P) = -27a^2d^2 + 18abcd + b^2c^2 - 4ac^3 - 4bd^3.$$

If P is an irreducible polynomial, then $\text{Discr}(P) \neq 0$.

The discriminant of the binary cubic form $F(X, Y)$ is, by definition, equal to the discriminant of $P(X) = F(X, 1)$:

$$D = \text{Discr}(F) = \text{Discr}(P).$$

Since F is irreducible, so is P, hence $D \neq 0$.

In the special case where $a = 1$, $b = 0$, that is,

$$F(X, Y) = X^3 + cXY^2 + dY^3,$$

then $D = -(27d^2 + 4c^3)$. If $F(X,Y) = aX^3 + dY^3$, then $D = -27a^2d^2$.

It is convenient to say that the discriminant of the equation $F(X,Y) = k$ is $D = \text{Discr}(F)$.

In 1909, Thue proved a very important theorem about diophantine approximation, more specifically, about the approximation of algebraic numbers by rational numbers. From this theorem, he derived general results about diophantine equations. This will be discussed in §C1.

In particular, for every irreducible binary cubic form $F(X,Y)$, the equation (15.2) has only finitely many solutions in integers.

It should be stressed that the proof of Thue's theorem does not lead to an estimate of the number or size of the solutions. In fact, for many equations it turns out that there is no solution at all.

The number of solutions is not bounded in the following sense (see Mahler, 1934):

(A15.1). *Let $F(X,Y)$ be an irreducible binary cubic form*

i) There exist only finitely many cube-free integers k (that is, no cube $h^3 > 1$ divides k) such that the equation $F(X,Y) = k$ has at least one solution.

ii) On the other hand, for every natural number n, there exists an integer $k \neq 0$ (but not cube-free in general), such that the equation $F(X,Y) = k$ has at least n solutions.

Now I indicate results about upper bounds for the number of solutions.

Delone proved (in 1922), see also Delone and Faddeev (1940):

(A15.2). *If d is not the cube of an integer, then the equation*

$$X^3 + dY^3 = 1, \qquad (15.3)$$

which has discriminant $D = -27d^2 < 0$, has the trivial solution $(1,0)$ and, in some cases, still another solution. The non-trivial solution exists if and only if the fundamental unit ϵ of the number field $\mathbb{Q}(\sqrt[3]{d})$, such that $0 < \epsilon < 1$, is of the form $\epsilon = x + y\sqrt[3]{d}$; in this situation (x,y) is the non-trivial solution.

In 1925b, Nagell extended the preceding result:

(A15.3). *If $k = 1$ or 3, if $a > 0$, d are integers, which are not cubes, then the equation*

$$aX^3 + dY^3 = k, \tag{15.4}$$

which has discriminant $D = -27a^3d^2 < 0$, has at most one solution in non-zero integers, except that

$$2X^3 + Y^3 = 3 \tag{15.5}$$

has the solutions $(x,y) = (1,1)$ and $(4,-5)$.

A supplementary result was given by Ljunggren (1953).

Delone (1922), and later Nagell (1925), studied the more general cubic diophantine equation

$$\langle a,b,c,d \rangle (X,Y) = 1, \tag{15.6}$$

when the discriminant D is negative.

(A15.4). *Assuming $D < 0$, apart from the exceptions listed below, the equation (15.6) has at most three solutions in integers. The exceptions are:*

a) *If $\langle a,b,c,d \rangle$ is equivalent to $\langle 1,0,-1,1 \rangle$ or to $\langle 1,-1,1,1 \rangle$, then the equation has exactly four solutions.*

b) *If $\langle a,b,c,d \rangle$ is equivalent to $\langle 1,0,-1,1 \rangle$, then the equation has exactly five solutions, namely*

$$(x,y) = (0,1), (1,0), (1,1), (-1,1), (4,-3).$$

It should be noted that the theorem cannot be sharpened, because for every integer $m > 1$, the equation

$$X^3 + mXY^2 + Y^3 = 1 \qquad (15.7)$$

has exactly three solutions, namely $(x, y) = (1, 0), (0, 1), (1, -m)$.

The same method yields:

(A15.5). *Assuming $D < 0$, then for every integer k, the equation*

$$\langle a, b, c, d \rangle (X, Y) = k$$

has at most $5k$ solutions (x, y) in relatively prime integers.

Now I turn to binary cubic forms with positive discriminant. This case is much more difficult to handle. Siegel showed in 1929:

(A15.6). *Let k be an integer. There exists a positive integer $D(k)$ such that if $F(X, Y)$ is any binary cubic form with discriminant $D > D(k)$, then the equation $F(X, Y) = k$ has at most 18 solutions.*

It is worth noting that the number $D(k)$ is not effectively determined.

In his thesis (1983) Evertse showed:

(A15.7). *Let $F(X, Y)$ be a binary cubic form with positive discriminant. Then the equation $F(X, Y) = 1$ has at most twelve solutions in integers.*

In a series of papers (1933, 1934, 1935), Skolem developed a p-adic method to solve the equation $F(X, Y) = 1$, for some binary cubic forms.

Consider the equation

$$F(X, Y) = X^3 + bX^2Y + cXY^2 + dY^3 = 1, \qquad (15.8)$$

and assume that it has discriminant $D > 0$.

The equation

$$F(X, -1) = X^3 - bX^2 + cX - d = 1 \qquad (15.9)$$

has the same discriminant $D > 0$, so its three roots η, η', η'' are real, say

$$\eta'' < 0 < \eta' < \eta.$$

If (x, y) is a solution of (15.8), then

$$(x + \eta y)(x + \eta' y)(x + \eta'' y) = F(X, Y) = 1,$$

so $x + \eta y$ is a unit of norm 1 in the field $\mathbb{Q}(\eta)$. Since $\mathbb{Q}(\eta)$ has degree 3 and $\mathbb{Q}(\eta)$ is a real field, then it has two fundamental units ϵ_1, ϵ_2. So

$$x + \eta y = \pm \epsilon_1^m \epsilon_2^n,$$

where m, n are integers (note that $1, -1$ are the only roots of unity in the real field $\mathbb{Q}(\eta)$).

Taking the trace,

$$\begin{aligned}
&\mathrm{Tr}[(\eta' - \eta'')(x + \eta y)] \\
&= (\eta' - \eta'')(x + \eta y) + (\eta'' - \eta)(x + \eta' y) \\
&\quad + (\eta - \eta')(x + \eta'' y) \\
&= (\eta'\eta - \eta''\eta + \eta''\eta' - \eta\eta' + \eta\eta'' - \eta'\eta'')y = 0,
\end{aligned}$$

thus $\mathrm{Tr}[(\eta' - \eta'')\epsilon_1^m \epsilon_2^n] = 0$.

The above relation should be viewed as an exponential equation to determine m, n, hence also x, y. The determination of the fundamental units may be awkward, if not outright difficult. Moreover, one relation to find m, n is not sufficient and another has to be found, by whatever means.

In 1943, Ljunggren applied Skolem's method to deal with the specific equation

$$X^3 - 3XY^2 - Y^3 = 1. \qquad (15.1)$$

(A15.8). *The only solutions of* (15.1) *are* $(x,y) = (1,0)$, $(0,-1)$, $(-1,1)$, $(1,-3)$, $(-3,2)$, $(2,1)$.

Sketch of the proof:

$F(X,Y) = X^3 - 3XY^2 - Y^3$ is irreducible and it has discriminant $D = 81$. The polynomial

$$F(X,-1) = X^3 - 3X + 1$$

has three real roots η, η', η'', say,

$$\eta'' < 0 < \eta' < \eta.$$

These roots satisfy the following relations:

$$\begin{cases} \eta + \eta' + \eta'' = 0 \\ \eta\eta' + \eta'\eta'' + \eta''\eta = -3 \\ \eta\eta'\eta'' = -1, \end{cases}$$

and also

$$\begin{cases} \eta^2 = \eta' + 2 \\ \eta'^2 = \eta'' + 2 \\ \eta''^2 = \eta + 2, \end{cases}$$

from which it follows that

$$\begin{cases} \eta\eta' = \eta - 1 \\ \eta'\eta'' = \eta' - 1 \\ \eta''\eta = \eta'' - 1 \end{cases}$$

and also that

$$\begin{cases} \frac{\eta' - \eta''}{\eta - \eta'} = -\eta'' \\ \frac{\eta - \eta''}{\eta - \eta'} = -\eta\eta''. \end{cases}$$

Thus $\mathbb{Q}(\eta) = \mathbb{Q}(\eta') = \mathbb{Q}(\eta'')$, and this field has two fundamental units. Next, it may be shown that $\{\eta, \eta'\}$ is a fundamental system of units.

If (x, y) is a solution, then $x + \eta y$ is a unit of norm 1, so there exist integers m, n, such that

$$x + \eta y = \pm \eta^m \eta'^n.$$

Therefore

$$(\eta' - \eta'')\eta^m \eta'^n + (\eta'' - \eta)\eta'^m \eta''^n + (\eta - \eta')\eta''^m \eta^n = 0,$$

hence

$$\frac{\eta' - \eta''}{\eta - \eta'} \eta^m \eta'^n + \frac{\eta'' - \eta}{\eta - \eta'} \eta'^m \eta''^n + \eta''^m \eta^n = 0,$$

and so

$$-\eta^m \eta'^n \eta'' + \eta \eta'^m \eta''^{n+1} + \eta''^m \eta^n = 0.$$

Multiplying with $\eta^{m-n} \eta'^m$ gives

$$\eta^{2m-n-1} \eta'^{n+m-1} + (-1)^{n+1} \eta^{m-2n} \eta'^{2m-n-1} = (-1)^{m+1}. \tag{15.10}$$

If m is even and n is odd, the two sides would have distinct signs.

Note that if (m, n) is a solution of (15.10) then $(-n+1, m-n)$ and $(n+1-m, -m+1)$ are also solutions of (15.10), as follows by conjugation and from the various relations satisfied by η, η', η''. This allows us to organize the set of solutions in groups of three solutions:

$$\{(m, n), (-n+1, m-n), (n+1-m, -m+1)\}.$$

In each group, there is one solution with both components even. Thus, it suffices to determine all solutions (m, n) of (15.10) with m, n even.

Then equation (15.10) takes the form

$$\eta^{2m-n-1}\eta'^{m+n-1} - \eta^{m-2n}\eta'^{2m-n-1} = -1.$$

Multiplying with $\eta\eta'\eta'' = -1$:

$$\eta^{2m-n}\eta'^{m+n}\eta'' - \eta^{m-2n}\eta'^{2m-n}\eta'' = 1,$$

which may be written as

$$M^2\eta'' - N^2\eta\eta'' = 1,$$

where

$$\begin{cases} M = \eta^{m-\frac{n}{2}}\eta'^{\frac{n+m}{2}} \\ N = \eta^{\frac{m-2n}{2}}\eta'^{m-\frac{n}{2}}. \end{cases}$$

Hence

$$\left(M\sqrt{\eta''} + N\sqrt{\eta\eta''}\right)^2 \left(M\sqrt{\eta''} - N\sqrt{\eta\eta''}\right)^2$$
$$= \left(M^2\eta'' - N^2\eta\eta''\right)^2 = 1,$$

so

$$\left(M\sqrt{\eta''} + N\sqrt{\eta\eta''}\right)^2 = M^2\eta'' + N^2\eta\eta'' + 2MN\eta''\sqrt{\eta}$$

$\in \mathbb{Q}(\sqrt{\eta})$, and it is a unit, with relative norm (over $\mathbb{Q}(\sqrt{\eta})$), which is equal to 1.

According to Dirichlet's theorem, $\mathbb{Q}(\sqrt{\eta})$ has a system of four fundamental units, because four conjugates of $\mathbb{Q}(\sqrt{\eta})$ are real and two are non-real. Of these fundamental units, two may be taken as η, η'. Two others were laboriously determined by Ljunggren:

$$\begin{cases} \lambda_1 = (\sqrt{\eta''} + \sqrt{\eta\eta''})^2 \\ \lambda_3 = \frac{1}{2}(\eta'' + \sqrt{\eta})^2. \end{cases}$$

Two other related units are
$$\begin{cases} \lambda_2 = (\eta'\sqrt{\eta''} + \eta'\eta''\sqrt{\eta\eta''})^2 \\ \lambda_4 = (\eta\eta''^2\sqrt{\eta''} + \eta^2\eta''\sqrt{\eta\eta''})^2. \end{cases}$$

They satisfy the relations
$$\begin{cases} \lambda_1\lambda_4 = \lambda_3^2 \\ \lambda_1^2\lambda_4 = \lambda_2, \end{cases}$$

so $\lambda_1\lambda_3^2 = \lambda_2$.

The unit $(M\sqrt{\eta''} + N\sqrt{\eta\eta''})^2$ is expressible in terms of λ_1, λ_3; that is, there exist integers p, q such that $(M\sqrt{\eta''} + N\sqrt{\eta\eta''})^2 = \lambda_1^p\lambda_3^q$. Explicitly,

$$(M + N\sqrt{\eta})^2 = \eta''^{p-1}(1+\sqrt{\eta})^{2p}\frac{1}{2^q}(\eta'' + \sqrt{\eta})^{2q},$$

or also

$$\frac{M + N\sqrt{\eta}}{(1+\sqrt{\eta})^p(\eta'' + \sqrt{\eta})^q} = \pm\frac{\eta''^{\frac{p-1}{2}}}{2^{\frac{q}{2}}}.$$

If p is even, then $\eta'' \in \mathbb{Q}(\sqrt{\eta})$ when q is even, or $\eta'' \in \mathbb{Q}(\sqrt{2\eta})$ when q is odd. But this is not true, thus p is odd. If q is odd, then $\sqrt{2} \in \mathbb{Q}(\eta) = \mathbb{Q}(\eta'')$, which is again not true. Thus, p is odd and q is even. Writing $q = 2q_1$, then

$$(M\sqrt{\eta''} + N\sqrt{\eta\eta''})^2 = \lambda_1^p(\lambda_3^2)^{q_1} = \lambda_1^p(\lambda_1^{-1}\lambda_2)^{q_1} = \lambda_1^{p-q_1}\lambda_2^{q_1}.$$

Hence
$$M\sqrt{\eta''} + N\sqrt{\eta\eta''}$$
$$= \pm(\sqrt{\eta''} + \sqrt{\eta\eta''})^h(\eta'\sqrt{\eta''} + \eta'\eta''\sqrt{\eta\eta''})^k,$$

where $h = p - q_1$ and $k = q_1$; therefore h, k have different parity.

Representation of Integers by Binary Cubic Forms

Let
$$\begin{cases} \epsilon_1 = \sqrt{\eta''} + \sqrt{\eta\eta''} \\ \epsilon_2 = \eta'\sqrt{\eta''} + \eta'\eta''\sqrt{\eta\eta''}, \end{cases}$$

and their conjugates

$$\begin{cases} \epsilon_1' = \sqrt{\eta''} - \sqrt{\eta\eta''} & \epsilon_2' = \eta'\sqrt{\eta''} - \eta'\eta''\sqrt{\eta\eta''} \\ \epsilon_1'' = \sqrt{\eta} + \sqrt{\eta'\eta} & \epsilon_2'' = \eta''\sqrt{\eta} + \eta''\eta\sqrt{\eta'\eta} \\ \epsilon_1''' = \sqrt{\eta} - \sqrt{\eta'\eta} & \epsilon_2''' = \eta''\sqrt{\eta} - \eta''\eta\sqrt{\eta'\eta} \\ \epsilon_1^{iv} = \sqrt{\eta'} + \sqrt{\eta''\eta'} & \epsilon_2^{iv} = \eta\sqrt{\eta'} + \eta'\sqrt{\eta''\eta'} \\ \epsilon_1^{v} = \sqrt{\eta'} - \sqrt{\eta''\eta'} & \epsilon_2^{v} = \eta\sqrt{\eta'} - \eta'\sqrt{\eta''\eta'}. \end{cases}$$

Note that $\epsilon_1^{-1} = \sqrt{\eta''} - \sqrt{\eta\eta''}$ and $\epsilon_2^{-1} = \sqrt{\eta''} - \eta'\eta''\sqrt{\eta\eta''}$, due to the relations previously indicated and those satisfied by η, η', η''.

Then
$$M\sqrt{\eta''} + N\sqrt{\eta\eta''} = \pm\epsilon_1^h \epsilon_2^k.$$

Considering the conjugates

$$\begin{cases} M' = \eta'^{m-\frac{n}{2}} \eta''^{\frac{n+m}{2}} \\ M'' = \eta''^{m-\frac{n}{2}} \eta^{\frac{n+m}{2}} \end{cases}$$

and

$$\begin{cases} N' = \eta'^{\frac{m-2n}{2}} \eta''^{m-\frac{n}{2}} \\ N'' = \eta''^{\frac{m-2n}{2}} \eta^{m-\frac{n}{2}}, \end{cases}$$

then
$$MM'M'' = \pm 1, \quad NN'N'' = \pm 1.$$

Therefore,
$$\begin{cases} \epsilon_1^h \epsilon_2^k + \epsilon_1'^h \epsilon_2'^k = \pm 2M\sqrt{\eta''} \\ \epsilon_1^h \epsilon_2^k - \epsilon_1'^h \epsilon_2'^k = \pm 2N\sqrt{\eta\eta''}, \end{cases}$$

and similarly for the conjugates. Hence

$$\begin{cases} (\epsilon_1^h \epsilon_2^k + \epsilon_1'^h \epsilon_2'^k) \times (\epsilon_1''^h \epsilon_2''^k + \epsilon_1'''^h \epsilon_2'''^k) \\ \times ((\epsilon_1^{iv})^h (\epsilon_2^{iv})^k + (\epsilon_1^v)^h (\epsilon_2^v)^k) = \pm 8i \\ ((\epsilon_1)^h (\epsilon_2)^k - (\epsilon_1')^h (\epsilon_2')^k) \times (\epsilon_1''^h \epsilon_2''^k - \epsilon_1'''^h \epsilon_2'''^k) \\ \times ((\epsilon_1^{iv})^h \epsilon_2^{iv})^k - \epsilon_1^v)^h \epsilon_2^v)^k)) = \pm 8. \end{cases} \quad (15.11)$$

Thus, h, k have to satisfy the exponential equations (15.11).

It is easy to see that $(h, k) = (1, 0)$ and $(0, 1)$ are solutions. Moreover, if (h, k) is a solution, then $(-h, -k)$ is also a solution. Since $h \not\equiv k \pmod 2$, by the above remark there are four possible cases:

$$1° \quad \begin{cases} h \equiv 1 \pmod 4 \\ k \equiv 0 \pmod 4 \end{cases}$$

$$2° \quad \begin{cases} h \equiv 1 \pmod 4 \\ k \equiv 2 \pmod 4 \end{cases}$$

$$3° \quad \begin{cases} h \equiv 0 \pmod 4 \\ k \equiv 1 \pmod 4 \end{cases}$$

$$4° \quad \begin{cases} h \equiv 2 \pmod 4 \\ k \equiv 1 \pmod 4 \end{cases}$$

After long calculations and the use of Skolem's p-adic method, Ljunggren concluded, in the various cases:

1°). $h = 1, \ k = 0$,

2°). No possible values for h, k,

3°). $h = 0, \ k = 1$,

4°). $h = -2, k = 1$.

Taking into account the implicit change in signs of h, k (to reduce to four cases) and that M, N are positive, then:

1°). $M\sqrt{\eta''} + N\sqrt{\eta\eta''} = \sqrt{\eta''} + \sqrt{\eta\eta''}$, hence $M = 1, N = 1$ and so $m = n = 0$.

3°). $M\sqrt{\eta''} + N\sqrt{\eta\eta''} = \pm\epsilon_2$ or $\pm\epsilon_2^{-1}$;

and since M, N are positive, then

$$M\sqrt{\eta''} + N\sqrt{\eta\eta''} = \eta'\sqrt{\eta''} - \eta'\eta''\sqrt{\eta\eta''};$$

hence

$$\begin{cases} \eta^{\frac{m-2n}{2}}\eta'^{\frac{n+m}{2}} = \eta' \\ \eta^{\frac{m-2n}{2}}\eta'^{m-\frac{n}{2}} = -\eta'\eta'' = \frac{1}{\eta}, \end{cases}$$

and this implies that $m = \frac{2}{3}, n = \frac{4}{3}$, a contradiction.

4°). $M\sqrt{\eta''} + N\sqrt{\eta\eta''} = \pm\epsilon_1^{-2}\epsilon_2$ or $\pm\epsilon_1^2\epsilon_2^{-1}$, and a similar calculation gives $m = n = 2$. Therefore the solutions, (x, y) when $(m, n) = (0, 0)$, are $(1, 0)$ and $(1, 1)$, and thus

$$\begin{aligned} x + y\eta &= 1, & \text{so } (x, y) = (1, 0), \text{ or} \\ x + y\eta &= -\eta, & \text{so } (x, y) = (0, -1), \text{ or} \\ x + y\eta &= \eta\eta' = \eta - 1, & \text{so } (x, y) = (-1, 1). \end{aligned}$$

The solutions (x, y) when $(m, n) = (-2, -2)$, are $(3, 0)$ and (1.3) and then

$$x + y\eta = \eta^{-2}\eta'^{-2} = \eta''^2 = 2 + \eta, \text{ so } (x, y) = (2, 1); \text{ or}$$
$$x + y\eta = \eta^3 = 1 - 3\eta, \text{ so } (x, y) = (1, -3); \text{ or}$$
$$x + y\eta = \eta\eta'^3 = -3 + 2\eta, \text{ so } (x, y) = (-3, 2).$$

Except for the steps which have not been developed in detail, this concludes the sketch of the proof. ∎

Recently, practical methods have been devised to solve Thue's equation (15.2). See, for example, the papers by Blass, Glass, Meronk and Steiner (1987), Tzanakis and de Weger (1989a, 1989b) and the thesis of de Weger (1987).

For example, Tzanakis and de Weger solved the Thue-Mahler equation

$$X^3 - 3XY^2 - Y^3 = \pm 2^{n_0} \times 17^{n_1} \times 19^{n_2}$$

(where $0 \leq n_0, n_1, n_2$). They found all of the 156 solutions $\pm(x, y)$ with $gcd(x, y) = 1$.

In particular, the solutions of

$$X^3 - 3XY^2 - Y^3 = \pm 3 \times 17 \times 19$$

are $(\pm 10, \pm 1), (\pm 1, \mp 11), (\pm 11, \mp 10)$.

The solutions of $X^3 - 3XY^2 - Y^3 = 1$, given in (A15.8), were also found by this method.

16. Some Quartic Equations

In §A12, I indicated that Størmer attributed erroneously to Lagrange the statement that $(1,1)$ and $(239, 13)$ are the only solutions in positive integers of the equation

$$X^2 - 2Y^4 = -1. \tag{16.1}$$

However, Lagrange made no such claim in his paper of 1777, nor is it possible to prove it using Lagrange's or Lebesgue's method of §A14 and §A13. This statement, which is true, had to wait until 1942 for a proof by Ljunggren.

Equation (16.1) is just an example in a class of quartic equations, which has been extensively studied and is relevant in many number theory problems, as I shall mention.

First of all, I should quote the following result of Thue (1917), proved with his method of diophantine approximation:

Some Quartic Equations

(**A16.1**). *Let A, B, C, D be integers, such that $B^2 - 4AC \neq 0$, and let $n \geq 3$. Then the equation*

$$AX^2 + BX + C = DY^n \qquad (16.2)$$

has only finitely many solutions in integers.

The proof given by Thue did not imply an algorithm to effectively compute the solutions, or even to bound their number. Another proof was given by Landau and Ostrowski (1920). Later, in 1933, Skolem proposed a new p-adic method to determine the eventual solutions. In a series of papers, beginning in 1938, Ljunggren studied the class of equations

$$AX^2 - BY^4 = C \qquad (16.3)$$

(with appropriate restrictions on A, B, C). By determining fundamental system of units in some fields (of degree 2, 4, 6 or 8—according to the equation studied) or certain rings of algebraic integers, Ljunggren was led to investigate exponential equations satisfied by the units, in many cases with more than one unknown exponent.

I shall report here the results of Ljunggren and Cohn relative to the equations

$$X^2 - DY^4 = \pm 1, \pm 4 \qquad (16.4)$$

and

$$Y^4 - DX^2 = \pm 1, \pm 4, \qquad (16.5)$$

where $D > 0$, D not a square.

Beginning in 1966, Cohn considered the above equations with the following supplementary hypothesis on D:

> *Cohn's hypothesis: the equation $X^2 - DY^2 = -4$ does not have a solution in odd integers x, y.*

In 1967, Cohn worked with another hypothesis for D.

The interest of Cohn's approach is the explicit relationship between the solutions of the above diophantine equations and the actual determination of the squares and double squares in some linear recurring sequences of second order. Thus, already in 1964, Cohn solved this problem for Fibonacci and Lucas numbers:

> *The only square Fibonacci numbers are 1 and 144. The only square Lucas numbers are 1 and 4. The only double-square Fibonacci numbers are 2 and 8. The only double-square Lucas numbers are 2 and 18.*

I proceed now to describe the results of Ljunggren (1938b):

(A16.2). i) *Let A, B be positive integers, and $C = 1, 2$ or 4. Then Equation (16.3) has at most two solutions in positive integers. These solutions may be effectively computed from a fundamental system of units of a certain biquadratic field.*

ii) *If $C = 1$ or $C = 4$, and $B \equiv 3 \pmod{4}$, then (16.3) has at most one solution in positive integers.*

In a previous paper of the same year (1938a), Ljunggren studied the more restrictive equations

$$AX^4 - BY^4 = \pm C \qquad (16.6)$$

where $A, B > 0$, and $C = 1, 2, 4$ or 8.

The equation (16.6) is said to be associated to the biquadratic field $\mathbb{Q}(\sqrt[4]{\frac{A}{B}})$.

(A16.3). *With above notations and hypotheses:*

i) *Among all equations of type (16.6) associated to a given biquadratic field $\mathbb{Q}(\sqrt[4]{d})$ (that is with $A = Bdm^4$, $m \neq 0$) and such that $d \neq 5$, there is at most one equation having some solution in non-zero integers. On the other hand, the two equations $X^4 - 5Y^4 = 1$ and $X^4 - 5Y^4 = -4$ have non-trivial solutions.*

ii) *The equation* (16.6) *has at most one solution in positive integers* x, y, *and*

$$\frac{1}{C}\left(x\sqrt[4]{A} + y\sqrt[4]{B}\right)\left(x^2\sqrt{A} + y^2\sqrt{B}\right) = \epsilon^k,$$

where ϵ *is a fundamental unit of relative norm* 1 *of the field* $\mathbb{Q}(\sqrt[4]{\frac{A}{B}})$, *and* $k = 1, 2$ *or* 4. *Moreover*, $k = 4$ *only for the equations*

$$X^4 - 5Y^4 = 1 \text{ and } X^4 - 3Y^4 = 1.$$

iii) *The equation* (16.6) *with* $C = 1$ *or* 2 *has at most one solution in positive integers, which may be effectively computed from a fundamental unit of the quadratic ring* $\mathbb{Z}(\sqrt{AB})$.

Even though an algorithm may, in principle, be derived from the proof of the above results, the computations are forbidding and require the explicit knowledge of fundamental systems of units. However, for special cases, the actual determination of the solutions may be carried out in full.

The first example had been dealt already by Tartakowsky (1926) and again, in a somewhat simpler way, by Ljunggren (1938a and 1942a).

(A16.4). *The equation*

$$X^4 - DY^4 = 1 \tag{16.7}$$

has at most one solution in positive integers x, y. *If* ϵ *is the fundamental unit of the ring* $\mathbb{Z}[\sqrt{D}]$ *and it has norm* -1, *then the only solution is* $x = 3, y = 2$, *and* $D = 5$. *If* ϵ *has norm* 1, *then* $x^2 + y^2\sqrt{D} = \epsilon$ *except when* $D = 7140$, *where* $(x, y) = (1239, 26)$ *and* $x^2 + y^2\sqrt{D} = \epsilon^2$, *with* $\epsilon = 169 + 2\sqrt{7140}$.

Ljunggren also proved a similar result, which I do not spell out, for the equation

$$X^2 - DY^4 = 4. \tag{16.8}$$

In (1942a), Ljunggren considered the equation

$$X^2 - DY^4 = -1 \qquad (16.9)$$

for which he proved:

(A16.5). *Assume that the fundamental unit of the field $\mathbb{Q}(\sqrt{D})$ is not equal to the fundamental unit of the ring $\mathbb{Z}(\sqrt{D})$. Then the equation (16.9) has at most two solutions in positive integers, and these may be effectively computed.*

In the same paper, Ljunggren applied his algorithm to the equation

$$X^2 - 2Y^4 = -1. \qquad (16.1)$$

After lengthy calculations he showed:

(A16.6). *The only solutions in positive integers of the equation (16.1) are $(1,1)$ and $(239, 13)$.*

This was the result required in §A12. Due to the complexity and intricacies of the proof, it has to be omitted. It is very desirable to discover an easier proof of this theorem.

In subsequent papers (1942b, 1951, 1967), Ljunggren treated the equations

$$X^2 - DY^4 = -4 \qquad (16.10)$$
$$AX^2 - BY^4 = -1 \qquad (16.11)$$
$$AX^2 - BY^4 = -4, \qquad (16.12)$$

obtaining results in the same vein.

Noteworthy was the paper of Bumby (1967), in which he found that $(1,1)$ and $(11,3)$ are the only solutions in positive integers of the equation $2X^2 - 3Y^4 = -1$.

Concerning the algorithms to actually determine the solutions, Cohn devised a simpler method, involving calculations with Jacobi symbols. It is applicable only for those values of D satisfying Cohn's hypothesis already indicated.

Some Quartic Equations

I indicate very succinctly how Cohn was led from these diophantine equations to linear recurring sequences of second order.

Let $D > 0$, D non-square, be such that the equation $X^2 - DY^2 = -4$ has a solution in odd positive integers. Let (a, b) be the fundamental solution in odd positive integers. Define

$$\begin{cases} \alpha = \frac{a+b\sqrt{D}}{2} \\ \beta = \frac{a-b\sqrt{D}}{2}, \end{cases}$$

hence $\alpha + \beta = a$, $\alpha - \beta = b\sqrt{D}$, $\alpha\beta = -1$.

For each integer n, let

$$\begin{cases} U_n = \frac{\alpha^n - \beta^n}{\sqrt{D}} \\ V_n = \alpha^n + \beta^n. \end{cases}$$

Then $U_0 = 0$, $U_1 = b$, $V_0 = 2$, $V_1 = a$, $U_{-n} = (-1)^{n-1}U_n$, $V_{-n} = (-1)^n V_n$, and

$$\begin{cases} U_n = aU_{n-1} + U_{n-2} \\ V_n = aV_{n-1} + V_{n-2}, \end{cases}$$

as is easily seen.

For each of the equations $X^2 - DY^2 = \pm 1, \pm 4$, the general solution may be expressed in terms of the above sequences, as indicated in Table 16.1.

Thus, the solutions of $X^2 - DY^4 = 1$ are given by the indices n such that $\frac{1}{2}U_{6n} = \square$ (a square), that is, $U_6 = 2\square$ (a double-square). Similarly, a solution of $X^4 - DY^2 = 1$ corresponds to the indices n such that $V_{6n} = 2\square$.

The same holds for each of the equations

$$X^2 - DY^4 = \pm 1, \pm 4, \qquad Y^4 - DX^2 = \pm 1, \pm 4.$$

Table 16.1

Equation	General Solution (x, y)
$X^2 - DY^2 = 1$	$\left(\frac{1}{2}V_{6n},\ \frac{1}{2}U_{6n}\right)$
$X^2 - DY^2 = 4$	$\left(V_{2n}, U_{2n}\right)$
$X^2 - DY^2 = -1$	$\left(\frac{1}{2}V_{6n-3},\ \frac{1}{2}V_{6n-3}\right)$
$X^2 - DY^2 = -4$	$\left(V_{2n-1}, U_{2n-1}\right)$

Cohn dealt with the sequences $(U_n)_{n \geq 0}$, $(V_n)_{n \geq 0}$ and showed that they have only one or two squares or double squares, managing in many cases to give an explicit determination. In a second paper (1967), he considered the values of D such that $X^2 - DY^2 = -4$ does not have a solution in odd integers but $X^2 - DY^2 = 4$ does have a solution in odd integers.

The reader will appreciate the elegance of Cohn's method. Recently, McDaniel and Ribenboim (1992), obtained the complete determination of squares and double squares in linear recurring sequences

$$U_0 = 0, \quad U_1 = 1, \quad U_n = PU_{n-1} - QU_{n-2} \quad \text{(for } n \geq 2\text{)}$$
$$V_0 = 1, \quad V_1 = P, \quad V_n = PV_{n-1} - QV_{n-2} \quad \text{(for } n \geq 2\text{)}$$

where P, Q are odd, relatively prime and $D = P^2 - 4Q > 0$.

Many results in the study of the above quartic equations (16.4), (16.5) concern the identification of values of D for which the equation has no solution in integers. Typical is the result of Mordell (1964), completed by Ljunggren (1966) and Mordell (1968):

Some Quartic Equations   199

(A16.7). i) *If p is a prime, $p \equiv 1 \pmod{4}$, then $X^2 - pY^4 = 1$ has no solution in positive integers.*

ii) *If $p \neq 2, 5, 29$, the equation $X^4 - pY^2 = 1$ has no solution in positive integers. For $p = 5$, the only solution is $(3, 4)$, and for $p = 29$, the only solution is $(99, 1820)$.*

See also Cohn's paper (1967) for a result of this kind.

Part B

DIVISIBILITY CONDITIONS

In this chapter, my purpose is to deduce divisibility conditions on the positive integers x, y, if they are assumed to satisfy $x^m - y^n = 1$, with $m, n \geq 2$. The idea behind this approach is to obtain conditions which are so restrictive that no numbers x, y may satisfy them.

1. Getting the Consecutive Powers 8 and 9

In this section I shall consider several special cases of Catalan's problem which admit as the only solution the consecutive powers 8 and 9.

Gérono proved in 1870 and 1871:

(B1.1). 1) *If q is a prime, $y \geq 1$ an integer, $m, n \geq 2$ and $q^m - y^n = 1$, then $q = 3$, $y = 2$, $m = 2$, and $n = 3$.*

2) *If p is a prime, $x \geq 1$ an integer, $m, n \geq 2$ and $x^m - p^n = 1$, then $p = 2$, $x = 3$, $m = 2$, and $n = 3$.*

Proof: 1) First I observe that n is odd. Indeed, if $n = 2u$ then $q^m - (y^u)^2 = 1$, which is impossible by (A3.1).

Let ℓ be a prime dividing n, and $n = \ell n'$, so $\ell \neq 2$. Since $m \geq 2$, then $y \neq 1$, hence $z = y^{n'} \geq 2$ and $q^m - z^\ell = 1$. Thus $q^m = (z+1)\frac{z^\ell+1}{z+1}$, where both factors are powers of q and not equal to 1 (by (B1.2)). Since q divides $\gcd\left(z+1, \frac{z^\ell+1}{z+1}\right) = 1$ or ℓ (by (P1.2)), then $q = \ell$, so q is odd. Again, by (P1.2), since q is odd, then q^2 does not divide $\frac{z^q+1}{z+1}$, so $\frac{z^q+1}{z+1} = q$. By (P1.2), $q = 3$ and $z = 2$, hence $y = 2$, $n = 3$.

2) The proof is similar. Let ℓ be a prime dividing m, $m = \ell m'$, $z = x^{m'}$, so $z \geq 2$. Hence $p^n = z^\ell - 1 = (z-1)\frac{z^\ell-1}{z-1}$ where both factors are powers of p. Clearly $\frac{z^\ell-1}{z-1} > 1$. If $z - 1 = 1$, then $z = 2$ (a prime), $2^\ell - p^n = 1$, and this is excluded by the first part of the proof. So p divides both factors. But $\gcd\left(z-1, \frac{z^\ell-1}{z-1}\right) = 1$ or ℓ, hence $p = \ell$. As before, if p is odd then p^2 does not divide $\frac{z^p-1}{z-1}$, so $\frac{z^p-1}{z-1} = p$, which is clearly impossible because $z \geq 2$. So $p = 2$, $2^n = z^2 - 1$, and

$$\begin{cases} z - 1 = 2^{n_1} \\ z + 1 = 2^{n_2} \end{cases}$$

with $0 \leq n_1 < n_2$, and $n_1 + n_2 = n$. Therefore $2 = 2^{n_1}(2^{n_2 - n_1} - 1)$, $n_1 = 1$, $n_2 = 2$, $n = 2$, and finally, $x = 3$ and $m = 2$. ∎

Many special cases of the above results of Gérono may be found in the literature: see Catalan (1885), Carmichael (1909), Goheen's problem solved by Hausmann (1941), Cassels (1953), Wall (1957).

Due to its simplicity, I include Wall's proof that if two prime powers are consecutive, then they must be 8 and 9. Indeed, one of the prime powers must be even, so assume that p is an odd prime, $m, n \geq 2$ and $2^m - p^n = \pm 1$.

If n is odd, then

$$2^m = p^n \pm 1 = (p \pm 1)\frac{p^n \pm 1}{p \pm 1}.$$

The second factor in the right-hand side is greater than 1 and odd (by (P1.2)), which is impossible. So n is even, say $n = 2k$. Let $p^k = 2h + 1$, hence

$$2^m = (2h+1)^2 \pm 1.$$

With the $+$ sign, $2^m = 4h^2 + 4h + 2 = 2(2h^2 + 2h + 1)$, which is impossible.

With the $-$ sign, $2^m = 4h^2 + 4h$, hence dividing by 4, $2^{m-2} = h(h+1)$. This is possible only if $m = 3$, from which $h = 1$, $p = 3$, $n = 2$.

The following result extends the one of Moret-Blanc (A6.3):

(B1.2). 1) If $x^m - y^x = 1$ with $m, x \geq 2$, then $x = 3$, $m = 2$, $y = 2$.

2) If $x^y - y^n = 1$ with $n, y \geq 2$, then $x = 3$, $n = 3$, and $y = 2$.

Proof: 1) Let p be a prime factor of m, $m = pm'$ and $z = x^{m'}$. Let q be the largest prime factor of x, and $x = qx'$, $t = y^{x'}$. Then $z^p - t^q = 1$. By (A3.1), $q \geq 3$ and if p is odd, then by Lemma (A1.1), either

$$\begin{cases} t + 1 = b^p \\ \frac{t^q+1}{t+1} = v^p \end{cases} \text{ with } x = bv,\ q \nmid bv,\ \gcd(b,v) = 1$$

or

$$\begin{cases} t + 1 = q^{p-1} b^p \\ \frac{t^q+1}{t+1} = qv^p \end{cases} \text{ with } x = qbv,\ q \nmid v,\ \gcd(b,v) = 1.$$

By (B1.1), $t \neq 2$, hence by (P1.11), the binomial $t^q + 1$ has a primitive factor ℓ, and $\ell \equiv 1 \pmod{q}$, so $q < \ell$. Also, $\ell \nmid t + 1$, hence ℓ divides $\frac{t^q+1}{t+1}$. So, in both cases, $\ell | v$, hence $\ell | x$, and therefore $\ell \leq q$, which is a contradiction. It follows that $p = 2$ and by (A6.2), necessarily $z = 3$, $t = 2$, $q = 3$, and $x = 3$, $m = 2$, $y = 2$.

2) The proof is similar. ∎

Now, it will be shown that 8, 9 are the only consecutive powers of consecutive integers. This was proved by Hampel (1956), and Obláth gave a simpler proof in 1954 (even before Hampel's paper was published). Another straightforward proof was given by Schinzel (1956). In 1956, Rotkiewicz generalized the result (the case $a = 1$ is Hampel's theorem).

(B1.3). *Let* $a \geq 1$, $x, y, m, n \geq 2$ *be integers such that* $gcd(x, y) = 1$ *and* $|x - y| = a$, $x^m - y^n = a^n$. *Then* $x = 3$, $y = 2$, $m = 2$, $n = 3$, *and* $a = 1$.

Proof: By hypothesis $x = y \pm a$ and $gcd(y, a) = 1$, $y \neq a$.

If $x = y + a$, then $y^n + a^n = (y + a)^m$. By (P1.7), $n = 3$, $y = 2$, $a = 1$, so $x = 3$, $m = 2$ (or $n = 3$, $y = 1$, $a = 2$, but this was excluded).

If $x = y - a$ and if p is a primitive factor of $y^{2n} - a^{2n}$, then $p \nmid y^n - a^n$, hence p divides $y^n + a^n = (y - a)^m$, so $p | y - a$, which is impossible. By (P1.7), $n = 3$, $a = 1$, $y = 2$ so $x = 1$ (which is excluded), or $y + a$ is a power of 2 and $n = 1$ (which is also excluded). ∎

2. The Theorem of Cassels and First Consequences

The theorem of Cassels plays a central role in the study of Catalan's problem. Its proof will require three lemmas.

At the end of the section, I shall give an immediate but striking consequence: the non-existence of three consecutive powers.

(B2.1). Lemma. *Let* a, b, t *be real numbers, such that* $b > 0$, $t > 1$, *and* $a + b^t > 0$. *Let* $f_{a,b}(t) = (a + b^t)^{1/t}$. *Then* $f'_{a,b}(t) > 0$ *if and only if* $b^t \log b^t > (a + b^t) \log(a + b^t)$.

In particular, if $m > n > 1$ *and* $z > 1$, *then* $(z^n - 1)^m < (z^m - 1)^n$, *and* $(z^m + 1)^n < (z^n + 1)^m$.

The Theorem of Cassels and First Consequences

Proof: For simplicity, let $f(t) = f_{a,b}(t)$. Then

$$f'(t) = \frac{(a+b^t)^{\frac{1}{t}}}{t}\left[\frac{b^t \log b}{a+b^t} - \frac{1}{t}\log(a+b^t)\right].$$

Hence $f'(t) > 0$ if and only if $\frac{b^t \log b}{a+b^t} > \frac{1}{t}\log(a+b^t)$, or equivalently $b^t \log b^t > (a+b^t)\log(a+b^t)$.

Taking $a = -1$, $b = z > 1$, $t > 1$, then $z^t > z^t - 1 > 0$, $\log z^t > \log(z^t - 1)$, therefore $z^t \log z^t > (z^t - 1)\log(z^t - 1)$. So $f'_{-1,z}(t) > 0$, and if $m > n > 1$, then

$$\left(z^n - 1\right)^{\frac{1}{n}} < \left(z^m - 1\right)^{\frac{1}{m}},$$

so $(z^n - 1)^m < (z^m - 1)^n$.

Similarly, taking $a = 1$, $b = \frac{1}{z}$, $t > 1$, then $0 < \frac{1}{z^t} < 1$, $\frac{1}{z^t}\log\frac{1}{z^t} < 0 < \left(1 + \frac{1}{z^t}\right)\log\left(1 + \frac{1}{z^t}\right)$. So $f'_{1,\frac{1}{z}}(t) < 0$, and if $m > n > 1$, then $\left(1 + \frac{1}{z^m}\right)^{\frac{1}{m}} < \left(1 + \frac{1}{z^n}\right)^{\frac{1}{n}}$, hence $(z^m + 1)^n < (z^n + 1)^m$. ∎

The next lemma gives an upper bound for the ℓ-adic value of a factorial:

(B2.2). Lemma. *Let r, m, n be positive integers. Let ℓ be a prime number, not dividing n. Then*

$$v_\ell(r!) \leq v_\ell\left[\frac{m}{n}\left(\frac{m}{n} - 1\right)\cdots\left(\frac{m}{n} - (r-1)\right)\right].$$

Proof: Let $a = \frac{m}{n}\left(\frac{m}{n} - 1\right)\cdots\left(\frac{m}{n} - (r-1)\right)$ and let $v_\ell(a) = e < \infty$; since $\ell \nmid n$, then $e \geq 0$. Moreover there exists $n' \geq 1$ such that $nn' \equiv 1 \pmod{\ell^{e+1}}$. Let $m' = mn'$ then $\frac{m}{n} - m' = \frac{m}{n}(1 - nn') \equiv 0 \pmod{\ell^{e+1}}$. So $\frac{m}{n} - k \equiv m' - k \pmod{\ell^{e+1}}$ for $k = 0, 1, \ldots, r-1$.

Let $a' = m'(m'-1)\ldots(m'-(r-1))$, then $a' \equiv a \pmod{\ell^{e+1}}$, that is, $v_\ell(a'-a) \geq e+1$. Since $\frac{a'}{r!} = \binom{m'}{r}$, then $v_\ell(a') \geq v_\ell(r!)$. If $v_\ell(r!) > v_\ell(a)$, then $v_\ell(a-a') = v_\ell(a) = e$, which is a contradiction, proving that $v_\ell(a) \geq v_\ell(r!)$. ∎

(B2.3). Lemma. *If $p > q$ are odd primes, x, y are integers, $x, y \geq 2$, and $x^p - y^q = \pm 1$, then*

$$(x \mp 1)^p q^{(p-1)q} > (y \pm 1)^q.$$

Proof: $x \mp 1 \geq \frac{x}{2}$, $x^p = y^q \pm 1 > \frac{y^q}{2}$, $y > \frac{y\pm 1}{2}$, so

$$(x \mp 1)^p \geq \left(\frac{x}{2}\right)^p > \frac{y^q}{2^{p+1}} > \frac{(y \pm 1)^q}{2^{q+p+1}}.$$

But $q^{(p-1)q} > 2^{p+q+1}$. Indeed, $(p-1)(q-1) \geq (q+1)(q-1) = q^2 - 1 > 2 + q$ since $q \geq 3$, so $(p-1)q > p+q+1$. Therefore $(x \mp 1)^p > \frac{(y\pm 1)^q}{q^{(p-1)q}}$, concluding the proof. ∎

Now I prove Cassels' theorem (1953, 1961):

(B2.4). *Let p, q be odd primes, and x, y be positive integers such that $x^p - y^q = \pm 1$. Then p divides y and q divides x.*

Proof: Without loss of generality, I may assume that $p > q$, and I note also that $x \geq 2$, $y \geq 2$.

1°). First I show that $q|x$.

If $q \nmid x$ then $q \nmid y^q \pm 1$. From $x^p = y^q \pm 1 = (y \pm 1)\frac{y^q \pm 1}{y \pm 1}$ and (P1.2), it follows that $\gcd\left(y \pm 1, \frac{y^q \pm 1}{y \pm 1}\right) = 1$, and so there exists an integer $b \geq 1$ such that $y \pm 1 = b^p$.

<u>Case 1.</u> If $y + 1 = b^p$ then $b \geq 2$, and

$$x^p = y^q + 1 = (b^p - 1)^q + 1 < b^{pq},$$

so $x < b^q$, hence $x \leq b^q - 1$. It follows from Lemma (B2.1) that $(b^q - 1)^p < (b^p - 1)^q$ because $q < p$. Hence

$$y^q + 1 = x^p \leq (b^q - 1)^p < (b^p - 1)^q = y^q,$$

which is an absurdity.

Case 2. If $y - 1 = b^p$ then $x^p = y^q - 1$, and from $q < p$ it follows that $x < y$; in particular $y \geq 3$ and $b \geq 2$. Now I have $x^p = (b^p + 1)^q - 1 > b^{pq}$, hence $x > b^q$, so $x \geq b^q + 1$. It follows from Lemma (B2.1) that $\left(1 + \frac{1}{b^p}\right)^q < \left(1 + \frac{1}{b^p}\right)^p$, so $y^q - 1 = x^p \geq (b^q + 1)^p > (b^p + 1)^q = y^q$, which is an absurdity.

2°). I shall show now that $p|y$.

Clearly $y^q \geq 8$. Since $q|x$ then $q \leq x$; from $x^p = y^q \pm 1 = (y \pm 1)\frac{y^q \pm 1}{y \pm 1}$ it follows that $\gcd\left(y \pm 1, \frac{y^q \pm 1}{y \pm 1}\right) = q$.

By (P1.2), there exist integers $b, c > 0$ such that

$$\begin{cases} y \pm 1 = q^{p-1} b^p \\ \frac{y^q \pm 1}{y \pm 1} = q c^p \end{cases}$$

and $q \nmid c$, $x = qbc$.

I deduce also from (P1.2) that $c \neq 1$, otherwise $y = 2$, $q = 3$ and $x^p = y^q + 1 = 9$, so $x = 3$, $p = 2 < q$, which is contrary to the hypothesis.

Next, I show that $c \equiv 1 \pmod{q^{p-1}}$.

Indeed, by (P1.2), $qc^p = \frac{y^q \pm 1}{y \pm 1} = k(y \pm 1) + q$, where k is an integer multiple of q.

Since q^{p-1} divides $y \pm 1$, then q^p divides $q(c^p - 1)$, hence q^{p-1} divides $c^p - 1$. If $c \not\equiv 1 \pmod{q^{p-1}}$ then c modulo q^{p-1} has order p, thus p divides $\varphi(q^{p-1}) = q^{p-2}(q-1)$, hence $p|q-1$ so $p < q$, which is a contradiction. Therefore $c \equiv 1 \pmod{q^{p-1}}$.

So $x \neq qb$ (since $c \neq 1$) and $x \equiv qb \pmod{q^p}$, because $c \equiv 1 \pmod{q^{p-1}}$.

I shall now assume that $p \nmid y$ in order to derive a contradiction.

From $y^q = x^p \mp 1 = (x \mp 1)\frac{x^p \mp 1}{x \mp 1}$ and Lemma (A1.1) it follows from $p \nmid y$ that there exists an integer $a \geq 1$ such that $x \mp 1 = a^q$. Now, I compare the integers a, b:

First, I show that $a > b$. Indeed, by Lemma (B2.3)

$$a^{pq} = (x \mp 1)^p > \frac{(y \pm 1)^q}{q^{(p-1)q}} = b^{pq},$$

hence $a > b$.

Next, I show that $a^q \geq \frac{1}{2}q^p$. As already seen, $x \neq qb$ and $x \equiv qb \pmod{q^p}$, so if I assume that $a^q < \frac{1}{2}q^p$, then

$$q^p \leq |x - qb| = |a^q \pm 1 - qb| \leq a^q + qb \pm 1 < \frac{1}{2}q^p + qb \pm 1,$$

hence $qb \mp 1 > \frac{1}{2}q^p$. But $b \geq 2$, $q \geq 3$, hence $a^q > b^q \geq qb + 1$, so $a^q \geq \frac{1}{2}q^p$.

Now I give lower bounds for x^p, y^q. First, I have $x^p = (a^q \pm 1)^p \geq (a^q - 1)^p$, $y^q = x^p \mp 1 = (a^q \mp 1)^p \mp 1 \geq (a^q - 1)^p$; noting that

$$\left(1 - \frac{2}{q^p}\right)^p \geq \left(1 - \frac{2}{3^p}\right)^p > \frac{1}{3} \geq \frac{1}{q}$$

it follows that:

$$\min\{x^p, y^q\} \geq (a^q - 1)^p = a^{pq}\left(1 - \frac{1}{a^q}\right)^p \geq a^{pq}\left(1 - \frac{2}{q^p}\right)^p > a^{pq}\frac{1}{q}, \quad (2.1)$$

because $a^q \geq \frac{1}{2}q^p$.

Next, I give an upper bound for $|x^{\frac{p}{q}} - y|$. Since

$$\left(x^{\frac{p}{q}} - y\right)\frac{\left(x^{\frac{p}{q}}\right)^q - y^q}{x^{\frac{p}{q}} - y} = x^p - y^q = \pm 1,$$

then

$$\left|x^{\frac{p}{q}} - y\right| = \frac{1}{\left|\sum_{i=0}^{q-1} x^{\frac{pi}{q}} y^{q-1-i}\right|}. \quad (2.2)$$

The Theorem of Cassels and First Consequences

But for each $i = 0, 1, \ldots, q-1$,

$$x^{\frac{pi}{q}} y^{q-1-i} > \left(a^{pq}\frac{1}{q}\right)^{\frac{i}{q}+\frac{q-1-i}{q}} = a^{p(q-1)} \frac{1}{q^{\frac{q-1}{q}}} > a^{p(q-1)}\frac{1}{q}.$$

Therefore

$$\left|x^{\frac{p}{q}} - y\right| < \frac{1}{a^{p(q-1)}}. \tag{2.3}$$

I write:

$$x^{\frac{p}{q}} = (a^q \pm 1)^{\frac{p}{q}} = \sum_{r=0}^{\infty} t_r, \tag{2.4}$$

where

$$t_r = (\pm 1)^r \frac{\frac{p}{q}\left(\frac{p}{q} - 1\right)\left(\frac{p}{q} - r + 1\right)}{r!} a^{p-rq} \neq 0; \tag{2.5}$$

in particular, $t_0 = a^p$. If ℓ is any prime, $\ell \neq q$ and $r \geq 1$, then by Lemma (B2.2)

$$v_\ell(r!) \leq v_\ell\left[\frac{p}{q}\left(\frac{p}{q} - 1\right) \cdots \left(\frac{p}{q} - r + 1\right)\right],$$

so $v_\ell(t_r) \geq v_\ell(a^{p-nq}) \geq 0$, for $p \geq nq$. Let

$$R = \left[\frac{p}{q}\right] + 1, \quad \rho = \left[\frac{R}{q-1}\right], \quad \text{so } Rq > p.$$

As is well known, $v_q(R!) = \frac{R-s}{q-1}$, where s is the sum of the digits of R in its q-adic expansion: $R = R_0 + R_1 q + \ldots + R_m q^m$, $0 \leq R_i \leq q - 1$, $s = R_0 + R_1 + \ldots + R_m$. Since $\frac{R-s}{q-1} < \frac{R}{q-1}$ then $v_q(R!) \leq \rho$.

For every $r < R$, if $\ell \neq q$ then

$$v_\ell\left(t_r q^{R+\rho} a^{R_q - p}\right) \geq v_\ell\left(a^{(R-r)q}\right) \geq 0$$

and
$$v_\ell\left(t_R q^{R+\rho} a^{Rq-p}\right) = 0.$$

Also
$$v_q\left(t_r q^{R+\rho} a^{Rq-p}\right) = -r - v_q(r!) + R + \rho + (R-r)q v_q(a) \geq 0,$$

thus $t_r q^{R+\rho} a^{Rq-p}$ is an integer, for every $r = 0, 1, \ldots, R$.

So the number

$$I = a^{Rq-p} q^{R+\rho}\left((y - x^{\frac{p}{q}}) + \sum_{r \geq R+1} t_r\right) = a^{Rq-p} q^{R+\rho}\left(y - \sum_{r=0}^{R} t_r\right) \tag{2.6}$$

is an integer, since $Rq - p > 0$.

I shall show that $I \neq 0$.

Write $I = I_1 + I_2 + I_3$, where

$$\begin{cases} I_1 = a^{Rq-p} q^{R+\rho}\left(y - x^{\frac{p}{q}}\right) \\ I_2 = a^{Rq-p} q^{R+\rho} t_{R+1} \neq 0 \\ I_3 = a^{Rq-p} q^{R+\rho} \sum_{r > R+1} t_r. \end{cases} \tag{2.7}$$

If $r > R$ then $\left|\frac{t_{r+1}}{t_r}\right| = \left|\frac{\frac{p}{q} - r}{r+1}\right| \frac{1}{a^q} < \frac{1}{a^q} \leq \frac{2}{q^p}$, since $\left|\frac{p}{q} - r\right| = r - \frac{p}{q} < r + 1$. Hence

$$\left|\frac{I_3}{I_2}\right| = \left|\sum_{r > R+1} \frac{t_r}{t_{R+1}}\right| \leq \sum_{r > R+1} \left|\frac{t_r}{t_{R+1}}\right|$$
$$= \left|\frac{t_{R+2}}{t_{R+1}}\right| + \left|\frac{t_{R+3}}{t_{R+1}}\right| + \cdots$$
$$< \frac{2}{q^p} + \left(\frac{2}{q^p}\right)^2 + \left(\frac{2}{q^p}\right)^3 + \cdots$$
$$= \frac{2}{q^p} \cdot \frac{1}{1 - \frac{2}{q^p}} = \frac{2}{q^p - 1} \leq \frac{2}{3^5 - 2} < \frac{1}{10}.$$

Next I show that

$$\frac{1}{q^2(R+1)^2} \le \left|a^{(R+1)q-p}t_{R+1}\right| \le \frac{1}{4}. \qquad (2.8)$$

Indeed,

$$\left|\frac{p}{q}\left(\frac{p}{q}-1\right)\cdots\left(\frac{p}{q}-R\right)\right| \le R(R-1)\cdots 2\left|\frac{p}{q}-R+1\right|\left|\frac{p}{q}-R\right| \le R!\frac{1}{4},$$

because $\left(R - \frac{p}{q}\right) + \left(\frac{p}{q} + 1 - R\right) = 1$, so their product is at most equal to $\frac{1}{4}$. By (2.5), $|t_{R+1}| \le \frac{a^{p-(R+1)q}}{4(R+1)}$.

On the other hand

$$\left|\frac{p}{q}\left(\frac{p}{q}-1\right)\cdots\left(\frac{p}{q}-R\right)\right| \ge (R-1)(R-2)$$

$$\cdots 1\left|\frac{p}{q}-R+1\right|\left|\frac{p}{q}-R\right| \ge \frac{(R-1)!}{q^2},$$

since $R - \frac{p}{q} \ge \frac{1}{q}$ and $\frac{p}{q} - (R-1) \ge \frac{1}{q}$. Therefore by (2.5), $|t_{R+1}| \ge \frac{a^{p-(R+1)q}}{q^2 R(R+1)}$, so

$$\frac{1}{q^2(R+1)^2} \le \frac{1}{q^2(R+1)R} \le \left|a^{(R+1)q-p}t_{R+1}\right| \le \frac{1}{4(R+1)} \le \frac{1}{4}. \qquad (2.9)$$

I use these estimates to show that $\left|\frac{I_1}{I_2}\right| < \frac{1}{10}$. Indeed by (2.9) and (2.3),

$$\left|\frac{I_1}{I_2}\right| = \left|\frac{y - x^{\frac{p}{q}}}{t_{R+1}}\right| \le a^{(R+1)q-p}\left|y - x^{\frac{p}{q}}\right|q^2(R+1)^2$$

$$< \frac{a^{(R+1)q-p}q^2(R+1)^2}{a^{p(q-1)}} = \frac{q^2(R+1)^2}{a^{q(p-R-1)}}.$$

Since $p > q \ge 3$, then $p \ge 5$, hence

$$p - R - 1 = p - \left[\frac{p}{q}\right] - 2 \ge 2;$$

indeed, this is true if $p = 5$, and if $p \geq 7$ then $p\left(\frac{q-1}{q}\right) \geq p^{\frac{2}{3}} \geq 4$ so $p - \left[\frac{p}{q}\right] \geq p - \frac{p}{q} \geq 4$. Then $R + 1 \leq p - 2 \leq p$. Therefore

$$\frac{q^2(R+1)^2}{a^{q(p-R-1)}} \leq \frac{q^2(R+1)^2}{\left(\frac{1}{2}q^p\right)^2} \leq \left(\frac{2p}{q^{p-1}}\right)^2 \leq \left(\frac{2p}{3^{p-1}}\right)^2$$

$$\leq \left(\frac{2 \times 5}{3^4}\right)^2 \leq \frac{1}{10}.$$

I deduce that

$$|I| = |I_2|\left|1 + \frac{I_1}{I_2} + \frac{I_3}{I_2}\right| \geq |I_2|\left(1 - \frac{1}{10} - \frac{1}{10}\right) \neq 0.$$

Hence $I \neq 0$ and since I is an integer, then $|I| \geq 1$.

Now, I shall derive a contradiction, by obtaining an upper estimate for $|I|$.

The following holds by (2.8),

$$|I_2| = \left|\frac{q^{R+\rho}a^{(R+1)q-p}t_{R+1}}{a^q}\right| \leq \frac{q^{R+\rho}}{4a^q} \leq \frac{1}{2}q^{R+\rho-p}.$$

Hence

$$1 \leq |I| = |I_2|\left|1 + \frac{I_1}{I_2} + \frac{I_3}{I_2}\right| \leq \frac{1}{2}q^{R+\rho-p}\left(1 + \frac{1}{10} + \frac{1}{10}\right) < q^{R+\rho-p},$$

therefore $R + \rho - p > 0$. But

$$R + \rho \leq R\left(1 + \frac{1}{q-1}\right) \leq \left(\frac{p}{q} + 1\right)\frac{q}{q-1} = \frac{p+q}{q-1} < \frac{2p}{q-1} \leq p.$$

I conclude that $R + \rho - p < 0$ and this is a contradiction. ∎

Another proof of the theorem of Cassels was given by Hyyrö (1964b).

Using Cassels' theorem Mąkowski solved in 1962 a problem by LeVeque (1956) and Sierpiński (1960). The same proof was also found by Hyyrö in 1963 (unaware of Mąkowski's result), and published in Finnish.

(B2.5). *Three consecutive integers cannot be proper powers.*

Proof: Suppose that three consecutive integers are proper powers, so there exist prime numbers ℓ, p, q and integers x, y, z such that
$$x^\ell - y^p = 1, \qquad y^p - z^q = 1.$$

By Cassels' theorem, $p|x$ and $p|z$, hence $p|x^\ell - z^q = 2$. So $x^\ell - y^2 = 1$. By (A3.1) this is impossible. ∎

Another consequence of Cassels' theorem concerns the numbers $F_{a,n} = a^{a^n} + 1$, with $a \geq 2$, $n \geq 0$, studied by Ferentinou-Nicolacopoulou in 1963 and by Ribenboim in 1979. I show now:

(B2.6). $F_{a,n}$ *is not a proper power (for $n \geq 1$).*

Proof: If $F_{a,n}$ is a proper power, write $a^{a^n} + 1 = m^p$ with some prime p and $m \geq 2$. If q is a prime dividing a and $a^n = qa'$, then $m^p - (a^{a'})^q = 1$. By Cassels' theorem, $q|m$; but $q|a$ and this is impossible. ∎

To end this section, I use Cassels' theorem, to rephrase Lemma (A1.1):

(B2.7). Lemma. *If p, q are odd primes, and if $x, y \geq 1$ are such that $x^p - y^q = 1$, then there exist natural numbers a, b, u, v such that*

$$\begin{cases} x - 1 = p^{q-1}a^q \\ \frac{x^p - 1}{x - 1} = pu^q \end{cases} \quad p \nmid u, \; gcd(a, u) = 1, \; y = pau,$$

$$\begin{cases} y + 1 = q^{p-1}b^p \\ \frac{y^q + 1}{y + 1} = qv^p \end{cases} \quad q \nmid v, \; gcd(b, v) = 1, \; x = qbv.$$

Proof: By Cassels' theorem, $p|y$ and $q|x$, hence only the alternatives indicated above can occur. ∎

3. Prime Factors of Solutions of Catalan's Equation

In this section, I shall indicate conditions which must be satisfied by the prime factors of eventual solutions x, y of Catalan's equation. Without loss of generality, I may assume that the exponents are primes and consider the equation $X^p - Y^q = 1$; moreover by (A3.1), (A6.2), and (A7.3), $p, q > 3$.

According to Gérono's results (B1.1), x and y cannot be primes.

There have been many extensions of Gérono's result, concerning the kind of prime factors of the solutions x, y.

It is best to prove first a result of Rotkiewicz (1960), from which it is immediate to deduce statements first proved by Obláth and Hampel.

(B3.1). 1) *If q is a prime, m is odd and $m \geq 3$, if $x, y \geq 1$ and $x^m - y^q = 1$, then there exists a prime ℓ such that $\ell \equiv 1 \pmod{q}$ and $q\ell$ divides x.*

2) *If p is an odd prime, $n \geq 2$, if $x, y \geq 1$ and $x^p - y^n = 1$, then there exists a prime h such that $h \equiv 1 \pmod{p}$ and ph divides y.*

Proof: 1) Let p be a prime dividing m, $m = bm'$ and $z = x^{m'}$, so $z^p - y^q = 1$. By the hypothesis and (A3.1) p, q are odd. By Lemma (B2.7)
$$\begin{cases} y + 1 = q^{p-1}b^p \\ \frac{y^q + 1}{y + 1} = qv^p, \end{cases}$$
where $gcd(b, v) = 1$ and $qbv = z$.

Since $y \neq 2$ by (B1.1), it follows from (P1.7) that $y^q + 1$ has a primitive factor ℓ, thus $\ell \nmid y + 1$, hence ℓ divides $\frac{y^q+1}{y+1}$. Also $\ell \equiv 1 \pmod{q}$ by (P1.4): it follows that $\ell | v$, hence $\ell | z$, so $\ell | x$. Since $\ell \neq q$ and $q | x$ by (B2.4), then $q\ell | x$.

2) The proof is similar. ∎

Obláth's result of 1941 is a simple corollary. Obláth proved the following result in 1941:

(B3.2). *Assume that $p, q > 3$, $x, y \geq 1$ and that $x^p - y^q = 1$. Then the prime factors of xy cannot all be of the form $2^a 3^b + 1$ (with $a, b \geq 0$).*

Proof: If $x^p - y^q = 1$, then $p, q > 3$ as already seen. By (B3.1), there exist prime factors ℓ of x and h of q, such that $\ell \equiv 1 \pmod{q}$ and $h \equiv 1 \pmod{p}$. So ℓ, h are not of the form $2^a 3^b + 1$ (with $a \geq 0$, $b \geq 0$). ∎

In particular, if $x^p - y^q = 1$ (with $p, q \geq 3$) then x, y have each at least two odd prime factors; another proof of this fact was given by Ribenboim (1979). Thus, x, y cannot be powers of 10 (proved in 1960 by Hampel). Also, x, y cannot be of the form $2^a \ell^b$, where ℓ is an odd prime and $a, b \geq 1$ (which is better than what Obláth proved in 1940).

Similarly, x, y cannot be of the form $2^a 3^b k^c$ where $a \geq 0$, $b \geq 1$, $c \geq 1$ and k is a prime. Indeed, by Nagell's results (A7.3), $p, q > 3$. By (B3.1), there exists a prime $\ell > q$ such that ℓ divides $x = 2^a 3^b k^c$, so $\ell = k$. By (B2.4), $q | x$, so $q = k$ and this is a contradiction. Similarly, there exists a prime $h > p$ such that h divides $y = 2^a 3^b k^c$, so $h = k$; from $p | y$, then $p = k$, which is absurd.

For every integer $a \geq 1$, let $\omega(a)$ denote the number of distinct prime factors of a.

(B3.3). *If $x^m - y^n = 1$, with $m, n \geq 3$, $x, y \geq 1$, then $\omega(x) \geq 1 + \omega(n)$, $\omega(y) \geq 1 + \omega(m)$ and $\omega(xy) \geq 3 + \omega(m) + \omega(n)$.*

Proof: Let $p_1 < \ldots < p_{\omega(m)}$ be the primes dividing m, and let $q_1 < \ldots < q_{\omega(n)}$ be the primes dividing n. Clearly, $p_i \neq q_j$ and $p_1, q_1 \neq 2$, by (A3.1) and (A6.2). By (B2.4), each p_i divides x and each q_j divides y.

Let $n_1 = \frac{n}{q_{\omega(n)}}$, $y_1 = y^{n_1}$, hence $x^m - y_1^{q_{\omega(n)}} = 1$. By (B3.1) there exists a prime ℓ such that $\ell | x$ and $\ell \equiv 1 \pmod{q_{\omega(n)}}$. Then $q_{\omega(n)} < \ell$ and therefore $\omega(x) \geq 1 + \omega(n)$. Similarly $\omega(y) \geq 1 + \omega(m)$.

Finally, since no prime can divide both x, y, and $2 | xy$, it follows that $\omega(xy) \geq 3 + \omega(m) + \omega(n)$. ∎

In particular, if $x^p - y^q = 1$ with p, q odd primes, $x, y \geq 1$, then $\omega(xy) \geq 5$.

The above considerations also show that

$$x \geq q_1 \ldots q_{\omega(m)}(2q_{\omega(m)} + 1) \quad \text{and} \quad y \geq p_1 \ldots p_{\omega(n)}(2p_{\omega(n)} + 1).$$

Thus, if $x^p - y^q = 1$, $p, q \geq 3$, $x, y \geq 1$, then $x \geq q(2q+1)$, $y \geq p(2p+1)$. By the results of Part A $p, q \geq 5$, hence $x, y \geq 55$.

In 1961, Rotkiewicz showed with a finer analysis that $x, y > 10^6$. I omit this proof, since I will give in §C3 much larger lower bounds for x, y.

4. The Theorem of Hyyrö

I shall now indicate divisibility conditions which were given by Hyyrö and are sharper than the conditions in Cassels's theorem. I keep the notation of Lemma (B2.7).

Hyyrö proved (in 1964b):

(B4.1). *Let p, q be odd prime numbers. Let x, y be positive integers such that $x^p - y^q = 1$. Then:*

i) $a = qa_0 - 1$, $b = pb_0 + 1$ with integers $a_0, b_0 \geq 1$.

ii) $x \equiv 1 - p^{q-1} \pmod{q^2}$, $y \equiv -1 + q^{p-1} \pmod{p^2}$.

iii) $q^2 | x$ if and only if $p^{q-1} \equiv 1 \pmod{q^2}$, $p^2 | y$ if and only if $q^{p-1} \equiv 1 \pmod{p^2}$.

Proof: i) By Lemma (B2.7), $a \equiv a^q \equiv a^q p^{q-1} \equiv x - 1 \equiv -1 \pmod{q}$. Similarly, $b \equiv b^p \equiv b^p q^{p-1} \equiv y + 1 \equiv 1 \pmod{p}$.

Hence $a = qa_0 - 1$, with $a_0 \geq 1$ since $a > 0$, and $b = bp_0 + 1$, with $b_0 \geq 0$ since $b > 0$. But actually $b_0 \geq 1$. Otherwise $b = 1$, hence $x^p = y^q + 1 < (y+1)^q$, and this implies, as I explain below, that $x < y(y+1)^{\frac{q}{p}} = q^{\left(\frac{p-1}{p}\right)q} < q^q < 2^q(q-1)^q \leq p^{q-1}a^q = x - 1 < x$, an absurdity. I used the fact that $q - 1 \leq qa_0 - 1 = a$ and for $p, q \geq 3$, $2 < \left(\frac{p}{2}\right)^{q-1}$.

ii) I have $a^q \equiv -1 \pmod{q^2}$, so $x \equiv 1 + p^{q-1}a^q \equiv 1 - p^{q-1} \pmod{q^2}$.
Similarly $b^q \equiv 1 \pmod{p^2}$, so $y \equiv -1 + q^{p-1}b^p \equiv -1 + q^{p-1} \pmod{p^2}$.

iii) This follows at once from the above congruences. ∎

Now, I consider also the case where one of the exponents may be composite. Hyyrö (1964a) showed:

(B4.2). *Let $q \geq 3$ be a prime, let $m \geq 3$ and let x, y be positive integers such that $x^m - y^q = \pm 1$. Then:*

i) *If $e \geq 1$ and q^e divides x, then*

$$\begin{cases} y \pm 1 = q^{em-1}c^m \\ \frac{y^q \pm 1}{y \pm 1} = qv^m \end{cases}$$

with integers $c, v \geq 1$, $\gcd(c, v) = 1$, $q \nmid v$, $x = q^e cv$. Moreover, every prime divisor ℓ of v is such that $\ell \equiv 1 \pmod{q}$. Hence $v \equiv 1 \pmod{q}$. Also $y \pm 1$ divides $v^m - 1$, q^{em-1} divides $v - 1$ and $x > q^{e(m+1)-1}$, $x - 1 > q^{e(m+1)-1}c$.

ii) *If m is composite and if p is any prime dividing m, then $p^{q-1} \equiv 1 \pmod{q^2}$.*

iii) *If m is composite, then $q^2 | x$.*

Proof: i) I have $x^m = y^q \pm 1 = (y \pm 1)\frac{y^q \pm 1}{y \pm 1}$, and I note that $q \nmid m$. By (P1.2), $\frac{y^q \pm 1}{y \pm 1} = k(y \pm 1) + q = (y \pm 1)^{q-1} + k'q$ (with integers k, k'), so $q | y \pm 1$ if and only if q divides $\frac{y^q \pm 1}{y \pm 1}$. By hypothesis, q^e divides x, so from (P1.2), $\gcd(y \pm 1, \frac{y^q \pm 1}{y \pm 1}) = q$. Also, by the same result, q^2 does not divide $\frac{y^q \pm 1}{y \pm 1}$. Hence

$$\begin{cases} y \pm 1 = q^{em-1} c^m \\ \frac{y^q \pm 1}{y \pm 1} = q v^m \end{cases}$$

with $c, v \geq 1$, $\gcd(c, v) = 1$, $q \nmid v$, $x = q^e c v$.

From this, I deduce that

$$\left((\mp y)^{q-1} - 1 \right) + \left((\mp y)^{q-2} - 1 \right) + \ldots + \left((\mp y) - 1 \right)$$
$$= \frac{(\mp y)^q - q}{(\mp y) - 1} - q = q(v^m - 1), \quad \text{that is,}$$
$$\left[\frac{(\mp y)^{q-1} - 1}{(\mp y) - 1} + \ldots + \frac{(\mp y)^2 - 1}{(\mp y) - 1} + 1 \right] \left((\mp y) - 1 \right) = q(v^m - 1).$$

Since $\mp y \equiv 1 \pmod{q}$, then the number in brackets is

$$\sum_{j=0}^{q-2} (\mp y)^j + \sum_{j=0}^{q-3} (\mp y)^j + \ldots + \left((\mp y) + 1 \right) + 1$$
$$\equiv (q-1) + (q-2) + \ldots + 2 + 1 \equiv \frac{q(q-1)}{2} \equiv 0 \pmod{q}.$$

It follows that $(\mp y) - 1$ divides $v^m - 1$, hence q^{em-1} divides $v^m - 1$.

Now I show that if ℓ is any prime dividing v, then $\ell \equiv 1 \pmod{q}$. Indeed $\ell \neq q$, $\ell \nmid c$, so $\ell \nmid y \pm 1$. On the other hand, ℓ divides $\frac{(\mp y)^q - 1}{(\mp y) - 1}$, hence divides $y^q \pm 1$. By (P1.4), $\ell \equiv 1 \pmod{q}$. This implies that $v \equiv 1 \pmod{q}$. But by (P1.2), $\frac{v^m - 1}{v - 1} = (v - 1)f + m$ (with some integer f), since q divides $v - 1$, and $q \nmid m$, then $q \nmid \frac{v^m - 1}{v - 1}$. Since $q^{em-1} | v^m - 1$, then

$q^{em-1}|v-1$ and so $x = q^e cv > q^e(v-1) \geq q^{e(m+1)-1}$. Actually, I have shown also that $x - 1 > q^e c(v-1) \geq q^{e(m+1)-1}c$.

ii) By (A3.1) and (A6.2), m is odd. Let p be any prime dividing m, so $m = pm'$, with $m' > 2$. If $z = x^{m'}$, then $z^p - y^q = \pm 1$ and so by Cassels' theorem, $q|z$ and $q|x$. Therefore, $q^2|x^{m'} = z$, so by (B4.1), $p^{q-1} \equiv 1 \pmod{q^2}$.

iii) In order to show the third assertion, I write $m = p_1 p_2 \ldots p_s$ with each p_i an odd prime (not necessarily distinct), so $s \geq 2$. I define inductively $x_s = x$, $x_{s-1} = x_s^{p_s}$, $x_{s-2} = x_{s-1}^{p_{s-1}}, \ldots, x_0 = x_1^{p_1}$.

I show that for every $i = 0, 1, \ldots, s$, there exist integers $a_i, h_i \geq 1$ such that all prime factors of h_i are among $p_1, \ldots, p_i$ and $x_i \mp 1 = h_i a_i^q$.

Indeed, if $i = 0$ then $x_0 \mp 1 = x_1^{p_1} \mp 1 = y^q$ (since $x_1 = x^{p_2 \cdots p_s}$), thus, I take $h_0 = 1$, $a_0 = y$.

Proceeding by induction on i, and noting that each p_i is odd,

$$h_{i-1} a_{i-1}^q = x_{i-1} \mp 1 = x_i^{p_i} \mp 1 = (x_i \mp 1) \frac{x_i^{p_i} \mp 1}{x_i \mp 1}.$$

Since $gcd\left(x_i \mp 1, \frac{x_i^{p_i} \mp 1}{x_i \mp 1}\right) = 1$ or p_i, it follows that $x_i \mp 1$ must be equal to $h_i a_i^q$, where the prime factors of h_i are among $p_1, \ldots, p_{i-1}, p_i$. For $i = s$, I obtain $x \mp 1 = h_s a_s^q$.

Since each p_i satisfies $p_i^{q-1} \equiv 1 \pmod{q^2}$, by the second part of the proof, then $h_s^{q-1} \equiv 1 \pmod{q^2}$, thus $h_s^q \equiv h_s \pmod{q^2}$. Since $q|z$, hence $q|x$, then $\mp 1 \equiv h_s a_s^q \equiv h_s a_s \pmod{q}$, so $\mp 1 \equiv (h_s a_s)^q \equiv h_s a_s^q \pmod{q^2}$, therefore $q^2|x$. ∎

5. The Theorems of Inkeri

I shall present some theorems of Inkeri which give many pairs of prime exponents (p, q) for which the equation $X^p - Y^q = 1$ has no solution in non-zero integers.

The hypothesis of Inkeri's theorems involve the class number h_p of the cyclotomic field $\mathbb{Q}(\zeta_p)$ and the class number $H(-p)$ of the imaginary quadratic field $\mathbb{Q}(\sqrt{-p})$, where p is an odd prime.

In §P2 I have recalled the basic facts about the arithmetic of the cyclotomic field $\mathbb{Q}(\zeta)$, where $\zeta = \cos\frac{2\pi}{p} + i\sin\frac{2\pi}{p}$ is a primitive p^{th} root of 1.

Concerning the class number $H(-p)$, I shall need the following facts:

Let $K = \mathbb{Q}(\sqrt{-d})$, with $d > 1$, d square-free.
Its discriminant is

$$D = \begin{cases} -d & \text{if } d \equiv 3 \pmod{4} \\ -4d & \text{if } d \equiv 1 \text{ or } 2 \pmod{4}, \end{cases}$$

so

$$K = \mathbb{Q}(\sqrt{D}).$$

For the following result, see Narkiewicz (1974), page 389:

(B5.1). *The class number $H(D)$ satisfies the following inequality:*

$$H(D) \leq \frac{2}{\pi}\sqrt{|D|}\left(1 + \log\frac{2}{\pi}\sqrt{|D|}\right). \tag{5.1}$$

For $|D| > e^{24}$, the simpler estimate holds:

$$H(D) \leq \frac{1}{3}\sqrt{|D|}\log|D|. \tag{5.2}$$

In particular, if p is a prime, $p \equiv 3 \pmod{4}$, then $-p$ is the discriminant of $\mathbb{Q}(\sqrt{-p})$ and

$$H(-p) \leq \frac{1}{2}\sqrt{p}\log p; \tag{5.3}$$

this weaker estimate holds however for every $p \geq 7$.

The Theorems of Inkeri

In 1963, Gut gave an elementary proof of the estimate

$$H(D) < \frac{|D|}{4} \tag{5.4}$$

which will suffice in many of the proofs to follow.

I shall also need some easy properties of the *Fermat quotient* with base $a \geq 1$:

$$\psi_p(a) = \frac{a^{p-1} - 1}{p} \tag{5.5}$$

(where p is an odd prime and $p \nmid a$).

(B5.2). Lemma. *If p is an odd prime, if p does not divide ad, then*

$$d\psi_p(p \pm d) \equiv d\psi_p(d) \mp 1 \pmod{p}.$$

Proof:

$$(p \pm d)^p \equiv \pm d^p \pmod{p^2}, \quad \text{so}$$
$$(p \pm d)^p - (p \pm d) \equiv \pm(d^p - d) - p \pmod{p^2},$$

and dividing by p,

$$(p \pm d)\psi_p(p \pm d) \equiv \pm d\psi_p(d) - 1 \pmod{p},$$

so finally

$$d\psi_p(p \pm d) \equiv d\psi_p(d) \mp 1 \pmod{p}. \quad \blacksquare$$

In particular,

$$\psi_p(p \pm 1) \equiv \mp 1 \pmod{p},$$
$$2\psi_p(p \pm 2) \equiv 2\psi_p(2) \mp 1 \pmod{p}.$$

The applications of Inkeri's theorems will require tables of residues of q^{p-1} modulo p^2 (where p, q are distinct primes), tables of the first factor h_p of the class number of the pth-cyclotomic field, as well as tables of $H(-p)$.

Now I prove Inkeri's first theorem (1964):

(B5.3). *Let p, q be odd primes, $p, q > 3$, and let x, y be non-zero integers such that $x^p - y^q = 1$.*

i) *If $p \equiv 3 \pmod{4}$ and $q \nmid H(-p)$, then $p^{q-1} \equiv 1 \pmod{q^2}$, q^2 divides x and $y \equiv -1 \pmod{q^{2p-1}}$.*

ii) *If $q \equiv 3 \pmod{4}$ and $p \nmid H(-q)$, then $q^{p-1} \equiv 1 \pmod{p^2}$, p^2 divides y and $x \equiv 1 \pmod{p^{2q-1}}$.*

iii) *If $p > q > 3$, $p \equiv q \equiv 3 \pmod{4}$ and $q \nmid H(-p)$, then $p^{q-1} \equiv 1 \pmod{q^2}$, $q^{p-1} \equiv 1 \pmod{p^2}$, $p^2 | x$, $q^2 | y$, $x \equiv 1 \pmod{p^{2q-1}}$ and $y \equiv -1 \pmod{q^{2p-1}}$.*

Proof: i) By Lemma (B2.7)

$$\begin{cases} x - 1 = p^{q-1} a^q \\ \frac{x^p - 1}{x - 1} = p u^q, \end{cases} \tag{5.6}$$

with $p \nmid u$, $\gcd(a, u) = 1$, $y = pau$; moreover, u is odd, by (P1.2), part (vi).

I shall use (P1.10). Noting that $p \equiv 3 \pmod{4}$, then $p^* = (-1)^{\frac{p-1}{2}} p = -p$, hence

$$pu^q = \frac{x^p - 1}{x - 1} = F_1(x)^2 + pG_1(x)^2, \tag{5.7}$$

where $F_1(X) = \frac{F(X)}{2}$, $G_1(X) = \frac{G(X)}{2}$ and

$$\begin{cases} F(X) = A(X) + B(X) \\ G(X) = -\frac{\tau}{p}[A(X) - B(X)] \end{cases}$$

The Theorems of Inkeri

are polynomials, with integral coefficients. Recall from (P1.10) that

$$A(X) = \prod_a (X - \zeta^a)$$

(a runs over the quadratic residues mod p, $1 \le a \le p-1$),

$$B(X) = \prod_b (X - \zeta^b)$$

(b runs over the quadratic non-residues mod p, $1 \le b \le p-1$),

and τ is the principal Gaussian sum

$$\tau = \sum_{m=1}^{p-1} \left(\frac{m}{p}\right)\zeta^m = \sum_a \zeta^a - \sum_b \zeta^b.$$

Note also that $\tau^2 = -p$. Now I observe that $F_1(x)$, $G_1(x)$ are integers. Indeed, if x is even by (P1.12) $G(x)$ is even, so $G_1(x)$ is an integer, hence $F_1(x)$ is an integer. If x is odd, $G(x) \equiv G(1) \equiv 1 + a_2 + \ldots + a_{\frac{p-5}{2}} + 1 \pmod{2}$. Since $G(X) = X^{\frac{p-1}{2}} G\left(\frac{1}{X}\right)$, then $a_2 = a_{\frac{p-5}{2}}$, $a_3 = a_{\frac{p-7}{2}}, \ldots$, hence $G(x) \equiv 0 \pmod{2}$, thus $G_1(x)$ is an integer and again $F_1(x)$ is an integer, by (5.7).

From (5.7), $p | F_1(x)$. Writing $F_1(x) = ps$, $G_1(x) = t$, then from (5.7) it follows that

$$a^q = ps^2 + t^2 = (t + \sqrt{-p}s)(t - \sqrt{-p}s). \tag{5.8}$$

Now I show that $gcd(s,t) = 1$. Indeed, if a prime ℓ divides s and t, then ℓ divides u and $\ell \ne 2$, since u is odd. But ℓ divides $F(x)$, $G(x)$, so it divides $A(x) = \frac{1}{2}[F(x) + G(x)]$ and $B(x) = \frac{1}{2}[F(x) - G(x)]$ in the cyclotomic field $\mathbb{Q}(\zeta)$.

Let L be a prime ideal of $\mathbb{Q}(\zeta)$ which divides ℓ; then there exist a, b, with $1 \le a, b \le p-1$, such that $\left(\frac{a}{p}\right) = 1$, $\left(\frac{b}{p}\right) = -1$, and L divides the factor $x - \zeta^a$ of $A(x)$ and the factor $x - \zeta^b$ of $B(x)$. Therefore L divides $\zeta^a - \zeta^b = \zeta^a(1 - \zeta^{b-a})$. Since $a \not\equiv b \pmod{p}$, then $1 - \zeta^{b-a}$ is associated with $1 - \zeta$. But p is associated with $(1 - \zeta)^{p-1}$, so L divides p, that is $\ell = p$, hence $p|u$, which is a contradiction.

In view of $gcd(s, t) = 1$, it follows that the principal ideals of $\mathbb{Q}(\sqrt{-p})$ generated by $t + \sqrt{-p}s$ and $t - \sqrt{-p}s$ are relatively prime. Indeed, if a prime ideal Q of $\mathbb{Q}(\sqrt{-p})$ divides the above ideals, then Q divides $2t$; but $Q|u$ and u is odd, so $Q \nmid 2$, hence $Q|t$, so Q divides $\sqrt{-p}s$. But $gcd(s, t) = 1$ hence Q must divide $\sqrt{-p}$, so $Q|p$. The norm of Q is a power of p, not equal to 1, so p divides the norm of Q, which divides u^2, so p divides u, which is contrary to the hypothesis.

From (5.8), I conclude that there exists an ideal I of $\mathbb{Q}(\sqrt{-p})$ such that the principal ideal generated by $t + \sqrt{-p}s$ is $\left(t + \sqrt{-p}s\right) = I^q$. Thus, in the class group of $Q(\sqrt{-p})$, the order of the class of the ideal I divides q; but it also divides the class number $H(-p)$ of $\mathbb{Q}(\sqrt{-p})$. Since $q \nmid H(-p)$, then I is itself a principal ideal. Recalling that the units of $\mathbb{Q}(\sqrt{-p})$ are $1, -1$ when $p > 3$, I may write:

$$t + \sqrt{-p}s = \left(\frac{m + \sqrt{-p}n}{2}\right)^q,$$

where m, n are integers, $m \equiv n \pmod{2}$. Comparing the two sides of the above relation

$$2^{q-1}G(x) = 2^q t = m^q - \binom{q}{2}m^{q-2}n^2 p \pm \ldots \pm qmn^{q-1}p^{\frac{q-1}{2}}$$

$$\equiv m^q \pmod{qm}.$$

By (P1.10) $G(x) = x(1 + a_2 x + \ldots + x^{\frac{p-5}{2}})$. Since $q|x$, the above congruence implies that $q|m$. Therefore $q^2|m^q$, $q^2|mq$, hence $q^2|G(x)$ and so $q^2|x$ as I wanted to prove.

Since $x^p - y^q = 1$, by (B4.1) $p^{q-1} \equiv 1 \pmod{q^2}$. By Lemma (B2.7) there exist positive integers b, v such that

$$\begin{cases} y + 1 = q^{p-1} b^p \\ \frac{y^p + 1}{y + 1} = qv^q \end{cases}$$

with $q \nmid v$, $x = qbvc$, and $\gcd(b, v) = 1$. Since $q^2 | x$, then $q | b$ and therefore q^{2p-1} divides $y + 1$, that is $y \equiv -1 \pmod{q^{2p-1}}$.

ii) If $x^p - y^q = 1$, then $(-y)^q - (-x)^p = 1$.

The statement ii) follows at once from i), interchanging p and q.

iii) Since $p > q > \frac{q}{4} > H(-1)$, by Gut's result (5.4) quoted above, then $p \nmid H(-q)$. The result now follows from i) and ii). ∎

Here are simple consequences, given by Aaltonen and Inkeri (1990):

(B5.4). *Let p, q be primes such that $p = q + d$, where $-3p < d < 3q$. Assume that the equation $X^p - Y^q = 1$ has non-trivial solutions in integers.*

i) *If $p \equiv 3 \pmod{4}$, then $p^{q-1} \equiv 1 \pmod{q^2}$ and $\frac{d^q - d}{q} \equiv 1 \pmod{q}$*

ii) *If $q \equiv 3 \pmod{4}$, then $q^{p-1} \equiv 1 \pmod{p^2}$ and $\frac{d^p - d}{p} \equiv -1 \pmod{p}$.*

Proof: i) By hypothesis, $d < 3q$, hence $\frac{p}{4} = \frac{q+d}{4} < q$. Since $H(-p) < \frac{p}{4}$ as already indicated, by (B5.1), $p^{q-1} \equiv 1 \pmod{q^2}$. Hence the Fermat quotient satisfies $\psi_q(p) \equiv 0 \pmod{q}$. From Lemma (B5.2),

$$d\psi_q(q + d) \equiv d\psi_q(d) - 1 \equiv \pmod{q},$$

so
$$\frac{d^q - d}{q} \equiv 1 \pmod{q}.$$

ii) The proof is similar and therefore omitted. ∎

The above result applies in particular for twin primes $p = q+2$. The only pair of twin primes less than 10^4 satisfying the congruence
$$\begin{cases} p^{q-1} \equiv 1 \pmod{q^2}, & \text{if } p \equiv 3 \pmod 4, \text{ or} \\ q^{p-1} \equiv 1 \pmod{p^2}, & \text{if } q \equiv 3 \pmod 4 \end{cases}$$
is (5.7). Thus for all twin primes in the above range, different from (5.7), Catalan's equation has only trivial solutions. The equations $X^5 - Y^7 = \pm 1$ will be considered afterwards.

Here is another application of (B5.3):

(B5.5). *Let p, q be primes such that $q = kp + r$, with $1 \le k$, r odd, $|r| < p$, and also $r^q \equiv r \pmod{q^2}$. Assume that $p \equiv 3 \pmod 4$ and $X^p - Y^q = 1$ has a solution in non-zero integers. Then:*
$$r\psi_q(k) \equiv 1 \pmod{q}.$$

Proof: k is even and $r > -p$, hence $q > (k-1)p \ge p$. Therefore $q \nmid H(-p)$ and also $q \nmid rk$. Since $p \equiv 3 \pmod 4$, by (B5.3), $p^q \equiv p \pmod{q^2}$. Hence $k^q p \equiv (kp)^q \equiv (q-r)^q \equiv -r^q \equiv -r \pmod{q}$. Thus $0 \equiv k^q p + r = (k^q - k)p + q \pmod{q^2}$ and dividing by q,
$$r\psi_q(k) \equiv r\frac{k^{q-1} - 1}{q} \equiv -p\frac{k^q - k}{q} \equiv 1 \pmod{q}. \qquad \blacksquare$$

Here it should be noted that for each $m = 0, 1, \ldots, q-1$, the congruence $X^q - X \equiv mq \pmod{q^2}$ has at least $q-1$ pairwise incongruent solutions, modulo q^2. Indeed, first let

$m = 0$. If $j = 1, 2, \ldots, q - 1$, then $j^q \equiv j \pmod{q}$, so $j^q \equiv j + h_j q \pmod{q^2}$ with $0 \leq h_j \leq q - 1$, h_j uniquely defined by j. It follows that $(j + h_j q)^q \equiv j^q \equiv j + h_j q \pmod{q^2}$. This gives rise to $q - 1$ solutions of $X^q - X \equiv 0 \pmod{q^2}$, which are pairwise incongruent modulo q^2. Moreover, if $m = 1, 2, \ldots, q-1$, and $x^q \equiv x \pmod{q^2}$, then $(x - mq)^q \equiv x^q \equiv x \pmod{q^2}$, so $(x - mq)^q - (x - mq) \equiv mq \pmod{q^2}$. This gives rise to $q-1$ pairwise incongruent solutions of $X^q - X \equiv mq \pmod{q^2}$.

(B5.6). Lemma. *Let p, q be odd primes. Assume that $p = 2kq + e$, with $q \nmid k$, $e^q \equiv e \pmod{q^2}$. Then $p^{q-1} \not\equiv 1 \pmod{q^2}$.*

Proof:
$$p^q \equiv (2kq + e)^q \equiv e^q \equiv e \pmod{q^2}.$$

If $p^q \equiv p \pmod{q^2}$, then $e \equiv p \equiv 2kq + e \pmod{q^2}$, hence $q|k$, which is contrary to the hypothesis. ∎

This result holds, in particular, when $e = \pm 1$.
Thus if $p = 2q + 1$, or $4q - 1$, or $6q + 1$ (with $q > 3$), then $p^{q-1} \not\equiv 1 \pmod{q^2}$.

(B5.7). Lemma. *Let p, q be primes and assume one of the following conditions is satisfied:*

i) $p = 2q + 1$.

ii) $p = 4q - 1$.

iii) $p < q^{1.462}$ and $p \equiv 3 \pmod 4$.

iv) $p = 6q + 1$.

Then $H(-p) < q$.

Proof: i) As indicated above, $H(-p) < \frac{p}{4} = \frac{2q+1}{4} < q$.

ii) Again, $H(-p) < \frac{p}{4} = \frac{4q-1}{4} < q$.

iii) It is trivial if $p = 3$ or 7. If $p \geq 11$, then necessarily $q \geq 7$.

Let $2t = 1.462$. By the estimate for $H(-p)$ given in (b):

$$\frac{H(-p)}{q} < \frac{1}{2q}\sqrt{p}\log p < \frac{\log q^t}{q^{1-t}} = \frac{t}{1-t}\frac{\log q^{1-t}}{q^{1-t}}.$$

The function $f(x) = \frac{\log x^{1-t}}{x^{1-t}}$ is decreasing for $x^{1-t} > e$. Also $\frac{0.731}{0.269} \times \frac{\log 7^{0.269}}{7^{0.269}} < 1$, hence for every prime $q \geq 7$, $H(-p) < q$.

iv) If $q \geq 53$, then

$$\frac{p}{q} = 6 + \frac{1}{q} \leq 6.2 < q^{0.462}, \text{ so } p < q^{1.462}.$$

But $p \equiv 3 \pmod 4$, hence by (iii), $H(-p) < q$.

If $5 \leq q < 53$, then since $p = 6q + 1$ is a prime then, $q = 5, 7, 11, 13, 17, 23, 37, 47$. Therefore $p = 31, 43, 67, 79, 103, 139, 223, 283$. According to the tables, $H(-p) = 3, 1, 1, 5, 5, 3, 7, 3$, hence $H(-p) < q$. ∎

(B5.8). *Let p, q be primes and assume that one of the following conditions is satisfied:*

i) $p = 2q + 1$

ii) $p = 4q - 1$

iii) $p = 6q + 1$

iv) $p = 2kq + e$, with $q \nmid k$, $e^q \equiv e \pmod{q^2}$ and $p < q^{1.462}$, $p \equiv 3 \pmod 4$.

Then the equation $X^p - Y^q = 1$ has no solution in non-zero integers.

Proof: Note that in all cases $p \equiv 3 \pmod 4$ and also $H(-p) < q$, so $q \nmid H(-p)$ by Lemma (B5.7). By Lemma (B5.6), $p^{q-1} \not\equiv 1 \pmod{q^2}$. By (B5.1), $X^p - Y^q = 1$ has only trivial solutions. ∎

The Theorems of Inkeri

In 1990, Inkeri proved another theorem in which the class number of the cyclotomic field intervenes.

First, it is convenient to establish the following result:

(B5.9). *Let $p, q \geq 3$ be distinct primes, such that q does not divide the class number h_p of the pth cyclotomic field $\mathbb{Q}(\zeta)$. Assume that there exist non-zero integers x, y, such that $x^p - y^q = 1$. Then there exist real units ϵ, η of $\mathbb{Q}(\zeta)$, such that*

$$\begin{cases} \epsilon^p = \alpha^q + \bar{\alpha}^q \\ \eta x = \beta^q + \bar{\beta}^q, \end{cases}$$

where $\alpha, \beta \in \mathbb{Z}[\zeta]$ and α, β are not units.

Proof: By Lemma (B2.7),

$$pu^q = \frac{x^p - 1}{x - 1} = (x - \zeta)(x - \zeta^2)\ldots(x - \zeta^{p-1}).$$

Since $q | x$, then $3 \leq q \leq |x|$, so

$$pu^q = x^{p-1} + x^{p-2} + \ldots + x + 1 \geq |x|^{p-1}$$
$$- |x|^{p-2} + \ldots - |x| + 1$$
$$\geq |x|^{p-2}(|x| - 1) + 1 \geq 3^{p-2} + 1 > p.$$

Hence $u \neq 1$. Next $p = \Phi_p(1) = (1 - \zeta)(1 - \zeta^2)\ldots(1 - \zeta^{p-1})$, hence

$$u^q = \prod_{i=1}^{p-1} \delta_i, \text{ with} \tag{5.9}$$

$$\delta_i = \frac{x - \zeta^i}{1 - \zeta} = \frac{x - 1}{1 - \zeta^i} + 1, \quad \text{for } i = 1, \ldots, p - 1. \tag{5.10}$$

But $p | x - 1$ (by Lemma (B2.7)) and by §P2, $1 - \zeta^i$, $1 - \zeta$ are associate and $(p) = (1 - \zeta)^{p-1}$. Hence $\delta \in \mathbb{Z}[\zeta]$ for every $i = 1, \ldots, p - 1$.

Now I observe that δ_i, δ_j are relatively prime when $i < j$. Indeed, if P is a prime ideal of $\mathbb{Z}[\zeta]$ such that P divides δ_i, δ_j, then P divides $\zeta^i(1 - \zeta^{j-1})$; since $1 - \zeta^{j-1}$ and $1 - \zeta$ are associate, then P divides the prime ideal $(1 - \zeta)$, so $P = (1 - \zeta)$. From Lemma (B2.7), $x - 1 = p^{q-1}a^q$, so $p^2|x-1$, hence $p^{2(p-1)}$ divides $x - 1$. By (5.10), $\delta_i \equiv 1 \pmod{P}$, and this is a contradiction.

From (5.9), $(\delta_i) = J_i^q$, where J_i is an ideal of $\mathbb{Z}[\zeta]$. Since $u \neq 1$, δ_i is not a unit, so $J_i \neq \mathbb{Z}[\zeta]$.

From the hypothesis that $q \nmid h_p$, each ideal J_i is principal. So for every $i = 1, \ldots, p - 1$, $\delta_i = \epsilon_i \alpha_i^q$, where ϵ_i is a unit, $\alpha_i \in \mathbb{Z}[\zeta]$, α_i not a unit. Therefore, $x - \zeta^i = \epsilon_i \alpha_i^q(1 - \zeta^i)$.

As it was indicated in §P2, $\epsilon_i = \zeta^{k_i} \eta_i$ where $0 \leq k_i \leq p-1$, and η_i is a real unit. So

$$x - \zeta^i = \zeta^{k_i} \eta_i \alpha_i^q (1 - \zeta^i). \tag{5.11}$$

In particular, for $i = 2$,

$$x - \zeta^2 = \zeta^{k_2} \eta_2 \alpha_2^q (1 - \zeta^2) = \zeta^{k_2+1} \eta_2 \alpha_2^q (\zeta^{-1} - \zeta)$$

Since $p \neq q$, there exist integers e, f such that $ep + fq = 1$. Hence $\zeta^{k_2+1} = \zeta^{(k_2+1)ep}\zeta^{(k_2+1)fq} = \zeta^{(k_2+1)fq}$ and

$$x - \zeta^2 = \eta_2 \gamma^q (\zeta^{-1} - \zeta), \tag{5.12}$$

with $\gamma = \zeta^{(k_2+1)f}\alpha_2 \in \mathbb{Z}[\zeta]$, γ not a unit. Taking the complex conjugates

$$x - \zeta^{-2} = \eta_2 \bar{\gamma}^q (\zeta - \zeta^{-1}), \tag{5.13}$$

and subtracting

$$\zeta^2 - \zeta^{-2} = \eta_2(\gamma^q + \bar{\gamma}^q)(\zeta - \zeta^{-1}),$$

hence

$$\frac{\zeta + \zeta^{-1}}{\eta_2} = \gamma^q + \bar{\gamma}^q.$$

The Theorems of Inkeri

Now I observe that

$$\zeta + \zeta^{-1} = \frac{\zeta^2 - \zeta^{-2}}{\zeta - \zeta^{-1}} = \frac{\zeta^{-2}(1 - \zeta^4)}{\zeta^{-1}(1 - \zeta^2)};$$

since $1 - \zeta^2$, $1 - \zeta^4$ are associate, then $\zeta + \zeta^{-1}$ is a unit and therefore

$$\eta = \frac{\zeta + \zeta^{-1}}{\eta_2}$$

is a real unit. Let $\epsilon = -\eta^e$, so ϵ is a real unit, and $\alpha = \eta^{-f}\gamma \in \mathbb{Z}[\zeta]$, α not a unit. Then

$$\epsilon^p = \alpha^q + \bar{\alpha}^q.$$

From (5.7), $\zeta^{-2}x - 1 = \eta_2 \gamma^2 (\zeta^{-1} - \zeta)\zeta^{-2}$. Taking $\beta = \zeta^{-2f}\gamma \in \mathbb{Z}[\zeta]$, β not a unit, then

$$\zeta^{-2}x - 1 = \eta_2 \beta^q (\zeta^{-1} - \zeta).$$

Taking the conjugate,

$$\zeta^2 x - 1 = \eta_2 \bar{\beta}^q (\zeta - \zeta^{-1}),$$

and subtracting,

$$(\zeta^2 - \zeta^{-2})x = \eta_2 (\beta^q + \bar{\beta}^q)(\zeta - \zeta^{-1}),$$

so $\eta x = \beta^q + \bar{\beta}^q$, where

$$\eta = \frac{\zeta^2 - \zeta^{-2}}{\eta_2(\zeta - \zeta^{-1})} = \frac{\zeta + \zeta^{-1}}{\eta_2}. \qquad \blacksquare$$

From this proposition, Inkeri could derive another proof of Nagell's result (A7.3) about the equations $X^3 - Y^m = \pm 1$.

New proof that $X^3 - Y^q = \pm 1$ has only trivial solutions, when $q \geq 5$

Let $\zeta = \frac{-1+\sqrt{-3}}{2}$ be a primitive cubic root of 1, so $\zeta^2 = \frac{-1-\sqrt{-3}}{2}$ and $\mathbb{Q}(\zeta) = \mathbb{Q}(\sqrt{-3})$. The only real units of $\mathbb{Q}(\sqrt{-3})$ are $1, -1$, as well known.

If $x, y \neq 0$ and $x^3 - y^q = 1$, by (B5.9), $\pm 1 = \alpha^q + \bar{\alpha}^q$, where $\alpha \in \mathbb{Z}\left[\frac{-1+\sqrt{-3}}{2}\right]$, $\alpha \neq \pm 1$. Since

$$\frac{\pm 1}{\alpha + \bar{\alpha}} = \frac{\alpha^q + \bar{\alpha}^q}{\alpha + \bar{\alpha}} = \alpha^{q-1} - \alpha^{q-2}\bar{\alpha} + \ldots$$
$$+ \alpha\bar{\alpha}^{q-2} + \bar{\alpha}^{q-1} \in \mathbb{Z}\left[\frac{-1+\sqrt{-3}}{2}\right],$$

then $\alpha + \bar{\alpha}$ is real unit, so $\alpha + \bar{\alpha} = \pm 1$.

It follows from (P1.11), applied to $\alpha, \bar{\alpha}$, that

$$\alpha^q + \bar{\alpha}^q - (\alpha + \bar{\alpha})^q = \pm q(\alpha\bar{\alpha})N,$$

with $N = 1 + Mr$, $M \in \mathbb{Z}$ and r a prime dividing $\alpha\bar{\alpha}$ in $\mathbb{Q}(\sqrt{-3})$. Therefore, $\pm 1 - (\pm 1) = \pm q(\alpha\bar{\alpha})N$. The left-hand side is either 0 or 2, and it cannot be 2, since $q \neq 2$. Thus necessarily $N = 0$. But then, $1 + Mq = 0$, which is impossible.

Similarly, $x^q - y^3 = 1$ with $q \geq 5$, $x, y \neq 0$ is impossible, for it would imply that $(-y)^3 - (-x)^q = 1$. ∎

(B5.10). Let p, q be distinct odd primes. If there exist nonzero integers x, y such that $x^p - y^q = 1$, then:

i) If $q \nmid h_p$, then $q^2 | x$ and $p^{q-1} \equiv 1 \pmod{q^2}$.

ii) If $p \nmid h_q$, then $p^2 | y$ and $q^{p-1} \equiv 1 \pmod{p^2}$.

Proof: i) Assume that $q \nmid h_p$, so by (B5.9) there exist real units ϵ, η in $\mathbb{Q}(\zeta)$ (where ζ is a primitive p^{th} root of 1), such that

$$\begin{cases} \epsilon^p = \alpha^q + \bar{\alpha}^q \\ \eta x = \beta^q + \bar{\beta}^q \end{cases}$$

where $\alpha, \beta \in \mathbb{Z}[\zeta]$, α, β non-units.

Applying the identity (P1.11) for $\beta, \bar{\beta}$,

$$\eta x = \beta^q + \bar{\beta}^q = (\beta + \bar{\beta}^q)^q + q(\beta\bar{\beta})(\beta + \bar{\beta})\delta, \qquad (5.14)$$

with $\delta \in \mathbb{Z}[\zeta]$.

By Cassels' theorem, $q|x$, so $x = qx_1$. Then $q|(\beta+\bar{\beta})^q$. In the field $\mathbb{Q}(\zeta)$, the principal ideal (q) is the product of distinct prime ideals $Q_1, \ldots Q_f$ (as indicated in §P2). For each prime ideal Q_i, $Q_i|(\beta+\bar{\beta})^q$, hence $Q_i|\beta+\bar{\beta}$, so $Q_i^q|(\beta+\bar{\beta})^q$, and

$$Q_i^2|q(\beta\bar{\beta})(\beta+\bar{\beta}).$$

By (5.14), Q_i^2 divides $x = qx_1$. Hence $Q_i|x_1$ for every $i = 1, \ldots, f$. So from $(q) = Q_1 \ldots Q_f$, it follows that $q|x_1$, hence $q^2|x$.

From (B5.3), it follows that $p^{q-1} \equiv 1 \pmod{q^2}$.

ii) From $x^p - y^q = 1$ it follows that $(-y)^q - (-x)^p = 1$, and the result follows from i). ∎

This theorem applies in particular to show that the equations $X^5 - Y^7 = \pm 1$ have only trivial solutions in integers. Otherwise, since $7 \nmid h_5 = 1$, then $5^6 \equiv 1 \pmod{49}$, which is not true. This fact had not been established previously.

Here is a consequence of (B5.10):

(B5.11). *Let p, q be odd primes. Assume that $p = 2kq + e$, with $q \nmid k$, $e^q \equiv e \pmod{q^2}$. If $q \nmid h_p$, then the equation $X^p - Y^q = 1$ has no solution in non-zero integers.*

Proof: By Lemma (B5.6), $p^{q-1} \not\equiv 1 \pmod{q^2}$ and by (B5.10), the equation $X^p - Y^q = 1$ has only trivial solutions. ∎

The next results require the use of tables for the congruence $a^{q-1} \equiv 1 \pmod{q^2}$, where q is a prime not dividing a. The computations with base 2 for $q < 6 \times 10^9$ were due to

D.H. Lehmer. In 1971, Brillhart, Tonascia and Weinberger extended a previous table of Kloss, for $a \leq 2 < 100$ and various limits for q. These limits were increased by W. Keller and D. Clark, who pushed the calculation to the following limits: $q < 7.5 \times 10^9$ (for $a = 2$), $q < 5 \times 10^8$ (for $a < 100$, a odd prime), $q < 9.3 \times 10^7$ (for $a < 100$, a not a power and not a prime).

I shall only require solutions of the congruence $p^{q-1} \equiv 1 \pmod{q^2}$ where p is also a prime. In the past these computations have been done by many authors, each one extending the previous results. For example, Aaltonen and Inkeri (1991) listed all the solutions for $p < 10^3$ and $q < 10^4$. I give a more up-to-date list for the above range of p. The solutions marked with * were given by Keller, the solution with ** was found by Clark.

The results which follow will also use the class numbers $H(-q)$, h_q of the imaginary quadratic field $\mathbb{Q}(\sqrt{-q})$ and of the cyclotomic field of qth roots of 1; in particular, their factorization into primes will be needed. Tables (in the range required here) are found, for example, in the book of Borevich and Shafarevich.

I recall that the class number h_q is expressed as a product $h_q = h_q^- h_q^+$ of natural numbers; the second factor is the class number of the real cyclotomic field $\mathbb{Q}(\zeta_q + \zeta_q^{-1})$. Even though the tables for h_q^- are rather extensive, little computation has been done for the second factor h_q^+. For the present purpose, it is noted that $h_q^+ = 1$ for $q < 71$.

(B5.12). i) *If $p \equiv q \equiv 3 \pmod 4$ and $5 \leq p, q < 10^4$, then $X^p - Y^q = 1$ has only trivial solutions in integers, with the possible exceptions of $(p,q) = (83, 4871)$, $(4871, 83)$.*

ii) *If $p \equiv 3 \pmod 4$, $q \equiv 1 \pmod 4$ and $5 < p, q < 500$ then $X^p - Y^q = 1$ has only trivial solutions in integers with the possible exceptions of $(p,q) = (19, 137)$, $(107, 97)$, $(223, 349)$, $(251, 421)$, $(419, 173)$, $(419, 349)$, $(499, 109)$.*

Solutions of $p^{q-1} \equiv 1 \pmod{q^2}$

Base p	Solutions q	Base p	Solutions q
2	1093 3511	127	3 19 907
3	11 1006003	131	17
5	20771 40487 53471161	137	29 59 6733
7	5 491531	139	None
11	71	149	5
13	863 1747591	151	5 2251
17	3 46021 48947	157	5
19	3 13743 137 63061489	163	3
23	13 2481757 13703077	167	None
29	None	173	3079
31	7 79 6451 28606861*	179	3 17
37	3 77867	181	3 101
41	29 1025273 138200401*	191	13
43	5 103	193	5 4877
47	None	197	3 7 653
53	3 47 59 97	199	3 5
59	2777		
61	None		
67	7 47 268573		
71	3 47		
73	3		
79	7 263 3037	1012573*	
		6031284*	
83	4871 13691 315746063**		
89	3 13		
97	7 2914393*		
101	5	211	None
103	None	227	71 349
107	3 5 97	229	31
109	3	233	3 11 157
113	None	239	11 13

Base p	Solutions q	Base p	Solutions q
241	11 523 1163	449	3 5 1789
251	3 5 11 17 421	457	5 11 919
257	5 359	461	1697 5081
263	7 23 251	463	1667
269	3 11 83 8779	467	3 29 743 7393
271	3	479	47 2833
277	1993	487	3 11 23 41 1069
281	None	491	7 79
283	None	499	5 109
293	5 7 19 83	503	3 17 229 659 6761
307	3 5 19 487		
311	None	509	7 14
313	7 41 149 181	521	3 7 31 53
317	107 349	523	3 9907
333	211 359	541	3
337	13	547	31
347	None	557	3 5 7 23
353	8123	563	None
359	3 23 307	569	7 263
367	43 2213	571	23 29
373	7 113	577	3 13 17 71
379	3	587	7 13 31
383	None	593	3 5
389	19 373	599	5
397	3	601	5 61
401	5 83 347	607	5 7
409	None	613	3 4073
419	173 349 983 3257	617	101 1087 6077
421	101 1483	619	7 73
431	3 2393	631	3 1787 5741
433	3	641	43
439	31 79	643	5 17 307 859
443	5	647	3 23

The Theorems of Inkeri

Base p	Solutions q
653	13 17 19 1381
659	23 131 2221 9161
661	None
673	61
677	13 211
683	3 1279
691	37 509 1091 9157
701	3 5
709	None
719	None
727	None
733	17
739	3 9719
743	5
751	5 151 409
757	3 5 17 71

Base p	Solutions q
761	41 907
769	None
773	3
878	37 41
809	3 59
811	3 211
821	19 83 233 293 1229
823	13 2309
827	3 17 29 9323
829	3 17
839	5227
853	None
857	5 41 157 1697
863	3 7 23 467
877	None

Base p	Solutions q
881	3 7 23
883	3 7
887	11 607
907	5 17
911	127
919	3
929	None
937	3 41 113 853
941	11 1499
947	5021
953	3
967	11 19 4813
971	3 11 401 9257
977	11 17 109 239 401
983	None
991	3 13 431
997	197 1223

Proof: Assume that $X^p - Y^q = 1$ has a non-trivial solution in integers; so the same is true for $X^q - Y^p = 1$.

1°). First let $5 \leq p < 73$, $5 \leq q < 10^4$, with $p \equiv 31 \pmod 4$.

I show that $p^{q-1} \equiv 1 \pmod{q^2}$. Indeed, if $p^{q-1} \not\equiv 1 \pmod{q^2}$, by (B5.3) $q|H(-p)$. From the tables, it is seen that $(p,q) \in \{(47,5), (71,7)\}$. Then $p \nmid h_q$, hence by (B5.10), $q^{p-1} \equiv 1 \pmod{p^2}$, which is a contradiction.

From $p^{q-1} \equiv 1 \pmod{q^2}$, $p \equiv 3 \pmod 4$, $5 \leq p < 73$, $5 \leq q < 10^4$, the tables give:

$(p,q) \in \{$ (7,5), (11, 71), (19, 7), (19, 13),
(19, 43), (19, 137), (23, 13), (31, 7), (31, 79),
(31, 6451), (43,5), (43, 103), (59, 2777),
(67, 7), (67, 47), (71, 47), (71, 331)$\}$.

If $q \equiv 3 \pmod 4$, it is seen that $p \nmid H(-q)$, hence by (B5.3), $q^{p-1} \equiv 1 \pmod{p^2}$; but, as checked in the tables or by direct computation when $(p,q) = (31, 6451)$, this is not true. If $q \equiv 1 \pmod 4$, then with the possible exceptions of (19, 137), (59, 2777), $p \nmid h_q$. By (B5.10), $q^{p-1} \equiv 1 \pmod{p^2}$; but this is not true, as seen in the tables.

The same conclusion is reached if $q \equiv 3 \pmod 4$ and $5 \leq q < 73$, $5 \leq p < 10^4$.

2°). Now let $73 \leq p$, $q < 10^4$, $p \equiv q \equiv 3 \pmod 4$.

I show that $q \nmid H(-p)$. Indeed, if $q|H(-p)$, then from the tables it is seen that

$(p,q) \in \begin{Bmatrix} (4391, 79), & (5399, 79), & (7127, 79), & (3911, 83), \\ (5039, 83), & (8423, 83), & (8231, 107), & (9239, 139) \end{Bmatrix}$.

But $H(-q) < q < p$, so $p \nmid H(-q)$ and by (B5.3), $q^{p-1} \equiv 1 \pmod{p^2}$. However, according to the tables, this congruence is not satisfied by the above pairs (p,q).

This shows that $q \nmid H(-p)$, and by (B5.3) $p^{q-1} \equiv 1 \pmod{q^2}$. Since the equation $X^q - Y^p = 1$ would also have a non-trivial solution, then $q^{p-1} \equiv 1 \pmod{p^2}$. It suffices now to note that the only pair (p,q), with $73 \leq p, q < 10^4$ such that $p^{q-1} \equiv 1 \pmod{q^2}$, $q^{p-1} \equiv 1 \pmod{p^2}$ is $(p,q) = (83, 4871)$; this is determined by direct computation from the table.

3°). Now let $73 \leq p < 500$, $5 \leq q < 500$, and $p \equiv 3 \pmod 4$, $q \equiv 1 \pmod 4$. First I show that $q \nmid H(-p)$. Otherwise, from $q|H(-p)$, by the tables,

$$(p,q) \in \begin{Bmatrix} (79,5), & (103,5), & (127,5), & (131,5), (179,5), \\ (191,13), & (227,5), & (239,5), & (263,13), (347,5), \\ (383,17), & (439,5), & (443,5), & (479,5) \end{Bmatrix}.$$

Then $p \nmid h_q$ and by (B5.10), $q^{p-1} \equiv 1 \pmod{p^2}$, which is not true, according to the tables.

Hence $q \nmid H(-p)$ and by (B5.3), $p^{q-1} \equiv 1 \pmod{q^2}$. By the table,

$$(p,q) \in \{(107,5), \ (107, 97), \ (131, 17), \ (151, 5),$$
$$(179, 17), \ (191, 13), \ (199, 5), \ (223, 349), (239, 13),$$
$$(251, 5), \ (251, 17), \ (251, 421), (307, 5), (419, 173),$$
$$(419, 349), (443, 5), \ (467, 29), \ (487, 41),$$
$$(499, 5), \ (499, 109)\}.$$

With the possible exceptions of (107, 97), (223, 349), (251, 421), (419, 173), (419, 349), (499, 109), $p \nmid h_q$; hence by (B5.10), $q^{p-1} \equiv 1 \pmod{p^2}$. But according to the tables, this is not true.

4°). The facts established in 1°, 2°, 3° are enough to prove the proposition. ∎

Mignotte has recently performed (1992, graciously communicated to me) very extensive calculations of the Fermat quotient. For each prime $p < 700$, when $p \equiv 1 \pmod 4$ and for each prime $p < 3040$, when $p \equiv 3 \pmod 4$, he determined all the primes $q < 2^{30}$ such that $p^{q-1} \equiv 1 \pmod{q^2}$. These computations went on for 183 days. As a result, many more pairs (p,q) have been found such that $X^p - Y^q = \pm 1$ has only the trivial solution.

In §C10 I shall report how the present methods were combined with estimates of linear forms in logarithms to treat other pairs of exponents.

Part C

ANALYTICAL METHODS

My purpose is to indicate estimates for the number and size of solutions of Catalan's equation, assuming that non-trivial solutions exist.

First, I consider fixed exponents $m, n \geq 2$ and examine the possible solutions x, y of the equation $X^m - Y^n = 1$.

Second, given distinct integers $a, b \geq 2$, I look for solutions in natural numbers u, v of the equation $a^U - b^V = 1$.

Finally, I consider the solutions in natural numbers x, y, u, v of the exponential diophantine equation $X^U - Y^V = 1$.

Accordingly, this chapter is divided into three parts. They will be preceded by a section about some fundamental theorems on diophantine equations.

1. Some General Theorems for Diophantine Equations

In order to prove that an equation has only finitely many solutions $(x_1, \ldots, x_k)$, where each x_i is an integer, it suffices to attain successfully one of the following goals.

a) To show that the existence of infinitely many solutions leads to a contradiction.

b) To determine explicitly an integer $N \geq 1$ such that the number of solutions is at most equal to N.

c) To determine explicitly some integer $C \geq 1$ such that every solution $(x_1, \ldots, x_m)$ must satisfy $|x_i| \leq C$ for $i = 1, \ldots, k$. By trying all possible integers with absolute value up to C, it is possible in principle to identify all of the solutions.

In case a), there is no indication of how many solutions there are, or how large the solutions can be.

In case b), there is no indication of how large the solutions are; thus, even if $N - 1$ solutions are already known, nothing may be inferred about whether any other solution exists, or how large it may be.

Finally, case c) is the most satisfactory. Yet, if the constant C provided by the method of proof is much too large—as is often the case—it is not possible to identify all of the solutions in a reasonable time.

In this short section, I shall state explicitly the main general theorems which will be used to show that the equations being studied have finitely many solutions.

Definitely no attempt will be made to describe the beautiful theories which have been devised to ascertain that wide classes of diophantine equations have finitely many solutions. The reader interested in an accessible presentation, may consult my paper *Some fundamental methods in the theory of diophantine equations* (1986). There are, of course, many books which treat these questions extensively. Perhaps the closest to the topics presented here is the book of Shorey and Tijdeman (1986).

In 1909, Thue proved:

(C1.1). *Let $F(X,Y) = a_0 X^n + a_1 X^{n-1} Y + \ldots + a_{n-1} XY^{n-1} + a_n Y^n$, with $a_0, a_1, \ldots, a_n$ integers, $a_0 \neq 0$ and $n \geq 3$. If a is any non-zero integer, and if the roots of the polynomial $F(X, 1)$ are distinct, then the equation $F(X, Y) = a$ has only finitely many solutions in integers.*

Some General Theorems for Diophantine Equations

The proof of Thue's theorem depends on the approximation of algebraic numbers by rational numbers. The idea originated with Liouville, who proved the weak statement:

(C1.2). *Let α be a real algebraic number of degree $d \geq 2$. Then there exists an effectively computable number $C > 0$ (depending on α) such that if $\frac{a}{b}$ (with $b > 0$, $\gcd(a,b) = 1$) is any rational number, then*

$$\left|\alpha - \frac{a}{b}\right| > \frac{C}{b^d}. \tag{1.1}$$

The best possible approximation theorem was proved by Roth in 1955; it improved on previous work by Gel'fond, Dyson and Siegel:

(C1.3). *If α is a real irrational algebraic number (which may be assumed of degree greater than 2), then for every $\epsilon > 0$ there exists $C > 0$ (depending on α, ϵ) such that for every rational number $\frac{a}{b}$ ($b > 0$, $\gcd(a,b) = 1$),*

$$\left|\alpha - \frac{a}{b}\right| > \frac{C}{b^{2+\epsilon}}. \tag{1.2}$$

However, *the constant C is not effectively determined.*

On the basis of Roth's theorem, it is immediate that one can strenghten Thue's theorem but this matter is outside my purpose here.

On the other hand, concerning solutions of equations in integers, Siegel had already proved in 1929:

(C1.4). *Let $f(X,Y)$ be a polynomial of degree n, with integer coefficients, which is irreducible over the field of complex numbers. Let $F(X,Y,Z) = Z^n f\left(\frac{X}{Y}, \frac{Y}{Z}\right)$ be the homogenized polynomial, and let $\mathcal{C}$ be the projective plane curve with equation $F(X,Y,Z) = 0$. If the curve $\mathcal{C}$ has genus greater than zero, then the equation $f(X,Y) = 0$ has only finitely many solutions in integers.*

Here I am interested only in the applications to Catalan or similar equations. So it is enough to state an explicit consequence, as given in 1964 by Inkeri and Hyyrö (see also Leveque, 1964). By doing so, I need not explain any concept of algebraic geometry, like the genus, which appeared in the statement of (C1.4).

(**C1.5**). *Let $m, n \geq 2$ with $\max\{m, n\} \geq 3$. Assume that $f(X)$ is a polynomial of degree m with integer coefficients and with distinct roots. Then, for any non-zero integer a, the equation $f(X) = aY^n$ has at most finitely many solutions in integers.*

In 1976, Schinzel and Tijdeman proved the following theorem (which improves (C1.5), since the exponent n is not fixed a priori). The theory of linear forms in logarithms, to be discussed below, was an essential ingredient of the proof.

(**C1.6**). *Let $g(X)$ be a polynomial with rational coefficients.*
   i) *If $g(X)$ has at least three simple zeros, then there exists an effectively computable number $C(g) > 0$, such that if x, y, z are integers, $|y| \geq 2$, $z \geq 2$ and $g(x) = y^z$, then $|x|, |y|, z < C(g)$.*
   ii) *If $g(X)$ has exactly two simple zeros, then there exists an effectively computable number $C(g) > 0$, such that if x, y, z are integers, $|y| \geq 2$, $z \geq 3$ and $g(x) = y^z$, then $|x|, |y|, z < C(g)$.*

Concerning rational solutions of diophantine equations, the main conjecture was formulated by Mordell and proved by Faltings (1985). Here I quote only a special case of this fundamental theorem:

(**C1.7**). *Let $f(X, Y, Z) \in \mathbb{Z}[X, Y, Z]$ be a non-constant homogeneous polynomial such that the corresponding projective plane curve $\mathcal{C}$ is non-singular and has genus greater than one. Then there exist only finitely many points in the curve $\mathcal{C}$ having rational coordinates.*

Some General Theorems for Diophantine Equations

More explicitly, there exist only finitely many triples (x, y, z) of integers, with $gcd(x, y, z) = 1$ such that $f(x, y, z) = 0$.

It should be noted that up to now, there is no known effective bound on the number or size of the above integers x, y, z.

I recall that the curve $\mathcal{C}$ is non-singular if there does not exist any triple of integers $(x, y, z) \neq (0, 0, 0)$, such that simultaneously

$$\frac{\partial f}{\partial X}(x, y, z) = 0, \quad \frac{\partial f}{\partial Y}(x, y, z) = 0, \quad \frac{\partial f}{\partial Z}(x, y, z) = 0.$$

For non-singular plane curves $\mathcal{C}$, for which f has degree n, the genus is given by the formula

$$g = \frac{(n-1)(n-2)}{2}. \tag{1.3}$$

All the above results were not always sufficient to derive explicit upper bounds for the number or the size of solutions of Catalan's equation and many other diophantine equations.

Beginning in 1966, Baker studied linear forms in logarithms and was able to obtain fundamental theorems. His work, of the utmost importance in the theory of diophantine approximation and diophantine equations, has stimulated research by numerous mathematicians.

It is not my intention, nor do I have the competence, to enter here into the discussion of these developments. For a discussion of the history of the subject see Baker (1977 and 1994). The shape of the bound first described in Baker's 1973 paper (see also the other two papers in the same series) had special bearing in the proof of Tijdeman's theorem (see Section 10). For details and proofs in the theory of linear forms in logarithms, the reader may consult the following papers: Phillipon & Waldschmidt (1988), Wüstholz (1988), Blass, Glass, Mansky, Meronk and Steiner (1990) and Waldschmidt (1990a) and, especially for the latest results, Baker & Wüstholz (1993).

I shall quote the theorem of Baker—not in its original.
I introduce the following notation:

If α is any algebraic number of degree $d \geq 1$, let $F(X) = a_0 X^d + a_1 X^{d-1} + \ldots + a_d$ (with $a_0, a_1, \ldots, a_d$ integers, $a_0 \neq 0$; $gcd(a_0, a_1, \ldots, a_d) = 1$) be its minimal polynomial. So $F(X)$ is irreducible over $\mathbb{Q}$ and $F(\alpha) = 0$. The *height of* α is

$$H(\alpha) = \max\left\{|a_0|, |a_1|, \ldots, |a_d|\right\}. \tag{1.4}$$

Let log denote the principal determination of the logarithmic function.

Let positive integers n, d and real numbers $A \geq 1$, $B \geq e$ be given. Denote by $\mathcal{A}(A)$ the set of all n-tuples $(\alpha_1, \ldots, \alpha_n)$ of algebraic numbers $\alpha_i \neq 0, 1$, such that:

a) The degree of the field $\mathbb{Q}(\alpha_1, \ldots, \alpha_n)$ is at most equal to d;

b) If $A_i = \max\{H(\alpha_i), e\}$, then

$$(\log A_1)(\log A_2) \ldots (\log A_n) \leq A. \tag{1.5}$$

Let $\mathcal{B}(B)$ be the set of all n-tuples $(b_1, \ldots, b_n)$ of rational integers $b_i \neq 0$, such that

$$\max_{1 \leq i \leq n} \{|b_i|\} \leq B. \tag{1.6}$$

Let $S = S(n, d, A, B)$ be the set of all linear forms in logarithms

$$\Lambda = b_1 \log \alpha_1 + \ldots + b_n \log \alpha_n, \tag{1.7}$$

such that $\Lambda \neq 0$ and $(\alpha_1, \ldots, \alpha_n) \in \mathcal{A}(A)$, $(b_1, \ldots, b_n) \in \mathcal{B}(B)$.

I quote now the most recent theorem in the context, due to Baker & Wüstholz (*loc. cit.*) It is the fruit of extensive and laborious endeavours in the field over a period of some twenty-five years.

(C1.8). *If $\Lambda \in S$, then*

$$|\Lambda| > \exp(-CA \log B) \tag{1.8}$$

where

$$C = (16dn)^{2(n+2)}. \tag{1.9}$$

Baker used his method of estimating lower bounds for linear forms in logarithms in order to derive effective upper bounds for the size of solutions of wide classes of diophantine equations. I shall return to this matter later. Much of this work is already contained in Baker's classical book (1975). It has opened a Pandora's box and has since kept mathematicians busy in establishing effective bounds for solutions and constructing algorithms which produce these solutions.

I. The Equation $X^m - Y^n = 1$

Let $m, n > 3$ and consider the equation

$$X^m - Y^n = 1.$$

My aim is to discuss the following questions: Does this equation have solutions in non-zero integers? If so how many? I shall also indicate upper and lower bounds for the hypothetical solutions and an algorithm to determine the eventual solutions.

Besides, some estimates of the sizes of differences between m^{th} and n^{th} powers will be given.

2. Upper Bounds for the Number and Size of Solutions

I shall apply the results of §C1 to deduce upper bounds for the number and size of solutions of the equation $X^m - Y^n = 1$.

Even more generally:

(C2.1). *If $m, n \geq 2$ with $\max\{m, n\} \geq 3$, and if a, b, k are non-zero integers, then the equation $aX^m - bY^n = k$ has only finitely many solutions in integers.*

Proof: This result is a direct application of (C1.5) for the polynomial $f(X) = aX^m - k$, which clearly has distinct roots. ∎

In particular, taking $m, n \geq 2$, $\max\{m, n\} \geq 3$, and $k \geq 1$, the equation $X^m - Y^n = k$ has only finitely many solutions in integers.

An equivalent formulation for the above result is the following. Let

$$z_1 < z_2 < z_3 < \ldots$$

be the increasing sequence of all integers which are either an mth or an nth power.

(C2.2). *If $m, n \geq 2$, $\max\{m, n\} \geq 3$, then*

$$\lim_{i \to \infty} (z_{i+1} - z_i) = \infty.$$

Proof: For every $N \geq 1$ and every $k = 1, 2, \ldots, N$, each equation $X^m - Y^n = k$, $X^n - Y^m = k$, $X^m - Y^m = k$, $X^n - Y^n = k$ has only finitely many solutions. Hence there exists M such that if $i < j$ and $z_j - z_i \leq N$, then $j \leq M$. In other words, if $i \geq M$, then $z_{i+1} - z_i > N$. This means that $\lim_{i \to \infty} (z_{i+1} - z_i) = \infty$.

The special case of the equations $X^m - Y^2 = k$, or $X^2 - Y^n = k$, was obtained in 1917 by Thue (as quoted in (A16.1) and again in 1920 by Landau and Ostrowski). ∎

The fact that $X^m - Y^n = k$ has finitely many solutions also follows from Mahler's theorem (1953) on the growth of the largest prime factor of $x^m - y^n$ as $\max\{x, y\}$ tends to infinity; see (C8.4).

The sequence of squares or cubes has been the object of particular attention. Concerning the gaps, Stark proved in 1973, by refining the original Baker's bounds in linear forms in logarithms:

(C2.3). *For every $\epsilon > 0$, there exists a number $C(\epsilon) > 0$, such that if $x, y > 0$ and $x^3 \neq y^2$, then $|x^3 - y^2| > C(\epsilon)(\log x)^{1-\epsilon}$.*

This result falls short of what might be expected and has been conjectured by Hall (1971):

Conjecture of Hall. There exists a number $C > 0$, such that if $x, y > 0$ and $x^3 \neq y^2$, then $|x^3 - y^2| > Cx^{\frac{1}{2}}$.

Even the following weaker statement is still unproven:

Weaker conjecture of Hall. There exist numbers $C > 0$ and $\delta > 0$, such that if $x, y > 0$, $x^3 \neq y^2$, then $|x^3 - y^2| > Cx^\delta$.

These conjectures have been discussed in a paper by Nair (1978). In their favour, there is the result of Birch, Chowla, Hall and Schinzel (1965). There exist infinitely many integers $x, y > 0$ with $x^3 \neq y^2$ and $|x^3 - y^2| < \frac{1}{9}x^{\frac{3}{5}}$. Another problem of great interest is the study of Mordell's equation

$$X^3 = Y^2 - k$$

where $k \neq 0$. In view of (A2.2), it may be assumed that $k \neq \pm 1$.

The aim of this study is to determine, for each k, all of the solutions in positive integers of Mordell's equation. The seminal papers were by Mordell (1913) and Hemer (1952, 1954). The book of London and Finkelstein (1973) contains a good presentation of the methods of resolution and gives complete solutions for numerous values of k.

For each $k \neq 0$, let $N(k)$ denote the number of solutions in integers of Mordell's equation $X^3 = Y^2 - k$.

As shown in Mordell's book (1968), the equation $X^3 = Y^2 - 7$ has no solution in rational numbers; hence for every integer $t \geq 1$, the equation $X^3 = Y^2 - 7t^6$ has no solution in integers, that is, $N(7t^6) = 0$. This implies that $\liminf_{k \to \infty} N(k) = 0$.

On the other hand, in 1930, Fueter gave a criterion for an equation $X^3 = Y^2 - k$ (with k having no sixth power factor) to have infinitely many rational solutions (once it is known that it has a rational solution). A simpler proof was given by Mordell in 1966. Thus, $X^3 = Y^2 - 3$ has the solution $(1,2)$ and Fueter's criterion may be applied to this equation telling that it has infinitely many rational solutions. This fact was used by Mohanty (in 1973) to show:

$$\limsup_{k \to \infty} N(k) = \infty$$

Indeed, given $n \geq 1$, let $\left(\frac{a_i}{d}, \frac{b_i}{d}\right)$, $i = 1, 2, \ldots, n$, (with integers a_i, b_i, d, $d > 0$) be solutions of $X^3 = Y^2 - 3$. Then $(a_i d)^3 = (b_i d^2)^2 - 3d^6$. So $N(3d^6) \geq n$.

Now, let $N'(k)$ denote the number of solutions in relatively prime integers (x, y) of $X^3 = Y^2 - k$. Mohanty showed in the same paper (1973) that $\limsup_{k \to \infty} N'(k) \geq 6$ and he also commented that it would be difficult to establish that $\limsup_{k \to \infty} N'(k) = \infty$, were it true. This question hinges on ranks and generators of the Mordell-Weil group of the associated elliptic curve.

The subject of Mordell's equation is of great importance. It is vast and beyond my aim in this book.

The statement (C2.1) does not include any assertion about the number or size of the solutions. Applying his method of linear forms in logarithms, Baker obtained the following upper bound for the size of integers x, y such that $x^m - y^n = k$ (with $k \neq 0$):

(C2.4). *Let $m, n \geq 2$ with $\max\{m, n\} \geq 3$, let $k \neq 0$ and assume that $x^m - y^n = k$.*

  i) *If $x^3 - y^2 = k$, then $\max\{|x|, |y|\} < \exp((10^6 |k|)^{10^4})$.*
  ii) *If $m > 3$ and $x^m - y^2 = k$, then $\max\{|x|, |y|\} < \exp\exp\exp(m^{10m^3}|k|^{m^2})$.*
  iii) *If $n \geq 3$ and $x^2 - y^n = k$, then $\max\{|x|, |y|\} < \exp\exp((5n)^{10}(2^{20}|k|)^4)$.*
  iv) *If $m, n \geq 3$, and $x^m - y^n = k$, then*

$$\max\{|x|, |y|\} < \exp\exp\left((5m)^{10}(n^{10n}|k|)^{n^2}\right)$$

Upper Bounds for the Number and Size of Solutions

and also
$$\max\{|x|,|y|\} < \exp\exp\left((5n)^{10}(m^{10m}|k|)^{m^2}\right).$$

These bounds are enormous, but at present no method is known to reduce their size substantially.

However, concerning the number of solutions, a much smaller lower bound was given by Hyyrö, in 1964a, using the following general result of Davenport and Roth (1955).

(C2.5). *Let β be an algebraic integer of degree $d \geq 3$ and $H(\beta)$ its height (that is the maximum of the absolute values of the coefficients of its minimal polynomial). Let $C = 3 + \log(1 + |\beta|) + 2\log(1 + H(\beta))$. If $0 < k \leq \frac{1}{3}$, then the number of pairs of integers (c,b), with $b \geq 1$ and $\gcd(c,b) = 1$, such that*
$$\left|\beta - \frac{c}{b}\right| < \frac{1}{2^{b^{2+k}}}$$
is at most equal to
$$\frac{2}{k}\log C + \exp\left(\frac{70d^2}{k^2}\right).$$

Hyyrö showed the following lemma:

(C2.6). Lemma. *Assume that $x^p - y^q = 1$, with p, q odd primes and x, y positive integers. With the notations before Lemma (B2.7):*
$$0 < \frac{q^{\frac{p-1}{p}}}{p^{\frac{q-1}{q}}} - \frac{a}{b} < \frac{1}{2b^{\min\{p,q\}}}.$$

Proof: Let
$$\alpha = \frac{q^{\frac{p-1}{p}}}{p^{\frac{q-1}{q}}}.$$

Since
$$(1+y)^q = \frac{\left(1+\frac{1}{y}\right)^q \left(1+\frac{1}{x-1}\right)^p (x-1)^p}{1+\frac{1}{y^q}},$$

then
$$\alpha = \frac{a}{b} \cdot \frac{\left(1+\frac{1}{y}\right)^{\frac{1}{p}} \left(1+\frac{1}{x-1}\right)^{\frac{1}{q}}}{\left(1+\frac{1}{y^q}\right)^{\frac{1}{pq}}}.$$

But
$$\left(1+\frac{1}{y^q}\right)^{\frac{1}{q}} < 1+\frac{1}{y},$$

so
$$\left(1+\frac{1}{y^q}\right)^{\frac{1}{pq}} < \left(1+\frac{1}{y}\right)^{\frac{1}{p}} < \left(1+\frac{1}{y}\right)^{\frac{1}{p}} \left(1+\frac{1}{x-1}\right)^{\frac{1}{q}},$$

therefore $\frac{a}{b} < \alpha$.

On the other hand, since $1 < \left(1+\frac{1}{y^q}\right)^{\frac{1}{pq}}$, then

$$\alpha - \frac{a}{b} = \frac{a}{b}\left[\frac{\left(1+\frac{1}{y}\right)^{\frac{1}{p}} \left(1+\frac{1}{x-1}\right)^{\frac{1}{q}}}{\left(1+\frac{1}{y^q}\right)^{\frac{1}{pq}}} - 1\right]$$

$$< \frac{a}{b}\left[\left(1+\frac{1}{y}\right)^{\frac{1}{p}} \left(1+\frac{1}{x-1}\right)^{\frac{1}{q}} - 1\right].$$

But
$$\left(1+\frac{1}{x-1}\right)^{\frac{1}{q}} < 1+\frac{1}{q(x-1)}, \qquad \left(1+\frac{1}{y}\right)^{\frac{1}{p}} < 1+\frac{1}{py},$$

hence
$$\left(1+\frac{1}{y}\right)^{\frac{1}{p}} \left(1+\frac{1}{x-1}\right)^{\frac{1}{q}} - 1 < \left(1+\frac{1}{py}\right)\left(\frac{1}{q(x-1)}\right) - 1$$

$$= \frac{1}{py} + \frac{1}{q(x-1)} + \frac{1}{byq(x-1)}$$

$$< \frac{1}{py}\left[1+\frac{1}{2q(x-1)}\right] + \frac{1}{q(x-1)}\left[1+\frac{1}{2py}\right]$$

$$< \frac{2}{p(y+1)} + \frac{2}{q(x-1)},$$

because $\left(1 + \frac{1}{y}\right)\left(1 + \frac{1}{2q(x-1)}\right) < \frac{3}{2} \times \frac{13}{12} = \frac{39}{24} < 2$ (using $x - 1 \geq 2$, $y \geq 2$), hence $\frac{1}{y}\left[1 + \frac{1}{2q(x-1)}\right] < \frac{2}{y+1}$. Thus

$$\alpha - \frac{a}{b} < \frac{2a}{b}\left[\frac{1}{p(y+1)} + \frac{1}{q(x-1)}\right].$$

If $p < q$, then $x > y$; actually $x \neq y + 1$ (by (B1.3)). So

$$\alpha - \frac{a}{b} < \frac{2a}{b} \times \frac{2}{p(y+1)} = \frac{4a}{pq^{p-1}b^{p+1}};$$

but $\frac{a}{b} < \alpha = \frac{q^{\frac{p}{p}}}{p^{\frac{q-1}{q}}} < q$, hence $\alpha - \frac{a}{b} < \frac{4}{bq^p - 2b^p} < \frac{1}{2b^p}$.

If $q < p$, then $x < y$ and by Lemma (B2.3), $(x-1)^p q^{(p-1)q} > (y+1)^q$, i.e., $p^{p(q-1)}a^{pq}p^{(p-1)q} > q^{q(p-1)}b^{pq}$, hence $p^{\frac{q-1}{q}} > \frac{b}{a}$. Now

$$\alpha - \frac{a}{b} < \frac{2a}{b} \times \frac{2}{q(x-1)} = \frac{4a}{bqp^{q-1}a^q} < \frac{1}{2b^q}$$

because $\left(\frac{b}{a}\right)^{q-1} < p^{\frac{(q-1)^2}{q}} < \frac{qp^{q-1}}{8}$; the last inequality holds, because, by the results of Part A, $p \geq 5$, and then $8^9 < q^{2q-1} < q^q p^{q-1}$.

This completes the proof of the lemma. ∎

I now present the result of Hyyrö (1964a):

(C2.7). *If $m, n \geq 2$, then the equation $X^m - Y^n = 1$ has at most $\exp\{631m^2n^2\}$ solutions in integers.*

Proof: Let p, q be primes, such that $p|m$, $q|n$. Then the number of solutions of $X^m - Y^n = 1$ is at most equal to the number of solutions of $X^p - Y^q = 1$. It therefore suffices to show that for the equation $X^p - Y^q = 1$ this number is at most $\exp\{631p^2q^2\} \leq \exp\{631m^2n^2\}$. Moreover, I may assume that $r = \min\{p, q\} \geq 5$, by the results of Part A.

Using the previous notations, let $\beta = p\alpha$ with

$$\alpha = \frac{q^{\frac{p-1}{p}}}{p^{\frac{q-1}{q}}}.$$

If (x, y) is a solution in integers of $X^p - Y^q = 1$, by Lemma (C2.6), $0 < \alpha - \frac{a}{b} < \frac{1}{2b^r}$. Then $0 < \beta - \frac{pa}{b} < \frac{1}{2b^r} < \frac{1}{2b^{r-1}}$, since $p < b$ by (B4.1). I also note that $\gcd(pa, b) = 1$.

In order to apply the theorem of Davenport and Roth, observe that $\beta^{pq} = p^p q^{(p-1)q}$, so β is an algebraic integer. Since the polynomial $X^{pq} - p^p q^{(p-1)q}$ is irreducible over $\mathbb{Q}$ (because its constant term is not a pq-power), then β has degree $d = pq$ over $\mathbb{Q}$, and its height is $H(\beta) = p^p q^{(p-1)q}$.

The number of solutions (x, y) is at most the number of pairs of integers (c, b), with $\gcd(c, b) = 1$ and $\left|\beta - \frac{c}{b}\right| < \frac{1}{2b^{r-1}}$; but $r - 1 \geq 4 > 2 + \frac{1}{3} \geq 2 + k$, so $\frac{1}{2b^{r-1}} < \frac{1}{2b^{2+k}}$. Hence the number of solutions (x, y) is at most the number of solutions (c, b), as above, such that $\left|\beta - \frac{c}{b}\right| < \frac{1}{2b^{2+k}}$. By the theorem of Davenport and Roth this number is at most $\frac{2}{k} \log C + \exp\left\{\frac{70p^2 q^2}{k^2}\right\}$, that is, taking $k = \frac{1}{3}$; $6 \log C + \exp 630 p^2 q^2$. Here

$$C = 3 + \log(1 + \beta) + 2 \log(1 + p^p q^{(p-1)q}).$$

But

$$1 + \beta = 1 + \frac{pq^{\frac{p-1}{p}}}{p^{\frac{q-1}{q}}} = 1 + p^{\frac{1}{q}} q^{1-\frac{1}{p}} < 2 p^{\frac{1}{q}} q^{1-\frac{1}{p}} < p^{1-\frac{1}{q}} q^{1-\frac{1}{p}} < pq,$$

because $2 < p^{1-\frac{2}{q}}$, since $p^2 < \left(\frac{p}{2}\right)^5 \leq \left(\frac{p}{2}\right)^q$ when $p, q \geq 5$.

Also $1 + p^p q^{(p-1)q} = 1 + \beta^{pq} < (1 + \beta)^{pq} < (pq)^{pq}$. Thus $C < 3 + \log(pq) + 2pq \log(pq)$. Since $p, q \geq 5$, then $3 \leq$

$\log 35 \leq \log(pq)$, hence $c < 2(pq+1)\log(pq)$. But $\log(pq+1) < \log(2pq) + \log 2 + \log(pq) < \frac{4}{3}\log(pq)$, so

$$6\log C < 6\left(\log 2 + \frac{4}{3}\log(pq) + \log\log(pq)\right), \text{ and}$$

$$\log 2 + \log\log(pq) < 2\log\log(pq) < \log(pq),$$

hence $6\log C < 6 \times \frac{5}{3}\log(pq) = 10(\log(pq))$.

Therefore, $6\log C + \exp\{630p^2q^2\} < \exp\{631p^2q^2\}$, concluding the proof. ∎

In his paper (1964b), Hyyrö made a thorough study of the number and size of solutions of the equation $aX^n - bY^n = Z$, using methods of diophantine approximation.

In particular, Hyyrö proved the following theorem, which was delicate to establish:

(C2.8). *Let $n \geq 5$, $D \geq 2$ be integers. Then the exponential diophantine equations $X^n - D^U Y^n = \pm 1$ have at most one solution in integers u, x, y with $0 \leq u < n$, $x \geq 2$, $y \geq 1$ (and if $n = 5$ or 6, then $x \geq 3$).*

This result was used to show:

(C2.9). *If $m > 2$, if p, q are odd primes and $p^e | m$, where $e \geq q$, then the diophantine equations $X^m - Y^q = \pm 1$ have no solution in integers $x, y \geq 2$.*

Proof: It is clearly sufficient to show that the equations $X^{p^q} - Y^q = \pm 1$ have no solution in integers $x \geq 2$, $y \geq 1$.

Assume that there exist $x, y \geq 2$, such that $x^{p^q} - y^q = \pm 1$. Then $q \geq 5$ by the results of Part A.

From $(x^{p^{q-1}})^p - y^q = \pm 1$, by Cassels' theorem, $p|y$ and $q|x$, so $p \neq q$. Let r, s be defined by $p^q = rq + s$, with $0 < s < q$; from $p^q > q > s$ it follows that $r \neq 0$.

I shall define successively the integers $a_1, \ldots, a_q$. From

$$y^q = x^{p^q} \mp 1 = \left(x^{p^{q-1}} \mp 1\right)\frac{x^{p^q} \mp 1}{x^{p^{q-1}} \mp 1},$$

and since $p|y$ and $p^2 \nmid \frac{x^{p^q}\mp 1}{x^{p^{q-1}}\mp 1}$ (by (P1.2)), then $x^{p^{q-1}} \mp 1 = p^{q-1}a_1^q$ for some integer a_1. Again

$$x^{p^{q-1}} \mp 1 = \left(x^{p^{q-2}} \mp 1\right)\frac{x^{p^{q-1}}\mp 1}{x^{p^{q-2}}\mp 1}, \quad p^{q-1}|x^{p^{q-1}} \mp 1$$

but $p^2 \nmid \frac{x^{p^{q-1}}\mp 1}{x^{p^{q-2}}\mp 1}$, hence $x^{p^{q-1}} \mp 1 = p^{q-2}a_2^q$ for some integer a_2. Continuing in this way, $x^p \mp 1 = pa_{q-1}^q$ and $x \mp 1 = a_q^q$.

Now, consider the exponential diophantine equations $X^q - x^U Y^q = \mp 1$, where $q \geq 5$, $x \geq 2$. These equations in the unknowns U, X, Y have the solutions (s, y, x^r), because $y^q - x^{s+rq} = \mp 1$, and $(1, a_q, 1)$, because $a_q^q - x = \mp 1$. This contradicts the preceding theorem. ∎

3. Lower Bounds for Solutions

In 1964a, Hyyrö calculated lower bounds for hypothetical positive integers x, y such that $x^m - y^n = 1$.

Assume that p, q are odd primes, x, y are positive integers and $x^p - y^q = 1$. For the convenience of the reader, I require Lemma (B2.7):

There exist positive integers a, b, u, v such that

$$\begin{cases} x - 1 = p^{q-1}a^q \\ \frac{x^p-1}{x-1} = pu^q \end{cases} \quad p \nmid u, \ y = pau, \ \gcd(a, u) = 1 \quad (3.1)$$

and

$$\begin{cases} y + 1 = q^{p-1}b^p \\ \frac{y^q+1}{y+1} = qv^p \end{cases} \quad q \nmid v, \ x = qbv, \ \gcd(b, v) = 1 \quad (3.2)$$

Moreover, $a = qa_0 - 1$, $b = pb_0 + 1$ with a_0, b_0 positive integers, by (B4.1).

Lower Bounds for Solutions

Let a_1 (resp. b_1) be the largest integer dividing a (resp. b) and such that all of its prime factors are not congruent to 1 modulo q (resp. p).

(C3.1). Lemma. *With the above notations*

i) $$a \equiv \frac{q^{p-1} - 1}{p} \pmod{p}$$
$$b \equiv -\frac{p^{q-1} - 1}{q} \pmod{q}.$$

ii) $u = p^{q-1} a_1^q u_1 + 1$, with $a_1 \geq q - 1$, $u_1 \geq 1$,
$2 | a_1 u_1, q | a_1 + 1$, and
$v = q^{p-1} b_1^p v_1 + 1$, with $b_1 \geq 1$, $v_1 \geq 1$,
$2 | b_1 v_1, p | b_1 - 1$.

iii) Either $2 | a_0$, and $2 | b_1$, or $2 | b_0$ and $2 | a_1$.

iv) $u \equiv p^{q-2} \pmod q$, $u_1 \equiv 1 - p^{q-2} \pmod q$, and
$v \equiv q^{p-2} \pmod p$, $v_1 \equiv q^{p-2} - 1 \pmod p$.

Proof: i) By Cassels' theorem (B2.4), $q | x$. By (B4.2) part i), $v \equiv 1 \pmod q$. Similarly $u \equiv 1 \pmod p$. It follows from (B4.1) that

$$a \equiv au \equiv \frac{y}{p} \equiv \frac{q^{p-1} - 1}{p} \pmod p, \text{ and}$$

$$b \equiv bv \equiv \frac{x}{q} \equiv -\frac{p^{q-1} - 1}{q} \pmod q.$$

ii) By Cassels' theorem, $p | y$. By (B4.2), part i), $x - 1$ divides $u^q - 1$, p divides $u - 1$, so $u > 1$.

Now I show that $\gcd(a_1, \frac{u^q - 1}{u - 1}) = 1$. Indeed, if ℓ is a prime such that $\ell | \frac{u^q - 1}{u - 1}$, but $\ell \nmid u - 1$, then ℓ is a primitive factor of $u^q - 1$ and by (P1.4), $\ell \equiv 1 \pmod q$, hence $\ell \nmid a_1$. If $\ell | a_1$,

$\ell | u - 1$, then $\ell | \gcd(u - 1, \frac{u^q-1}{u-1})$ so $\ell = q$ by (P1.2), part ii); so $q|a$ hence $q|x - 1$, but $q \nmid x$ by Cassels' theorem, and this is absurd. Hence $p^{q-1}a_1^q$ divides $p^{q-1}a^q = x - 1$, which divides $u^q - 1 = \frac{u^q-1}{u-1}(u - 1)$, and therefore $p^{q-1}a_1^q$ divides $u - 1$. So $u = p^{q-1}a_1^q u_1 + 1$, with $u_1 \geq 1$. But $\frac{x^p-1}{x-1}$ is always odd, so u is odd, hence $2|a_1 u_1$. Moreover $a = a_1 a_2$, where $a_2 \equiv 1 \pmod{q}$, hence $a_1 \equiv a \equiv -1 \pmod{q}$ by (B4.1); thus $q|a_1 + 1$.

Now I consider the integer v. Since $\frac{(-y)^q-1}{(-y)-1} = \frac{y^q+1}{y+1} = qv^p$, I prove in the same way that $v = q^{p-1}b_1^p v_1 + 1$, with $v_1 \geq 1$, $2|b_1 v_1$ and $p|b_1 - 1$. Writing $b = b_1 b_2$, with $b_2 \equiv 1 \pmod{p}$ by (B4.1), $b_1 \equiv b \equiv 1 \pmod{p}$, hence $p|b_1 - 1$, $b_1 \geq 1$.

iii) Assume that $2 \nmid a_0$ or $2 \nmid b_1$. If $2 \nmid a_0$ then by (B4.1) $2|a$, hence $2|a_1$; also $2|y$ (since $y = pau$), hence $2 \nmid x$, so $2 \nmid b$ and therefore, by (B4.1), $2|b_0$.

If $2 \nmid b_1$ then $2 \nmid b$ so $2|b_0$; also by ii) $2|v_1$, so $2 \nmid v$, hence $2 \nmid x$, so $2|y$, hence $2|au$, but since $2|a_1 u_1$ by ii), then $2 \nmid u$. Thus $2|a$, so $2|a_1$.

iv) $pu \equiv pu^q \equiv \frac{x^p-1}{x-1} \equiv 1 \pmod{q}$, since $q|x$ by Cassels' theorem. Then $u \equiv p^{q-2} \pmod{q}$. Also, $u \equiv a_1 u_1 + 1 \pmod{q}$ and $a_1 \equiv -1 \pmod{q}$, hence $u_1 \equiv 1 - u \equiv 1 - p^{q-2} \pmod{q}$.

Similarly, $qv \equiv qv^p \equiv \frac{y^q+1}{y+1} \equiv 1 \pmod{p}$, since $y \equiv 0 \pmod{p}$, by Cassels' theorem. Then $v \equiv q^{p-2} \pmod{p}$. From $v \equiv b_1 v_1 + 1 \pmod{p}$ and $p_1 \equiv 1 \pmod{p}$, it follows that $v_1 \equiv v - 1 \equiv q^{p-2} - 1 \pmod{p}$. ∎

Note in particular, that $u_1 \geq 2$.

(C3.2). Lemma. *If $k, z > 1$, then*

i) $\left(1 + \frac{1}{z}\right)^{1/k} > 1 + \frac{1}{2kz}$.

ii) $\left(1 - \frac{1}{z}\right)^{1/k} > 1 - \frac{1}{k(z-1)}$.

Proof: This is a very simple exercise.

i) Writing $t = \frac{1}{z}$, then $0 < t < 1$. Let $f(t) = (1+t)^{1/k} - (1 + \frac{t}{2k})$. Then $f(0) = 0$ and $f'(t) = \frac{1}{k}(1+t)^{\frac{1}{k}-1} - \frac{1}{2k}$.
Since $t < 1$ then $(1+t)^{k-1} < 2^k$, hence $\frac{1}{(1+t)^{(k-1)/k}} > \frac{1}{2}$, showing that $f'(t) > 0$. Thus $f(t) > 0$ for every t, which proves (i).

ii) Let $g(t) = (1-t)^{1/k} - \left(1 - \frac{t}{k(1-t)}\right)$. Then $g(0) = 0$ and
$$g'(t) = \frac{1}{k}(1-t)^{\frac{1}{k}-1} + \frac{1}{k(1-t)^2} > 0.$$
Hence $g(t) > 0$, showing (ii). ∎

The following result gives lower bounds for positive integers x, y satisfying $x^p - y^q = 1$.

(C3.3). *With the above notations:*

i) $x \geq \max\left\{p^{q-1}(q-1)^q + 1, \ q(2p+1)(2q^{p-1}+1)\right\}$,

$y \geq \max\left\{q^{p-1}(p+1)^p - 1, \ p(q-1)(2p^{q-1}(q-1)^q+1)\right\}$.

ii) $x > (q^{1/p}v)^{q/(q-1)}$,
$$y > \left[p^{1/q}\left(1 - \frac{1}{q(x-1)}\right)u\right]^{p/(p-1)}.$$

Proof: i) First assume that x is even; then a is odd and by Lemma (C3.1) (iii), a_0 and b_1 are even, so $b_1 \geq p+1$, $b \geq 2p+1$, and $a \geq 2q - 1$.
From $x > p^{q-1}a^q \geq p^{q-1}(2q-1)^q$, then $x \geq p^{q-1}(2q-1)^q + 1$. On the other hand, $x = qbv \geq q(2p+1)[q^{p-1}(p+1)^p + 1]$.
Similarly, $y + 1 = q^{p-1}b^p \geq q^{p-1}(2p+1)^p$, hence $y \geq q^{p-1}(2p+1)^p - 1$. But $y = pau \geq p(2q-1)[2p^{q-1}(2q-1)^q+1]$, because $u_1 \geq 2$, a_1 is odd, so $2q | a_1 + 1$.

Now assume that x is odd, hence y is even. The same considerations show that
$$x \geq p^{q-1}(q-1)^q + 1,$$
$$x \geq q(2p+1)(2q^{p-1}+1)$$
and

$$y \geq q^{p-1}(2p+1)^p - 1,$$
$$y \geq p(q-1)[2p^{q-1}(q-1)^q + 1,$$

since $u_1 \geq 2$. Putting this together, it follows that

$$x \geq \max\left\{p^{q-1}(q-1)^q + 1,\ q(2p+1)(2q^{p-1}+1)\right\}$$
$$y \geq \max\left\{q^{p-1}(2p+1)^p - 1,\ p(q-1)[2p^{q-1}(q-1)^q + 1]\right\}.$$

ii) Since $x^p - y^q = 1$, then by (C3.2),

$$x = qbv = q^{1/p}y^{1/p}\left(1 + \frac{1}{y}\right)^{1/p} v$$
$$= q^{1/p}x^{1/q}\left(1 - \frac{1}{x^p}\right)^{1/pq}\left(1 + \frac{1}{y}\right)^{1/p} v.$$

Also

$$y = pau = p^{1/q}x^{1/q}\left(1 - \frac{1}{x}\right)^{1/q} u$$
$$= p^{1/q}y^{1/p}\left(1 + \frac{1}{y^q}\right)^{1/pq}\left(1 - \frac{1}{x}\right)^{1/q} u.$$

By Lemma (C3.2), it follows that

$$x > q^{1/p}x^{1/q}\left[1 - \frac{1}{pq(x^p - 1)}\right]\left[1 + \frac{1}{2py^q}\right] v$$
$$> q^{1/p}x^{1/q}v,$$

because

$$\frac{1}{pq(x^p-1)} + \frac{1}{2p^2q(x^p-1)y^q} = \frac{1}{pqy^q} + \frac{1}{2p^2qy^{2q}}$$
$$< \frac{1}{4py^q} + \frac{1}{4py^q} = \frac{1}{2py^q}.$$

Then $x^{\frac{q-1}{q}} > q^{1/p}v$ and so $x > (q^{1/p}v)^{q/(q-1)}$.

Similarly, by Lemma (C3.2)

$$y > p^{1/q} y^{1/p} \left[1 + \frac{1}{2pqy^q}\right] \left[1 - \frac{1}{q(x-1)}\right] u$$
$$> p^{1/q} y^{1/q} \left[1 - \frac{1}{q(x-1)}\right] u,$$

hence

$$y^{(p-1)/p} > p^{1/q} \left[1 - \frac{1}{q(x-1)}\right] u,$$

and finally

$$y > \left[p^{1/q} \left[1 - \frac{1}{q(x-1)}\right] u\right]^{p/(p-1)}. \qquad \blacksquare$$

More explicitly, Hyyrö showed (1964a):

(C3.4). *If $p, q > 3$ are primes, and if $x, y \geq 2$ satisfy $x^p - y^q = 1$, then $x, y > 10^{11}$.*

Proof: If $p < q$ then $p \geq 5$, $q \geq 7$, $x > y$, so by Lemma (C3.1),

$$x > y > \max\left\{7^4 \times 6^5 - 1,\ 5 \times 6(5^6 \times 6^7 + 1)\right\} > 10^{11}.$$

If $q < p$ then $q \geq 5$, $p \geq 7$ and $y > x$. First assume that x is even. In the proof of Lemma (C3.1) it was seen that

$$y > x \geq \max\left\{7^4 \times 9^5 + 1,\ 5 \times 8(5^6 \times 8^7 + 1)\right\} > 10^{11}.$$

Now assume that x is odd. From $x^p > y^q$ it follows that $(x+1)^p > (y+1)^q$. Indeed, from $x < y$, then

$$\frac{(y+1)^q}{y^q} = \left(1 + \frac{1}{y}\right)^q < \left(1 + \frac{1}{x}\right)^q < \left(1 + \frac{1}{x}\right)^p$$
$$= \frac{(x+1)^p}{x^p} < \frac{(x+1)^p}{y^q},$$

so $(y+1)^q < (x+1)^p$. Thus $x+1 > (y+1)^{\frac{q}{p}}$, and with the previous notations

$$y > x > (y+1)^{\frac{q}{p}} - 1 = q^{\frac{(p-1)q}{p}} b^q - 1 > q^{q-1} b^q - 1.$$

But, by (B4.1) and Lemma (C3.1),

$$\begin{cases} b \equiv 1 \pmod{p} \\ b \equiv -\frac{p^{q-1}-1}{p} \pmod{q}, \end{cases}$$

and

$$b \equiv 1 \pmod{2}, \quad \text{because } x \text{ is odd}.$$

I shall examine various cases, first for small values of p, q. Let $q = 5$. Then

P	7	11	13	17	19	≥ 23
$b \geq$	85	67	53	171	191	47

Indeed, if $p = 7$, then $b \equiv -\frac{7^4-1}{5} \equiv 0 \pmod{5}$, $b \equiv 1 \pmod{7}$ and $b \equiv 1 \pmod 2$, hence by the Chinese remainder theorem, $b \equiv 15 \pmod{70}$. If $b = 15$, then $b_1 = 15$ (recall that b_1 is the product of the factors of b which are not congruent to 1 modulo p); hence by (B4.1) and Lemma (C3.1) $x > v > 5^6 \times 15^7 = y+1$; however $y > x$, so this is absurd. This shows that $b \geq 85$.

Similarly, if $p = 11$ then $b \equiv -\frac{11^4-1}{5} \equiv 2 \pmod 5$, so $b \equiv 67 \pmod{110}$. If $p = 13$, then $b \equiv -\frac{13^4-1}{5} \equiv 3 \pmod 5$, so $b \equiv 53 \pmod{130}$. If $p = 17$, then $p \equiv -\frac{17^4-1}{5} \equiv 1 \pmod 5$, so $p \equiv 1 \pmod{170}$, and since $b \geq p+1$, so $b \geq 171$. If $p = 19$ then $p \equiv -\frac{19^4-1}{5} \equiv 1 \pmod 5$ so $p \equiv 1 \pmod{190}$, but $b \geq p+1$, so $b \geq 191$. Next, if $p \geq 23$, since $b = pb_0 + 1$ and b is odd, then $b \geq 2p + 1 \geq 47$.

So, always $b \geq 47$ and $y > x > 5^4 \times 47^5 - 1 > 10^{11}$.

Finally, let $q \geq 7$, then $p \geq 11$; b is odd so $b \geq 2p+1 \geq 23$ and therefore $x > 7^6 \times 23^7 - 1 > 10^{11}$. ∎

The next result of Hyyrö concerns lower bounds for x, y when one of the exponents m or n is composite, the other being possibly a prime.

(C3.5). *Let $m, n > 2$, and assume that x, y are positive integers such that $x^m - y^n = 1$. If m is composite then $x > 10^{84}$, while if n is composite then $y > 10^{84}$.*

Proof: With an appropriate change of notations, the two cases may be treated simultaneously. So assume that q is a prime, m is composite, x, y are positive integers such that $x^m - y^q = \pm 1$; by the results of Part A, necessarily $q \geq 5$. Let p be a prime dividing m, so again $p \geq 5$. By (B4.2) $p^{q-1} \equiv 1 \pmod{q^2}$.

Hyyrö used Riesel's table (1964) (which is contained in the table of §B5) to deduce that if $5 \leq p < 150$ and $5 \leq q < 150$, one of the possibilities shown in Table C3.5 holds.

Since m is composite, by (B4.2), $q^2 | x$, and by (B4.2) part (1), $x > q^{2(m+1)-1} = q^{2m+1}$, and also $x - 1 > q^{2m+1}c$, where

$$\begin{cases} y \pm 1 = q^{2m-1} c^m \\ \frac{y^q \pm 1}{y \pm 1} = qv^m \end{cases}$$

with $c, v \geq 1$, $\gcd(c, v) = 1$, $q \nmid v$, $x = q^2 cv$ (by (B4.2), part (1)).

I shall now show that $x > 10^{84}$, which is enough to prove the statement.

Let p be the smallest prime dividing m; hence by the results of Part A, $p \geq 5$ and $m \geq 25$.

I consider several cases in succession.

Table C3.5

q	p				
5	7	43	101	107	149
7	19	31	67	79	97
13	19	23	89		
17	131				
19	127				
29	41	137			
43	19				
47	53	67	71		
59	53	137			
71	11				
79	31				
97	53	107			
103	43				
137	19				

a) First let $p = 5$. By (B4.2), $5^{q-1} \equiv 1 \pmod{q^2}$. Using the above table, $q > 100$, hence $x > 100^{51} > 10^{100}$.

b) Now let $p = 7$ and $q = 5$. First assume that m is a power of 7, so that $m \geq 49$.

It was seen that $x - 1 > q^{2m+1}c$. From $x^m = y^q \pm 1$ it follows that $(x-1)^m < x^m - 1 \leq x^m \pm 1 = y^q$, and also $q^{2m-1}c^m \geq y - 1$. This gives

$$\begin{cases} x - 1 > 5^{99}c \\ y^s > (x-1)^{49} \\ 5^{97}c^{49} \geq y - 1, \end{cases}$$

hence $x - 1 > 5^{99}$, $y^5 > 5^{99 \times 49}$, so

$$c^{49} \geq \frac{y-1}{5^{97}} > \frac{5^{\frac{99 \times 49}{5}} - 1}{5^{97}} = 5^{\frac{99 \times 49}{5}} - \frac{1}{5^{97}} > 5^{873} - \frac{1}{5^{97}},$$

so $c^{49} \geq 5^{873}$, and hence $c > 5^{17}$. Therefore $x - 1 > 5^{99+17} = 5^{116}$ and $y^5 > 5^{116 \times 49}$, so $c^{49} \geq \frac{y-1}{5^{97}} > 5^{\frac{116 \times 49}{5}} - \frac{1}{5^{97}}$, hence $c^{49} > 5^{1039}$, and so $c > 5^{21.2}$. Therefore $x - 1 > 5^{99+21.2} = 5^{120.2} > 10^{84}$, thus $x > 10^{84}$. If $p = 7$, $q = 5$ and m is not a power of 7, let ℓ be a prime dividing m with $\ell > 7$. By (B4.2), $\ell^{q-1} \equiv 1 \pmod{\ell^2}$ with $q = 5$, so by the table $\ell \geq 43$. Thus $m \geq 7 \times 43 = 301$ and $x > q^{2m+1} > 5^{603} > 10^{100}$.

The remaining cases are the following:

c) If $p = 7$ and $7 \leq q < 150$, then by the table $7^{q-1} \not\equiv 1 \pmod{q^2}$, and by (B4.2) this case is impossible.

d) If $p > 7$ and $7 \leq q < 150$, by Riesel's table and the congruence $p^{q-1} \equiv 1 \pmod{q^2}$, either $p = 11$ or $p \geq 19$.

If $p = 11$ then $q = 71$, hence $m \geq 121$, so $x > q^{2m+1} \geq 71^{243} > 10^{100}$. On the other hand, if $p \geq 19$, then $m \geq 361$, so $x > q^{2m+1} \geq 5^{763} > 10^{100}$.

e) If $q > 150$ and $p \geq 7$, then $m \geq 49$ and $x > q^{2m+1} > 150^{99} > 10^{100}$.

∎

The next result of Hyyrö (1964a) indicated that consecutive powers with composite exponents must indeed be very large:

(C3.6). *If $x^m - y^n = 1$ where m, n are composite, then x^m, y^n have at least 10^9 digits.*

Proof: Let p be the smallest prime dividing m and q the smallest prime dividing n. By appropriate change of notation, and by allowing that $x^m - y^n = \pm 1$, it may be assumed without loss of generality that $p < q$.

Let $m = pm'$, $n = qn'$ and $x' = x^{m'}$, $y = y^{n'}$; so $x'^p - y^n = 1$, $x^m - y'^q = 1$.

By (B4.2), $p^{q-1} \equiv 1 \pmod{q^2}$ and $q^{p-1} \equiv 1 \pmod{p^2}$, simultaneously. By Riesel's table, this implies that $q > 150$.

If $p = 5$ or 7, the known tables indicate that $q > 200000$ (note that $5^{20770} \equiv 1 \pmod{20771^2}$). So $n > 4 \times 10^8$; by (B4.2) $y > p^{2m+1} > 5^{8 \times 10^8} > 10^{5 \times 10^8}$. Hence $y^n - 1 > 10^{3 \times 10^8 \times 2 \times 10^8} > 10^{10^9}$.

If $p \geq 11$ then $n \geq q^2 > 22500$, so $y > p^{2n+1} > 11^{45000}$, so $y^n - 1 > 11^{45000 \times 22500} > 10^{10^9}$. ∎

All the above lower bounds may be improved by extending the tables of congruences $p^{q-1} \equiv 1 \pmod{q^2}$ and $q^{p-1} \equiv 1 \pmod{p^2}$.

4. Algorithm to Determine the Eventual Solutions

Hyrrö indicated (in 1964a) a continued fraction algorithm to retrieve the solutions of $X^p - Y^q = 1$, if any exist.

I keep the notations of §C3.

Assume that p, q are distinct odd primes, $x, y \geq 1$ satisfy $x^p - y^q = 1$. Let

$$\alpha = \frac{q^{\frac{p-1}{p}}}{p^{\frac{q-1}{q}}}$$

First I give a lemma:

(C4.1). Lemma. *With the above notations:*

$$0 < \alpha - \frac{a}{b} < \frac{1}{2b^r} < \frac{1}{2b^2}, \qquad (4.1)$$

where $r = \min\{p, q\}$.

Proof: Clearly,

$$(x-1)^p < x^p - 1 < y^q + 1 < (y+1)^q,$$

so, by (3.1) and (3.2),

$$p^{(q-1)p} a^{qp} = (x-1)^p < (y+1)^q = q^{(p-1)q} b^{qp},$$

and therefore
$$\frac{a}{b} < \alpha.$$

Next $q^{\frac{p-1}{p}-1} < 1 < p^{\frac{q-1}{q}}$, hence
$$\frac{q^{\frac{p-1}{p}}}{p^{\frac{q-1}{q}}} = \alpha < q.$$

I show that $\frac{b}{a} < p$. By Lemma (B2.3)
$$p^{(q-1)p}a^{qp}q^{(p-1)q} = (x-1)^p q^{(p-1)q} > (y+1)^q = q^{(p-1)q}b^{pq},$$

hence
$$p^{(q-1)p}a^{qp} > b^{pq}$$

and so
$$\frac{b}{a} < p^{\frac{q-1}{q}} < p.$$

I now give an expression for α as a product with the factor $\frac{a}{b}$:

$$\alpha = \frac{q^{\frac{p-1}{p}}}{p^{\frac{q-1}{q}}} = \frac{a}{b} \times \frac{(y+1)^{\frac{1}{p}}}{(x-1)^{\frac{1}{q}}} = \frac{a}{b} \times \frac{\left(1+\frac{1}{y}\right)^{\frac{1}{p}}\left(1+\frac{1}{x-1}\right)^{\frac{1}{q}}}{\frac{x^{\frac{1}{q}}}{y^{\frac{1}{p}}}}$$

$$= \frac{a}{b} \times \frac{\left(1+\frac{1}{y}\right)^{\frac{1}{p}}\left(1+\frac{1}{x-1}\right)^{\frac{1}{q}}}{\left(1+\frac{1}{y^q}\right)^{\frac{1}{pq}}} < \frac{a}{b}\left(1+\frac{1}{y}\right)^{\frac{1}{p}}\left(1+\frac{1}{x-1}\right)^{\frac{1}{q}}.$$

Hence
$$\alpha - \frac{a}{b} < \frac{a}{b}\left[\left(1+\frac{1}{y}\right)^{\frac{1}{p}}\left(1+\frac{1}{x-1}\right)^{\frac{1}{q}} - 1\right],$$

and I estimate the function in the bracket. Let $x - 1 = \frac{1}{X}$, $y + 1 = \frac{1}{Y}$, so $y = \frac{1-Y}{Y}$. Then

$$\left(1+\frac{1}{x-1}\right)^{\frac{1}{q}}\left(1+\frac{1}{y}\right)^{\frac{1}{p}} = (1+X)^{\frac{1}{q}}\left(1+\frac{Y}{1-Y}\right)^{\frac{1}{p}} = F(X,Y)$$

has value $F(0,0) = 1$. By the mean value theorem, $F(X,Y) \leq F(0,0) + \frac{2}{q}X + \frac{2}{p}Y$, so

$$\alpha - \frac{a}{b} < \frac{a}{b}\left[\frac{2}{q(x-1)} + \frac{2}{p(y+1)}\right]$$

$$= \frac{a}{b}\left[\frac{2}{qp^{q-1}a^q} + \frac{1}{pq^{p-1}b^p}\right]$$

$$< \frac{a}{b}\left[\frac{2}{qab^{q-1}} + \frac{2}{pq^{p-1}b^p}\right]$$

$$< \frac{2}{qb^2} + \frac{1}{pq^{p-2}b^2} < \frac{1}{2b^r} < \frac{1}{2b^2},$$

because $5 \leq p, q$ (by Part A). ■

(C4.2). *Let p, q be distinct odd primes. If there exist integers $x, y \geq 2$ such that $x^p - y^q = 1$, they may be found by the following algorithm. Let $\alpha = \frac{q^{\frac{p-1}{p}}}{p^{\frac{q-1}{q}}} = [c_0, c_1, \ldots, c_n, \ldots]$ (the simple continued fraction expansion of α), and let $\frac{A_j}{B_j}$ (for $j \geq 0$) be the convergents of α. Then there exist an even index $i \geq 0$ such that $x = p^{q-1}A_i^q + 1$ and $y = q^{p-1}B_i^p - 1$. Moreover, for this index i, the following conditions hold.*

i) $A_i > 1$, $B_i > 1$.

ii) $A_i \equiv -1 \pmod{q}$, $B_i \equiv 1 \pmod{p}$.

iii) $A_i \equiv \dfrac{q^{p-1} - 1}{p} \pmod{p}$,

$B_i \equiv -\dfrac{p^{q-1} - 1}{q} \pmod{q}$.

iv) $c_{i+1} \geq -A_i^{r-2}$ *and* $c_{i+1} \geq B_i^{r-2}$, *where* $r = \min\{p, q\}$.

Proof: Suppose that $x, y \geq 2$ are integers such that $x^p - y^q = 1$. By the above lemma, $0 < \alpha - \frac{a}{b} < \frac{1}{2b^2}$. By (P4.9), there exists an index $i \geq 0$ such that $a = A_i$, $b = B_i$. Hence $x = p^{q-1}A_i^p + 1$

and $y = q^{p-1}B_i^q - 1$. By (P4.4), (P4.5) from $\frac{a}{b} < \alpha$ it follows that i is even. By (B4.1) $A_i > 1$, $B_i > 1$ and $A_i \equiv -1 \pmod{q}$, $B_i \equiv 1 \pmod{p}$. By Lemma (C3.1) $A_i \equiv \frac{q^{p-1}-1}{p} \pmod{p}$ and $B_i \equiv -\frac{p^{q-1}-1}{q} \pmod{q}$.

Hence

$$B_i + B_{i+1} = B_i + c_{i+1}B_i + B_{i-1} < 2(c_{i+1}B_i + B_i),$$

because $B_{i-1} < B_i$. It follows from (P4.7) and Lemma (C4.1) that

$$\frac{1}{2B_i^r} > \alpha - \frac{A_i}{B_i} > \frac{1}{B_i(B_i + B_{i+1})} > \frac{1}{2B_i(c_{i+1}B_i + B_i)},$$

so $B_i(c_{i+1}B_i + B_i) > B_i^r$, hence $B_i^2 c_{i+1} > B_i^r - B_i^2$, and therefore $c_{i+1} \geq B_i^{r-2}$.

To conclude the proof, I note that $c_{i+1} \geq -A_i^{r-2}$ because $\alpha > 0$ implies that $c_0 \geq 0$, and $A_i \geq 0$ and therefore $c_{i+1} > 0 \geq -A_i^{r-2}$. ∎

It should be added that the algorithm has never yet led to any solution of the equation.

II. The Equation $a^U - b^V = 1$

5. What Will Be Discussed

The study of the equation

$$a^U - b^V = 1 \tag{5.1}$$

may be approached from a somewhat higher standpoint, as I shall explain now.

Consider the following problems:

Problem 1. Given integers $A, B, k \geq 1$ and $M_1, \ldots, M_m$, $N_1, \ldots N_n > 1$ (with $m, n \geq 1$), to find all integers $u_1, \ldots, u_m$, $v_1, \ldots, v_n \geq 0$, such that

$$AM_1^{u_1} \ldots M_m^{u_m} - BN_1^{v_1} \ldots N_n^{v_n} = k \qquad (5.2)$$

Here is a special case of this problem:

Problem 1'. Given integers $A, B, k \geq 1$, $a, b, > 1$, to find all integers $u, v \geq 0$, such that

$$Aa^u - Bb^v = k. \qquad (5.3)$$

Now let E_1, E_2 be finite (non-empty) sets of prime numbers; denote by $E_1^\times$ (resp. $E_2^\times$) the set of all finite products of primes belonging to E_1 (resp. E_2). Consider also the problem:

Problem 2. Given integers $A, B, k \geq 1$, and the sets E_1, E_2 (as above), to find all integers $M \in E_1^\times$, $N \in E_2^\times$, such that

$$AM - BN = k. \qquad (5.4)$$

It is especially interesting to consider the above problems with $A = B = 1$.

In this section, I discuss whether these problems have only finitely many solutions.

A moment of reflection suffices to conclude that problem 1 has only finitely many solutions if and only if this is the case for problem 2. Indeed, one implication is obvious. For the other implication take

$$E_1 = \{p \text{ prime}|\ p \text{ divides } M_1 M_2 \ldots M_m\},$$

and

$$E_2 = \{p \text{ prime}|\ p \text{ divides } N_1 N_2 \ldots N_n\}.$$

In the next section, I prove that these problems have indeed only finitely many solutions. The proof will require Thue's theorem (C1.1).

Here is a chronology of the study of these problems.

1908—Størmer showed that problem 1, with $k = 1$ or 2, has only finitely many solutions, which may be effectively obtained in finitely many steps.

1908—Thue proved that problem 1' has only finitely many solutions; however, his proof did not allow one to determine these solutions in finitely many steps.

1918—Pólya showed that problem 2 has only finitely many solutions; his method does not lead to an effective procedure to determine the solutions.

1925a (resp. 1945a)—Nagell (resp. Skolem) extended Størmer's result for $k = 3$ (resp. $k = 4$), with an effective proof.

1931—Pillai gave an upper estimate for the number of pairs (u, v) of integers $u, v \geq 0$, such that $0 < a^u - b^v \leq k$, where $a, b > 1$, $k \geq 0$ are given integers and k tends to infinity.

1936—Herschfeld showed that for every sufficiently large $|k|$, the equation $2^U - 3^V = k$ has at most one solution. Pillai extended this result for the equation $a^U - b^V = k$.

1952, 1953—LeVeque established that $a^U - b^V = 1$ has at most one solution in integers $u, v \geq 2$, and Cassels gave an algorithm to determine the eventual solutions.

1958—Nagell gave another proof that problem 1 has only finitely many solutions, using Thue's result.

1960—Cassels proved that the solutions of problems 1 and 2 may be effectively computed in finitely many steps.

6. Finiteness of the Number of Solutions

I begin proving that problems 1 and 2 of the preceding sections have only finitely many solutions. The proof, based on Thue's theorem (C1.1), does not lead to the effective determination of the number or size of solutions.

The following proposition was known to Pólya (1918) and Pillai (1931) and also appears in Nagell (1958).

(C6.1). i) *Let E_1, E_2 be two finite non-empty sets of prime numbers, and let $A, B, k \geq 1$ be integers. Then, there exist only finitely many integers $M \in E_1^\times$, $N \in E_2^\times$, such that $AM - BN = k$.*

ii) *Let $M_1, \ldots, M_m$, $N_1, \ldots, N_n > 1$ (with $m \geq 1$, $n \geq 1$), and let $A, B, k \geq 1$. Then there exist only finitely many integers $u_1, \ldots, u_m, v_1, \ldots, v_n \geq 0$, such that*

$$AM_1^{u_1} \ldots M_n^{u_n} - BN_1^{v_1} \ldots N_n^{v_n} = k.$$

iii) *Let $A, B, k \geq 1$, $a, b > 1$. Then there exist only finitely many integers $u, v \geq 0$, such that $Aa^u - Bb^v = k$.*

Proof: i) Let $E_1 = \{p_1, \ldots, p_r\}$, $E_2 = \{q_1, \ldots, q_s\}$. Let C be the set of all pairs (a, b), where

$$\begin{cases} a = p_1^{e_1} \ldots p_r^{e_r} & \text{with } 0 \leq e_i \leq 2 \text{ for every } i, \\ b = q_1^{f_1} \ldots q_s^{f_s} & \text{with } 0 \leq f_j \leq 2 \text{ for every } j. \end{cases}$$

Thus, C is a finite set. For each $(a, b) \in C$ let $S_{(a,b)}$ be the set of solutions (u, v), in integers, of the equation

$$E_{(a,b)}: \quad AaU^3 - BbV^3 = k.$$

By (C1.1), each set $S_{(a,b)}$ is finite, hence $S = \bigcup_{(a,b) \in C} S_{(a,b)}$ is also finite.

Let T be the set of solutions $(x, y) \in E_1^\times \times E_2^\times$ of $AX - BY = k$.

Each $x \in E_1^\times$, $y^x \in E_2^\times$ may be written in unique way in the form
$$\begin{cases} x = au^3 \\ y = bv^3, \end{cases}$$
with $(a,b) \in C$, $u \in E_1^\times$, $v \in E_2^\times$.

If $(x,y) \in T$, then $Aau^3 - Bbv^3 = k$, so $(u,v) \in S_{(a,b)}$.

The mapping $\varphi : T \to C \times S$, given by $\varphi(x,y) = ((a,b), (u,v))$ is injective. Since $C \times S$ is finite, so is the set T.

ii) Let $E_1 = \{p\text{ prime}|\ p \text{ divides } M_1 M_2 \ldots M_m\}$, $E_2 = \{p \text{ prime}|\ p \text{ divides } N_1 N_2 \ldots N_n\}$. Then ii) follows at once from i).

iii) This is a special case of ii). ∎

The following result is an immediate consequence of (C6.1):

(C6.2). i) *Let* $E = \{p_1, p_2 \ldots p_r\}$ *be a non-empty set of prime numbers, and let*
$$E^\times : \quad z_1 < z_2 < z_3 < \ldots$$
be the sequence of elements of $E^\times$ in increasing order. Then
$$\lim_{i \to \infty} (z_{i+1} - z_i) = \infty.$$

ii) *Let* $a, b > 1$, $a \neq b$, *and let*
$$S : t_1 < t_2 < t_3 < \ldots$$
be the sequence of integers which are powers of a or of b. Then
$$\lim_{i \to \infty} (t_{i+1} - t_i) = \infty.$$

Proof: i) For every $k \geq 1$, there exist only finitely many integers $x, y \in E^\times$ such that $0 < x - y \leq k$, as was shown in (C6.1). Therefore $\lim_{i \to \infty} (z_{i+1} - z_i) = \infty$.

ii) This follows at once from i), by considering the set $E = \{p \text{ prime}| p \text{ divides } ab\}$. ∎

In §C9, I shall indicate effective versions of the preceding results.

The following result is similar to one given by Herschfeld in 1936.

(C6.3). *Given integers $A, B, k \geq 1$, $a, b \geq 2$, let $t \geq 0$ be the smallest integer such that $k \leq Aa^t$. For each $i = 0, 1, 2$ and $j = 0, 1$, consider the equation*

$$E_{i,j}: \quad X^3 - (Bb^j)Y^2 = (Aa^{t+i})^2 k,$$

and let N be the maximum of the number of solutions in integers $x, y \geq 1$ of the equations (E_{ij}) [note that by (C1.5) this number is finite].

Then the equation

$$E: \quad Aa^U - Bb^V = k$$

has at most $6N$ solutions in integers $u, v \geq 1$.

Proof: Let T be the set of all positive solutions (u, v) of equation (E). If $Aa^u - Bb^v = k$, then $k \leq Aa^u$, hence $t \leq u$. Let

$$\begin{cases} u - t = 3m + i & \text{with } 0 \leq i \leq 2 \\ v = 2n + j & \text{with } 0 \leq j \leq 1. \end{cases}$$

Then $Aa^{t+1}(a^m)^3 - Bb^j(b^n)^2 = k$, hence

$$[(Aa^{t+1})^2 a^m]^3 - Bb^j [Aa^{t+i} b^n]^2 = (Aa^{t+1})^2 k.$$

Let $S_{i,j}$ be the set of solutions (x, y), $x, y \geq 1$, of the equation $(E_{i,j})$, and let

$$S = \{(x, y, i, j) | 1 \leq x, y,\ 0 \leq i \leq 2,\ 0 \leq j \leq 1 \text{ and } (x, y) \in S_{i,j}\}.$$

It follows that $\#(S) \leq 6N$.

By the above considerations, the mapping $\varphi : T \to S$, given by
$$T(u,v) = ((Aa^{t+1})^2 a^m, \ b^n, \ i, j)$$
is injective, hence $\#(T) \leq \#(S) \leq 6N$. ∎

In 1936, Pillai studied the equations
$$a^U - b^V = k$$
for k sufficiently large. It should be noted that his handling of the case where $gcd(a,b) \neq 1$ is not acceptable.

In his study, Pillai used one of his earlier estimates (1931):

(C6.4). *Let $a, b \geq 2$ be integers such that $\frac{\log a}{\log b}$ is not a rational number.*

  i) *Let $A, B \geq 1$. For every $\delta > 0$, there exists an effectively computable positive integer u_0 (depending on δ, A, B, a, b) such that if $u > u_0$, $v \geq 1$ and $Aa^u > Bb^v$, then*
$$Aa^u - Bb^v \geq a^{u(1-\delta)}.$$

  ii) *For $k \geq 1$, let $N(k)$ denote the number of pairs of integers (u,v), with $u, v \geq 0$, such that*
$$0 < a^u - b^v \leq k.$$
Then, asymptotically as k tends to ∞,
$$N(k) \sim \frac{(\log k)^2}{2(\log a)(\log b)}.$$

Thus, for example, given $a, b \geq 2$, there exist u_0 such that if $u > u_0$, $v \geq 1$ and $a^u > b^v$, then $a^u - b^v > a^{\frac{1}{2}u}$.

Pillai proved in 1936:

(C6.5). Let $a, b \geq 2$ be relatively prime integers. Then there exists an effectively computable integer $k_0 \geq 1$, such that if $k \geq k_0$, then there exists at most one pair of positive integers (x, y) such that $a^x - b^y = k$.

Proof: The proof will be divided into several parts.

1°). There exists $n \geq 1$ such that $b^n = \ell a^m + 1$, where $m \geq 2$, and $gcd(a, \ell) = 1$. Moreover, m, ℓ are uniquely defined by n.

Indeed, since $gcd(a, b) = 1$, there exists $n \geq 1$ such that $b^n \equiv 1 \pmod{a^2}$; choose n smallest possible.

Letting $a = p_1^{e_1} \ldots p_r^{e_r}$ where $p_1, \ldots, p_r$ are distinct primes and
$e_1 \geq 1, \ldots, e_r \geq 1$, then I may write

$$b^n = 1 + h p_1^{f_1} \ldots p_r^{f_r} a^m,$$

where $m \geq 2$, $gcd(h, a) = 1$, $0 \leq f_i$, and for some j, $f_j \leq e_j - 1$.

It will be shown that there exists $n' \geq n$, $m' \geq m$, such that

$$b^{n'} = 1 + h' p_1^{f'_1} \ldots p_r^{f'_r} a^{m'},$$

where $gcd(h', a) = 1$, $0 \leq f'_i$, $f'_j \leq e_j - 1$ for some j, and moreover

$$f'_i = \begin{cases} 0 & \text{if } f_i \leq e_i \\ f_i - e_i & \text{if } e_i < f_i. \end{cases}$$

This process will be repeated. For every $i = 1, \ldots, r$ consider the sequence of exponents $f_i, f_i^{(1)}, f_i^{(2)}, f_i^{(3)}, \ldots$ thus obtained. Since there exist k such that $f_i < k e_i$ for every $j = 1, \ldots, r$, then $f_i^{(k+1)} = 0$. So, after finitely many repetitions, it follows that there exist n and $m \geq 2$ such that $b^n = 1 + \ell a^m$, where $gcd(a, \ell) = 1$.

Now I indicate how to determine n', m', h' and f'_i (for $i = 1, \ldots, r$).

Consider the set J of indices j such that $f_j \leq e_j - 1$, and let $t = \prod_{j \in J} p_j^{e_j - f_j}$ and $n' = nt$. Then, since $m + 2 \leq 2m$

$$b^{n'} = (1 + hp_1^{f_1} \ldots p_r^{f_r} a^m)^t$$
$$= 1 + htp_1^{f_1} \ldots p_r^{f_r} a^m + kp_1^{f_1} \ldots p_r^{f_r} a^{2m}$$
$$= 1 + hp_1^{f_1} \ldots p_r^{f_r} a^{m+1} + k' p_1^{f_1} \ldots p_r^{f_r} a^{m+2}$$
$$= 1 + h' p_1^{f_1'} \ldots p_r^{f_r'} a^{m'}$$

with $m' = m + 1$, $f_j' = 0$ if $f_j \leq e_j$, and $f_i' \leq f_i - e_i$ when $e_i + 1 \leq f_j$. This proves the assertion.

2°). Let $n \geq 1$ be the smallest integer such that $b^n = 1 + \ell a^m$, with $gcd(\ell, a)$ 1 and $m \geq 2$. Then n is the order of b mod a^m.

Indeed, let r be the order of b mod a^m. So $b^r = 1 + ha^m$ and $n = rt$. Then

$$1 + \ell a^m = b^{rt}$$
$$= (1 + ha^m)^t$$
$$= 1 + tha^m + \binom{t}{2} h^2 a^{2m} + \ldots$$
$$= 1 + h\left(t + \binom{t}{2} ha^m + \ldots\right) a^m.$$

So $\ell = h\left(t + \binom{t}{2} ha^m + \ldots\right)$, hence $gcd(h, a) = 1$. By the choice of n as minimal, $n = r$ is the order of b modulo a^m.

3°). With above notation, for every $i \geq 0$, the order of b mod a^{m+i} is na^i, and $b^{na^i} = 1 + \ell_i a^{m+i}$ with $gcd(\ell_i, a) = 1$.

This is shown by induction on i, being true for $i = 0$. Let it be assumed true for $i \geq 0$. If $j = kna^i + r$, with $0 \leq r < na^i$ and $b^j \equiv 1 \pmod{a^{m+i+1}}$, then $b^j \equiv 1 \pmod{a^{m+i}}$, so $b^r \equiv b^r (b^{na^i})^k \equiv b^j \equiv 1 \pmod{a^{m+i}}$ so by induction, $r = 0$. Now, let $1 \leq k$ then $b^{kna^i} = (1 + \ell_i a^{m+1})^k = 1 + \ell_i a^{m+i}(k + $

$\binom{k}{2}\ell_i a^{m+i} + \ldots)$. For $k = 1, \ldots, a-1$, $b^{kna^i} \not\equiv 1 \pmod{a^{m+i+1}}$, however

$$b^{na^{i+1}} = 1 + \ell_i\left(1 + \binom{a}{2}\ell_i a^{m+i-1} + \ldots\right) a^{m+i+1}$$
$$= 1 + \ell_{i+1} a^{m+i+1},$$

with $\gcd(\ell_{i+1}, a) = 1$. This completes the induction.

4°). If $N, M \geq 2$ are such that $b^N \equiv 1 \pmod{a^M}$, then na^{M-m} divides n.

Indeed, by (3°) the order of b modulo a^M is na^{M-m}, hence na^{M-m} divides N.

5°). If $\gcd(a, b) = 1$ and $a^x - b^y = a^X - b^Y$ with x, y, X, Y positive integers and $X > x$, then $Y \geq a^{x-m}$.

Indeed, first observe that $Y > y$. Then

$$a^x(a^{X-x} - 1) = b^y(B^{Y-y} - 1).$$

So $b^{Y-y} - 1$ is divisible by a^x. By (4°), na^{x-m} divides $Y - y$, so $a^{x-m} \leq na^{x-m} \leq Y - y \leq Y$.

6°). By (C6.3) there exists x_0, effectively computable, such that if $a^x > b^y$ and $x > x_0$, then $a^x - b^y \geq a^{x/2}$. By (1°), there exists m with the property indicated. Let $x_1 = \max\{x_0, 3(m+1)\}$ and $k > a^{x_1}$. If $a^x - b^y = a^X - b^Y$ with $x < X$, then $X < 2x$. Indeed, $a^x > k > a^{x_1}$, so $X > x > x_1 \geq x_0$. Then $a^x = k + b^y > k = a^X - b^Y \geq a^{X/2}$, and so $x > \frac{X}{2}$, as required.

On the other hand, by (5°)

$$a^X = k + b^Y > b^Y \geq b^{a^{x-m}} \geq 2^{a^{x-m}}$$
$$> 1 + a^{x-m} + \frac{a^{x-m}(a^{x-m} - 1)}{2}$$
$$+ \frac{a^{x-m}(a^{x-m} - 1)(a^{x-m} - 2)}{6}$$
$$> \frac{(a^{x-m} - 2)^3}{6} > a^{3(x-m-1)},$$

because $a \geq 2$. Hence $X > 3(x-m-1)$. But $x > x_1 \geq 3(m+1)$, so $X > 2x$, which is a contradiction, thus concluding the proof. ∎

Pillai stated that the above result is also valid for the equation $Aa^U - Bb^V = k$, when k is sufficiently large and further assuming that $gcd(a,b) = 1$.

Pillai also studied the equations $a^x - b^y = a^U - b^V$, $a^U + b^V = a^X - b^Y$, $a^U - b^V = b^Y - a^X$, when $gcd(a,b) = 1$. Note that $a^U + b^V = a^X + b^Y$ is the same as $a^U - b^Y = a^X - b^V$.

From (C6.1) and (C6.5), it follows at once that $a^U - b^V = a^X - b^Y$ has only finitely many solutions (u,v,x,y), with $(u,v) \neq (x,y)$. Indeed, by (C6.5), if $|k|$ is sufficiently large, there is at most one pair of integers (x,y) such that $a^x - b^y = k$; similarly, for any k, there are at most finitely many pairs (x,y) such that $a^x - b^y = k$.

Pillai has also shown (in 1944) that the equations $a^U + b^V a^X - b^Y$ and $a^U - b^V = b^Y - a^X$ have only finitely many solutions, when $gcd(a,b) = 1$. However, Pillai did not determine effectively the solutions.

In 1945, he studied the special equations

$$2^U \pm 3^V = 2^X - 3^Y, \qquad 2^U \pm 3^V = 3^Y - 2^X.$$

Pillai noted that the equation $2^X - 3^Y = 2^U - 3^V$ has the three solutions:

$$-1 = 2 - 3 = 2^3 - 3^2$$
$$5 = 2^3 - 3 = 2^5 - 3^3$$
$$18 = 2^4 - 3 = 2^8 - 3^5.$$

Pillai conjectured that there are no other solutions. In 1976, Chein announced that he had established this conjecture, but his proof has not appeared in print (at least, if it exists, it has not been reviewed in *Mathematical Reviews*).

Pillai could solve the other equations with elementary methods.

The only values $a = 2^u + 3^v = 2^x - 3^y$ are

a	$2^u + 3^v$	$2^x - 3^y$
5	$2 + 3$	$2^5 - 3^3 = 2^3 - 3$
7	$2^2 + 3$	$2^4 - 3^2$
13	$2^2 + 3^2$	$2^8 - 3^5 = 2^4 - 3$
29	$2 + 3^3$	$2^5 - 3$
247	$2^2 + 3^5$	$2^8 - 3^2$

The only values $a = 2^u - 3^v = 3^y - 2^x$ are

a	$2^u + 3^v$	$3^y - 2^x$
1	$2^2 - 3$	$3 - 2, \ 3^2 - 2^3$
5	$2^3 - 3, \ 2^5 - 3^3$	$3^2 - 2^2$
7	$2^4 - 3^2$	$3^2 - 2$
23	$2^5 - 3^2$	$3^3 - 2^2$

The only values $a = 2^u + 3^v = 3^y - 2^x$ are

a	$2^u + 3^v$	$3^y - 2^x$
5	$2 + 3$	$3^2 - 2^2$
7	$2^2 + 3$	$3^2 - 2$
11	$2 + 3^2, \ 2^3 + 3$	$3^3 - 2^4$
17	$2^3 + 3^2$	$3^4 - 2^6$
19	$2^4 + 3$	$3^3 - 2^3$
25	$2^4 + 3^2$	$3^3 - 2$
73	$2^6 + 3^2$	$3^4 - 2^3$

More has been found about the equation $a^U - b^V = 1$.

Supplementing the result of Pillai (C6.5), LeVeque proved in 1952:

(C6.6). *If $a, b \geq 2$, the equation $a^U - b^V = 1$ has at most one solution in positive integers u, v, unless $a = 3$, $b = 2$, where there are two solutions $(u, v) = (1, 1)$ and $(u, v) = (2, 3)$.*

Proof: Assume that $(u, v), (x, y)$ are solutions, with $u < x$. So $a^u - b^v = a^x - b^y = 1$. Then $v < y$ and $a^u(a^{x-u} - 1) =$

$b^v(b^{y-v} - 1)$. Since $a^u - b^v = 1$, then $\gcd(a,b) = 1$, hence $b^v = a^u - 1$ must be equal to $a^{x-u} - 1$, and also $a^u = b^v + 1$ must be equal to $b^{y-v} - 1$. Then $u = x - u$, so $x = 2u$, and also $b^{y-v} - b^v = 2$. Hence $y - v > v$ and $b^v(b^{y-2v} - 1) = 2$. Therefore $v = 1$, $b = 2$, $b^{y-2v} - 1 = 1$, thus $y - 2v = 1$ and $y = 3$. Therefore $u = 1$, $x = 3$, $a = 2$. ■

This simple result of LeVeque has an interesting corollary, concerning sums of powers of successive integers.

Let
$$S_1(n) = \sum_{j=1}^{n} j = \frac{n(n+1)}{2}$$

$$S_2(n) = \sum_{j=1}^{n} j^2 = \frac{n(n+1)(2n+1)}{6}$$

$$S_3(n) = \sum_{j=1}^{n} j^3 = \frac{n^2(n+1)^2}{4}, \text{ etc.}$$

It is therefore true that $S_3(n) = [S_1(n)]^2$ for every $n \geq 1$. The following shows that this is the only possible instance:

(C6.7). *If $t \geq 1$, $u \geq 2$, $v \geq 2$ are such that $S_v(n) = [S_t(n)]^u$ for every $n \geq 1$, then $t = 1$, $u = 2$, and $v = 3$.*

Proof: Taking $n = 2$, $1 + 2^v = (1 + 2^r)^u$. Let $1 + 2^t = a$, so $a^u - 2^v = 1$ and also $a - 2^t = 1$. Thus $(u, v), (1, t)$ are solutions of the equation $a^U - 2^V = 1$. By (C7.6), $a = 3$, $t = 1$, $u = 2$, and $v = 3$. ■

As a matter of fact, it is easy to prove a more general result, even without appealing to (C6.7) or any result in this book. Instead, it is required the fact that for every $k \geq 1$ there exists a polynomial $S_k(X) \in \mathbb{Q}[X]$, of degree $k + 1$ and leading coefficient $\frac{1}{k+1}$, such that

$$S_k(n) = \sum_{j=1}^{n} j^k, \quad \text{for every } n \geq 1.$$

This result was established first by Euler.

It follows as remarked by Allison in 1961: If t, u, v, w are positive integers with $t < v$ and $[S_v(n)]^w = [S_t(n)]^u$, for every $n \geq 1$, then $w = 1$, $t = 1$, $u = 2$, $v = 3$.

The proof is of course very easy.

7. Algorithm to Determine the Eventual Solutions

In 1953 Cassels indicated an algorithm to find the eventual solution of $a^U - b^V = 1$, thereby providing another proof of LeVeque's result:

(C7.1). *Let $a, b \geq 2$, and let $u, v \geq 2$ be such that $a^u - b^v = 1$. Either $a = 3$, $b = 2$, $u = 2$, and $v = 3$, or a and b are not powers of 2 and u, v are the smallest positive integers such that $a^u \equiv 1 \pmod{B}$ and $b^v \equiv -1 \pmod{A}$, where A (resp. B) is the product of the distinct odd prime factors of a (resp. b).*

Proof: Assume that there exist $u, v \geq 2$ such that $a^u - b^v = 1$. I exclude the case when $a = 3$, $b = 2$, $u = 2$, $v = 3$. By (B1.1), a and b cannot be powers of 2. So $A \neq 1$, $B \neq 1$.

Now let h, k be the smallest positive integers such that $a^h \equiv 1 \pmod{B}$, $b^k \equiv -1 \pmod{A}$.

First I show that $v = k$. From $b^v = -1 + a^u \equiv -1 \pmod{A}$, since $A|a$, it follows that k divides v. If $\frac{v}{k}$ is even then $-1 \equiv b^v \equiv (b^k)^{\frac{v}{k}} \equiv (-1)^{\frac{v}{k}} \equiv 1 \pmod{A}$, hence A divides 2; but $A \neq 2$, so $A = 2$, which is impossible, since A is odd.

Therefore $\frac{v}{k}$ is odd. If $v \neq k$, let p be an odd prime dividing $\frac{v}{k}$, so $v = pv_1$, with $k|v_1$ and $\frac{v_1}{k}$ is necessarily odd. Let $c = b^{v_1} \geq 2$. Then $a^u = b^v + 1 = (c+1)\frac{c^p+1}{c+1}$. But $\frac{c^p+1}{c+1} > p$, unless $c = 2$, $p = 3$ (by (P1.2)) hence $b = 2$, $v_1 = 1$, $v = 3$, which has been excluded. Moreover, by Lemma (A1.1), $p^2 \nmid \frac{c^p+1}{c+1}$ and also $\frac{c^p+1}{c+1}$ is odd, whether c is even or odd. Hence there exists an odd prime q, $q \neq p$, such that $q|\frac{c^p+1}{c+1}$;

then $q \nmid c+1$, because $\gcd\left(c+1, \frac{c^p+1}{c+1}\right) = 1$ or p. Then $q|a$, so $q|A$, $b^{v_1}+1 \not\equiv 0 \pmod{A}$, because $\frac{v_1}{k}$ is odd. This is a contradiction, proving that $v = k$.

Now I shall prove that $u = h$. From $a^u = 1 + b^v \equiv 1 \pmod{B}$, it follows that $h|u$. If $\frac{u}{h}$ is not a power of 2, let p be an odd prime dividing $\frac{u}{h}$, and proceed as before. Then $u = pu_1$, with $h|u_1$. Let $d = a^{u_1} \geq 2$, so $b^v = a^u - 1 = (d-1)\frac{d^p-1}{d-1}$. By the same argument, there exists an odd prime q such that $q \neq p$, q divides $\frac{d^p-1}{d-1}$ but $q \nmid d-1$; so $q|b$, hence $q|B$ and $a^{u_1} \not\equiv 1 \pmod{q}$, hence $a^{u_1} \not\equiv 1 \pmod{B}$, and a fortiori $a^k \not\equiv 1 \pmod{B}$, which is absurd.

This $\frac{u}{h}$ is a power of 2. I show now that if 2 divides $\frac{u}{h}$, then this leads to a contradiction, establishing that $u = h$.

Let 2 divide $\frac{u}{h}$, so $u = 2u_1$, with $h|u_1$. From $b^v = a^u - 1 = (a^{u_1}-1)(a^{u_1}+1)$, then either $a^{u_1}+1$ is a power of 2, or there exists an odd prime p dividing $a^{u_1}+1$ hence also b; then $p \nmid a^{u_1}-1$, hence $a^{u_1} \not\equiv 1 \pmod{p}$; thus $a^{u_1} \not\equiv 1 \pmod{B}$, and therefore $a^n \not\equiv 1 \pmod{B}$, an absurdity.

If $a^{u_1}+1 = 2^t$, then a is odd; if $u_1 \geq 2$, then $2^t \equiv 2 \pmod{4}$, so $t = 1$, $a^{u_1} = 1$, and $a = 1$, contrary to the hypothesis. If $u_1 = 1$, then $u = 2$, so $a^2 - b^v = 1$.

From $a+1 = 2^t$, it follows that $a-1 = 2^t - 2 = 2(2^{t-1}-1)$, so $t \geq 2$. From $b^v = a^2 - 1 = (a-1)(a+1)$, it follows that $a-1 = 2c^v$, since $\gcd(a+1, a-1) = 2$. Subtracting, $2 = 2^t - 2c^v$, then $2^{t-1} - c^v = 1$. By (B1.1), $t = 1$ or 2, hence $a = 1$ or 3 (hence $b = 2$, $u = 2$, $v = 3$). Since these values have been excluded by hypothesis, this completes the proof. ∎

8. The Largest Prime Factor of Values of Quadratic Polynomials

It is interesting to exhibit a connection between the problems considered above and the prime divisors of values of polynomials. Let $P[m]$ denote as before the largest prime factor of $m > 1$.

First, I give a very simple result, due to Schur (1912):

(C8.1). *If $f(X)$ is any non-constant polynomial with integral coefficients, then $\limsup_{n\to\infty} P[f(n)] = \infty$.*

Proof: It is equivalent to show that the set of primes which divide some value $f(n)$, with $n \geq 1$, is infinite.

Let $f(X) = a_0 X^d + a_1 X^{d-1} + \ldots + a_d$. It may be assumed, without loss of generality, that $a_0 > 0$.

If $a_d = 0$, then the statement is true, since there exist infinitely many primes.

Assume that $a_d \neq 0$ and that the set of primes p dividing some value $f(n) = (n \geq 1)$ is finite, say equal to $\{p_1, \ldots, p_r\}$. Let $a = |a_d|$ and let c be an integer such that $f(cap_1 \ldots p_r) > a$. Note that $b = \frac{f(cap_1 \ldots p_r)}{a}$ is an integer and that $b > 1$. It follows that $b \equiv 1 \pmod{p_1 p_2 \ldots p_r}$. From $b > 1$, there exists a prime p dividing b. Then $p | f(cap_1 \ldots p_r)$, hence $p \in \{p_1, \ldots p_r\}$. On the other hand, $b \equiv 0 \pmod{p_1 \ldots p_r}$ and this is a contradiction. ∎

Using Thue's result, Pólya showed in 1918:

(C8.2). *Let $f(X) = (aX + b)(cX + d)$, where a, b, c, d are integers, and $a, c \geq 1$, $\frac{b}{a} \neq \frac{d}{c}$. Then $\lim_{n\to\infty} P[f(n)] = \infty$.*

Proof: It suffices to show that $\liminf_{n\to\infty} P[f(n)] = \infty$.

Assume that $\liminf_{n\to\infty} P[f(n)] = k < \infty$. Hence there exists an infinite set S of integers n, such that if $p | f(n)$, then $n \leq k$. Let $E = \{p \text{ prime} | p \leq k\}$. Thus, for every $n \in S$, $an + b$, $cn + d \in E^x$. Note that $c(an + b) - a(cn + d) = bc - ad \neq 0$. However, by (C6.1), the equation $cX - aY = bc - ad$ has only finitely many solutions in E^x, and this is a contradiction. ∎

In particular, the above result also holds for $f(X) = X(X+1)$ or $X(X + 2)$; in this respect see another proof later (C9.3) given by Størmer.

Pólya has also shown:

(C8.3). *Let $f(X)$ be an irreducible quadratic polynomial with integral coefficients. Then*
$$\lim_{n\to\infty} P[f(n)] = \infty.$$

Proof: If $f(X) = aX^2 + Bx + c$, let $g(X) = X^2 + Bx + ac$, so $g(aX) = af(X)$. It suffices to show that $\lim_{n\to\infty} P[g(na)] = \infty$. So there exists n_0 such that if $n \geq n_0$, then $P[af(n)] = [g(an)] > a$. It follows, for $n \geq n_0$ that $P[f(n)] = P[g(na)]$, hence $\lim_{n\to\infty} P[f(n)] = \infty$. So I may assume, without loss of generality, that $f(X)$ is monic.

Let D be the discriminant of $f(X)$. Since $f[X]$ is irreducible, then $D \neq 0$ and the roots α, α' of $f(X)$ are algebraic integers of the field $K = \mathbb{Q}(\sqrt{D})$. For every $\gamma \in K$, denote its conjugate by γ'.

Let $\omega = \frac{1+\sqrt{D}}{2}$, so every algebraic integer of K is of the form $c + d\omega$ (where c, d are integers). Note that $\frac{\alpha'-\alpha}{\omega'-\omega}$ is an integer.

Denote by h the class number of K.

If the statement is false, by (C8.1), $\liminf_{n\to\infty} P[f(n)] = k < \infty$. So there exists an infinite set S of integers n such that if $p|f(n)$ then $p \leq k$. Let $E = \{p \text{ prime}|\ p \leq k\}$ and let $F = \{P|\ P \text{ prime ideal of the field } K \text{ such that } P \text{ divides some prime } p \in E\}$. So F is finite, say $F = \{P_1, P_2, \ldots, P_r\}$.

Let F^x denote the set of all ideals of K which are finite products of ideals belonging to F.

The idea of the proof is to define a mapping Φ from the set S to a set T of pairs of integers, which is known to be finite, and to show that for every $(x, y) \in T$ there are at most finitely many $n \in S$ such that $\Phi(n) = (x, y)$. This is an absurdity.

T is defined as follows.

Let U be the set of units of K. If $D < 0$, let $U_0 = U$, so U_0 has at most six elements. If $D > 0$, let ϵ be the fundamental unit and let $U_0 = \{\pm 1, \pm\epsilon \pm \epsilon^2\}$. Then every unit of K is of the form $\delta\gamma^3$, where $\gamma \in U$ and $\delta \in U_0$ are uniquely defined.

Let F_0^x be a set of algebraic integers β of K such that:
i) The principal ideal (β) is in F^x, and conversely, every principal ideal in F^x is equal to some (β), where $\beta \in F_0^x$;
ii) If $\beta_1, \beta_2 \in F_0^x$, $\beta_1 \neq \beta_2$, then $(\beta_1) \neq (\beta_2)$. Thus F_0^x, like F^x, is a finite set. For every $\delta \in U_0$ and $\beta \in F_0^x$, let

$$g_{\delta,\beta}(X,Y) = \frac{\delta\beta(X+\omega Y)^3 - \delta'\beta'(X+\omega' Y)^3}{\omega' - \omega}.$$

It is easy to see that $g_{\delta,\beta}(X,Y)$ is a binary cubic form with integral coefficients. Moreover, the polynomial $g_{\delta,\beta}(Z,1)$ is irreducible since its roots ρ are such that $\frac{\rho+\omega}{\rho+\omega'}$ are the cubic roots of $\frac{\delta'\beta'}{\delta\beta}$.

Let $T_{\delta,\beta}$ be the set of integral solutions (x,y) of the diophantine equation

$$g_{\delta,\beta}(X,Y) = \frac{\alpha' - \alpha}{\omega' - \omega}.$$

By Thue's theorem (C1.1), each set $T_{\delta,\beta}$ is finite. Let $T = U\{T_{\delta,\beta} | \delta_t U_0, \beta \in F_0^x\}$, so T is a finite set.

Now, I shall define a mapping $\Phi : S \to T$.

If $n \in S$, since $f(n) = (n-\alpha)(n-\alpha')$, then the principal ideal $(n - \alpha) \in F^x$. Hence

$$(n - \alpha) = P_1^{e_1} \ldots P_r^{e_r} I^{3h},$$

where I is an ideal of K, $0 \le e_i < 3h$ for every $i = 1, \ldots, r$.

But I is a principal ideal, say $I^h = (c + \omega d)$, with c, d integers. Then $P_1^{e_1} \ldots P_r^{e_r}$ is also a principal ideal, thus there exists a unique $\beta \in F_0^x$ such that $(n - \alpha) = (\beta) \cdot (c + \omega d)^3$.

Since every unit of K is of the form $\delta\gamma^3$ (with $\delta \in U_0$, $\gamma \in U$), then there exist $\delta \in U_0$ and integers x, y such that $n - \alpha = \delta\beta(x + \omega y)^3$. Taking the conjugate,

$$n - \alpha' = \delta'\beta'(x + \omega' y)^3.$$

Hence
$$\frac{\alpha' - \alpha}{\omega' - \omega} = \frac{\delta\beta(x+\omega y)^3 - \delta'\beta'(x+\omega' y)^3}{\omega' - \omega}.$$

With the previous notation, (x,y) is a solution of $g_{\delta,\beta}(X,Y) = \frac{\alpha'-\alpha}{\omega'-\omega}$, thus $(x,y) \in T_{\delta,\beta} \subseteq T$.

Define $\Phi(n) = (x,y)$.

If $(x,y) \in T$, since U_0, F_0^x are finite sets, there exist at most finitely many $n \in S$ such that $n - \alpha = \delta\beta(x+\omega y)^3$, hence $\Phi(n) = (x,y)$.

This concludes the proof. ∎

For the particular case of the polynomial $f(X) = X^2 + 1$, see another proof by Størmer, after (C9.3).

In 1933, Mahler showed even more:

$$\lim_{n \to \infty} \frac{P[n^2 - 1]}{\log\log n} \geq 1, \text{ and } \lim_{n \to \infty} \frac{P[n^2 + 1]}{\log\log n} \geq 2.$$

There are many more results in the literature concerning the growth of $P[f(n)]$, for various polynomials $f(X)$ with integral coefficients, but this subject is beyond the scope of this book. The reader may wish to consult Shorey and Tijdeman's book of 1986 (with special attention to pages 56–57, 124–137, 142, 149–150) and various papers of Langevin (1974, 1975), where further references to the extensive literature may be found.

I mention here that the finiteness of the number of solutions of $aX^m - bY^n = k$, as seen in (C2.1), may also be proved by using an interesting result of Mahler (1953), concerning the largest prime factor.

Mahler proved:

(C8.4). *Let $m \geq 2$, $n \geq 3$, let a, b be non-zero integers. For every integer $t \geq 1$, let*

$$M(t) = \min\{P[|ax^m - by^n|] \,|\, \max\{x,y\} = t, \, gcd(x,y) = 1\}.$$

Then $\lim_{t\to\infty} M(t) = \infty$.

In particular, for any $k \neq 0$, if $\max\{x,y\}$ is sufficiently large then
$$\left|x^m - y^n\right| \geq P\left[|x^m - y^n|\right] > k.$$

9. Effective Results

Here I shall give effective proofs of some of the propositions already established in the preceding sections.

I begin with the pioneering results of Størmer concerning sequences of integers with prime factors in a given finite set. Before starting, it is perhaps worthwhile to mention how Størmer was led to the study of these problems.

From the series expansion
$$\log(1+x) = x - \frac{x^2}{2} + \frac{x^3}{3} - \frac{x^4}{4} + \cdots$$

it follows easily, when N, h are positive integers, that

$$\log\left(\frac{2N+2h}{2N+h}\right)$$
$$= \log\left(1 + \frac{h}{2H+h}\right)$$
$$= \frac{h}{2N+h} - \frac{1}{2}\left(\frac{h}{2N+h}\right)^2 + \frac{1}{3}\left(\frac{h}{2H+h}\right)^3 - \cdots,$$
$$\log\left(\frac{2N}{2N+h}\right)$$
$$= \log\left(1 - \frac{h}{2N+h}\right)$$
$$= -\frac{h}{2N+h} - \frac{1}{2}\left(\frac{h}{2N+h}\right)^2 - \frac{1}{3}\left(\frac{h}{2N+h}\right)^3 - \cdots,$$
$$\log\left(\frac{N+h}{N}\right)$$
$$= 2\left(\frac{1}{1+\frac{2N}{h}} + \frac{1}{3}\left(\frac{1}{1+\frac{2N}{h}}\right)^3 + \frac{1}{5}\left(\frac{1}{1+\frac{2N}{h}}\right)^5 + \cdots\right).$$

Effective Results 289

If h is small, say $h = 1$ or 2, and N is large, then the above series converge very quickly.

Assume that the logarithms of the prime numbers $p_1, \ldots, p_{r-1}$ are already known, and the purpose is to calculate $\log p_r$. If N is a natural number having only the prime factors $p_1, \ldots, p_{r-1}$ and $N + h$ has only the prime factors $p_1, \ldots, p_{r-1}$ and p_r, the above formula allows one to calculate $\log p_r$. Moreover, the larger is N, the quicker the convergence.

As indicated in (C6.1), given $h \geq 1$ and $E = \{p_1, \ldots, p_r\}$, there exist only finitely many positive integers N such that $N(N + h) \in E^\times$; that result was not effective. Størmer took $h = 1$ or 2 and was able to determine effectively the integers N with the above property (see (C9.3)). His result defeated somewhat the original purpose of finding arbitrarily rapidly converging series to $\log p_r$, with the above method.

Let $E = \{p_1, \ldots, p_r\}$ ($r \geq 1$) be a set of prime numbers. As before, let $E^\times$ denote the set of natural numbers, all of whose prime factors belong to E.

Let $a \geq 1$ and consider the set

$$D_a(E) = \{D = ap_1^{e_1} \ldots p_r^{e_r} | D \text{ is not a square, each } e_i = 0, 1 \text{ or } 2\}.$$

Clearly $\#D_a(E) \leq 3^r$.

Størmer showed in 1897:

(C9.1). *Given $a \geq 1$ and E as above, if $x \geq 1$ and $x^2 \mp 1 \in aE^\times$, there exist $D \in D_a(E)$ and $y \geq 1$, such that (x, y) is the fundamental solution of $X^2 - DY^2 = \pm 1$. In particular, there exist at most 3^r such integers x, and they may be effectively computed in finitely many steps.*

Proof: Assume that $x^2 \mp 1 = au$ with $u \in E^\times$, so $u = \prod_{i=1}^{r} p_i^{d_i}$

and each $d_i \geq 0$. Let

$$e_i = \begin{cases} 0 & \text{if } d_i = 0 \\ 1 & \text{if } d_i \text{ is odd} \\ 2 & \text{if } d_i \text{ is even, } d_i \geq 2, \end{cases}$$

and $D = a \prod_{i=1}^{r} p_i^{e_i}$. Since $e_i \leq d_i$ and $d_i - e_i$ is even for every $i = 1, \ldots, r$, then $au = Dy^2$, for some integer $y \geq 1$; hence $x^2 - Dy^2 \mp 1$ and necessarily D is not a square, so $D \in D_a(E)$.

Note that $p_i | y$ if and only if $d_i \neq e_i$; this happens exactly when d_i is odd and greater than 1, or d_i is even and greater than 2. In turn, this implies that $e_i \geq 1$. Therefore, if $p_i | y$, then $p_i | D$. From $x^2 - Dy^2 = \pm 1$ it follows from (A4.2) that (x, y) is the fundamental solution of this equation.

If x, x' are such that $x^2 \mp 1$, $x'^2 \mp 1 \in aE^\times$ and they correspond to the same integer $D \in D_a(E)$, then $y = x'$, because both numbers are equal to the first integer of the fundamental solution of $X^2 - DY^2 = \pm 1$, so there exist at most 3^r such integers. Finally, as indicated in §P5, these integers may be computed in finitely many steps. ∎

It follows at once that if $a_1, \ldots, a_r \geq 1$, there exist only finitely many integers x such that $x^2 \mp 1 \in \{a_1^{e_1} \ldots a_r^{e_r} | e_1, \ldots, e_r \geq 0\}$, and these integers may be effectively computed in finitely many steps.

As a numerical illustration of his method, Størmer determined all the integers $x \geq 1$ such that all the prime divisors p of $1 + x^2$ are such that $p \leq 13$:

Effective Results

$$1 + 1^2 = 2$$
$$1 + 2^2 = 5$$
$$1 + 3^2 = 10 = 2 \times 5$$
$$1 + 5^2 = 26 = 2 \times 13$$
$$1 + 7^2 = 50 = 2 \times 5^2$$
$$1 + 8^2 = 65 = 5 \times 13$$
$$1 + 18^2 = 325 = 5^2 \times 13$$
$$1 + 57^2 = 3250 = 2 \times 5^3 \times 13$$
$$1 + 239^2 = 57122 = 2 \times 13^4.$$

Indeed, if $p | 1 + x^2$, then either $p = 2$ or $\left(\frac{-1}{p}\right) = +1$, so $p \equiv 1 \pmod 4$. Therefore $1 + x^2 = 2^a \times 5^b \times 13^c$ with $a, b, c \geq 0$ and $a = 0$ or 1. If $1 + x^2$ is of the above form, then there exists $D \in D_1(E) \cup D_2(E)$ where $E = \{5, 13\}$, such that (x, y) is the fundamental solution of $X^2 - DY^2 = -1$. But $D_1(E) \cup D_2(F) = \{2, 5, 13, 2 \times 5, 2 \times 13, 2 \times 5^2, 2 \times 13^2, 5 \times 13, 5^2 \times 13, 5 \times 13^2, 2 \times 5 \times 13, 2 \times 5^2 \times 13, 2 \times 5 \times 13^2, 2 \times 5^2 \times 13^2\}$. According to (P5.3), computing the continued faction expansion of $\sqrt{D}$, for each $D \in D_1(E) \cup D_2(E)$, yields the fundamental solution (x, y) of $X^2 - DY^2 = -1$:

D	x	y
2	1	1
5	2	1
10	3	1
13	18	5
26	5	1
50	7	1
65	8	1
130	57	5
325	18	1
338	239	13
650	–	–
845	12238	421
1690	–	–
8450	54608393	594061

Note that when $D = 650$ or 1690, the equation has no solution; also when $D = 845$ and 8450, y has a prime factor bigger than 13. Excluding these values of D, and observing that $D = 13$ and $D = 325$ have the same fundamental solution, there remain exactly the nine values of x indicated above.

Using the result (A4.3) of Skolem, the following may be shown, with the same proof as (C9.1):

(C9.2). *There exist at most finitely many integers $x \geq 1$ such that $x^2 \mp 4 \in aE^\times$, and they may be effectively computed in finitely many steps.*

Størmer deduced (1908):

(C9.3). *Let $a \geq 1$, $E = \{p_1, \ldots, p_r\}$, and $h = 1$ or 2. Then there are at most 3^r integers $x \geq 1$ such that $x(x+h) \in aE^\times$, and these integers may be effectively computed in finitely many steps.*

Proof: Let $h = 1$. Since $x(x+1) = x^2 + x = \frac{1}{4}[(2x+1)^2 - 1]$, then if $x(x+1) \in aE^\times$, $(2x+1)^2 - 1 \in 4aE^\times$. By (C9.1), there are at most 3^r such integers and they may be computed effectively in finitely many steps.

Let $h = 2$. Since $x(x+2) = x^2 + 2x = (x+1)^2 - 1$, then if $x(x+2) \in aE^\times$, $(x+1)^2 - 1 \in aE^\times$. Therefore, there are at most 3^r such integers $x \geq 1$ and they may be effectively computed in finitely many steps. ∎

The above results allows us to give a new proof that if $f(X) = X(X+1), (X+2), X^2+1$ or X^2-1, then $\lim_{n \to \infty} P[f(n)] = \infty$ (this is a special case of (C8.2) and (C8.3)).

Indeed, for the first two polynomials, by (C9.3), there exist only finitely many $n \geq 1$ such that $P[f(n)] \leq k$. Hence, there exists n_0 such that if $n \geq n_0$, then $P[f(n)] > k$, proving the statement. For the other two polynomials, the proof is similar and follows from (C9.1).

Now let $A, M_1, M_2, \ldots, M_m, B, N_1, N_2, \ldots, N_n (m \geq 1, n \geq 1)$ be integers, with $A, B > 0$ and each $M_i > 1$, $N_i > 1$.

Effective Results

Let E be the set of primes dividing the product $ABM_1 \ldots M_m N_1 \ldots N_n$, and let $\#(E) = r$.

In 1898, Størmer deduced from the above result:

(C9.4). *With the above notations, there exist only finitely many tuples $(e_1, \ldots e_m\ f_1, \ldots, f_n)$ of natural numbers such that*

$$AM_1^{e_1} \ldots M_m^{e_m} - BN_1^{f_1} \ldots N_n^{f_n} = 1 \text{ or } 2, \qquad (9.1)$$

and they may be effectively computed in finitely many steps.

Proof: Let $X = \{(e_1, \ldots, e_m,\ f_1, \ldots, f_n) |$ relation (9.1) holds$\}$ and $S' = \{(f_1, \ldots, f_n) |$ there exist $(e_1, \ldots, e_m)$ such that relation (9.1) holds$\}$.

I shall prove that S' is finite and may be effectively computed in finitely many steps. It follows at once that S has the same property.

For each $(f_1, \ldots, f_n) \in S'$, let $N = BN_n^{f_1} \ldots N_n^{f_n}$, so $N \in E^\times$ and also $N + h \in E^\times$ for $h = 1, 2$, as follows from relation (9.1). Hence $N(N + h) \in E^\times$.

By (C9.2), the set S'' of all integers N such that $N(N + h) \in E^\times$ is finite and effectively computable in finitely many steps. But for every $N \in S''$ there exist at most finitely many $(f_1, \ldots, f_n) \in S'$ such that $BN_1^{f_1} \ldots N_n^{f_n} = N$. Hence S' is finite and effectively computable in finitely many steps. ∎

Using (C9.1) and a proof similar to (C9.4), Skolem established in 1945a (see also Nagell, 1955):

(C9.5). *There exist only finitely many tuples $(e_1, \ldots, e_m, f_1, \ldots, f_n)$ of natural numbers, such that*

$$AM_1^{e_1} \ldots M_m^{e_m} - BN_1^{f_1} \ldots N_n^{f_n} = 4,$$

and they may be effectively computed in finitely many steps.

With the theory of cubic forms, Nagell proved in 1925a (see also Skolem 1945a):

(C9.6). *There exist only finitely many tuples* $(e_1, \ldots, e_m, f_1, \ldots, f_n)$ *of natural numbers, such that*

$$AM_1^{e_1} \ldots M_m^{e_m} - BN_1^{f_1} \ldots N_n^{f_n} = 3,$$

and they may be effectively computed in finitely many steps.

Using an effective theorem of diophantine approximation by Gel'fond, Cassels obtained an effective version of (C6.1) and an extension of (C9.1).

Gel'fond proved (see his book, 1960, page 174, or Gel'fond 1940):

(C9.7). *Let K be a number field and P a prime ideal in K. Let $\alpha, \beta \in K$, $\alpha, \beta \neq 0$, be multiplicatively independent (that is, $\alpha^m = \beta^n$ with m, n integers, implies that $m = n = 0$). Assume also that α, β are P-adic units (that is, $v_P(\alpha) = v_P(\beta) = 0$, where v_P denotes the P-adic valuation).*

Then there exists an effectively computable integer x_0 (depending on α, β, P), such that if there exist integers x, u, v, $m \geq 1$ such that $\alpha^u \equiv \beta^v \pmod{P^n}$, with $|u| + |v| \leq x$ and $m \geq [(\log x)^7]$, then $x < x_0$.

Cassels (1960) used this result to derive the following extension of (C9.1):

(C9.8). *Let E be a finite set of prime numbers, let $D > 0$, D square-free and $k \neq 0$, such that no prime factor of k belong to E. Then there are only finitely many solutions in integers (x, y), with $y \in E^\times$, of the equation*

$$X^2 - DY^2 = k,$$

and the solutions may be found in finitely many steps.

From this result, Cassels deduced the effective version of (C6.1).

(C9.9). *Let E_1, E_2 be finite sets of prime numbers. For any integer $k \neq 0$, there exist only finitely many integers $x \in E_1^\times$,*

$y \in E_2^\times$ such that $x - y = k$. Moreover, these integers x, y may be found in finitely many steps.

Proof: The existence of finitely many solutions was already established in (C6.1). The proof that the solutions may be effectively determined is based on (C9.8) and is similar to (C6.1).

To begin, I remark that it suffices to determine effectively the solutions (x, y) with $gcd(x, y, k) = 1$. These are exactly the solutions (x, y) such that x, y, k are pairwise relatively prime, because $x - y = k$.

By deleting, if necessary, primes from E_1, E_2 it may be assumed without loss of generality that $E_1 \cap E_2 = \emptyset$ and that k has no prime factor in $E_1 \cup E_2$.

Let T be the set of all pairs of integers (a, b) such that

$$a = \prod p_i^{e_i}, \text{ with each } p_i \in E_1 \text{ and } 0 \le e_i \le 1,$$
$$b = \prod p_j^{f_j}, \text{ with each } p_j \in E_2 \text{ and } 0 \le f_j \le 1.$$

Clearly, T is a finite set.

For each $(a, b) \in T$, consider the equation

$$F_{(a,b)} : U^2 - abV^2 = ak.$$

Denote by S the set of solutions $(x, y) \in E_1^\times \times E_2^\times$, of $X - Y = k$, such that x, y, k are pairwise relatively prime. Denote by $W_{(a,b)}$ the set of solutions (u, v) of $F_{(a,b)}$, such that $v \in E_2^\times$. Since every prime factor of ak does not belong to E_2, by (C9.8), $W_{(a,b)}$ is finite and may be effectively determined in finitely many steps. So $\bigcup_{(a,b) \in T} W_{(a,b)}$ is also finite and may be determined in finitely many steps.

To conclude the proof, it suffices to define an injective mapping $\varphi : S \to \bigcup_{(a,b) \in T} W_{(a,b)}$.

Let $(x, y) \in S$. Then write $x = ax_1^2$, $y = by_1^2$. Note that a, b, x_1, y_1 are completely determined by x, y. Then $(ax_1)^2 -$

$aby_1^2 = ak$ where no prime factor of ak belongs to $E_2^\times$. Hence $(ax_1, y_1) \in W_{(a,b)}$ and the required injective mapping φ is defined by $\varphi((x, y)) = (ax_1, y_1)$. ∎

Still on the same question, let $E = \{p_1, \ldots, p_r\}$ with $r \geq 2$, let $p = \max\{p_1, \ldots, p_r\}$, and as before, $E^\times : z_1 < z_2 < \ldots$.
Tijdeman showed in 1973:

(C9.10). *There exists an effectively computable number $C = C(p) > 0$ such that*

$$z_{i+1} - z_i \geq \frac{z_i}{(\log z_i)^C},$$

for $z_i \geq 3$.

The equations

$$aX^n - bY^n = \pm k \tag{9.2}$$

(a, b, k non-zero integers, $n \geq 3$) have at most finitely many solutions (see (C2.1)).

When $n = 3$, $a > 0$ and b is not a cube, then as indicated in (A15.3), the equations (9.2) have at most one solution in integers (with the exception of $2X^3 + Y^3 = 3$, which has two solutions).

For $n = 4$, as already indicated in (A16.3), Ljunggren showed that

$$aX^4 - bY^4 = \pm 1 \tag{9.3}$$

has at most one solution in positive integers.

In 1954, Domar showed that, for $n \geq 5$,

$$aX^n - bY^n = \pm 1 \tag{9.4}$$

has at most two solutions in integers.

In his thesis (1983), Evertse gave upper bounds for the number of solutions in integers of the equations (9.2).

Let $\omega(k)$ denote the number of distinct prime factors of k. Then Evertse showed that equations (9.2) have at most $2n^{\omega(k)} + 6$ solutions in integers.

To conclude this section, I wish to quote, still without proofs, several effective results about the equation

$$\frac{X^n - 1}{X - 1} = Y^m$$

studied in §A8, and similar equations.

In 1976, Shorey and Tijdeman proved:

(C9.11). *Let $x > 1$. There exists an effectively computable number $C = C(x) > 0$ such that, if $m \geq 2$, $n \geq 1$, $y \geq 1$ and $\frac{x^n - 1}{x - 1} = y^m$, then $m, n < C$.*

In other words, in base x, there is only an effectively bounded number of repunits.

In 1980, Balasubramanian and Shorey considered numbers which may be simultaneously written with repeated digits, in different bases. Precisely, they proved:

(C9.12). *Let E be a finite set of primes. There is an effectively computable number $C = C(E) > 0$ such that if $a, b \in E^\times$, $\gcd(a, b) = 1$, if $x, y > 1$ with $a(x - 1) \neq b(y - 1)$, if $m, n \geq 1$ and*

$$a\frac{x^m - 1}{x - 1} = b\frac{y^n - 1}{y - 1},$$

then $a, b, m, n, x, y < C$.

Special cases had been obtained by various authors, as indicated in the paper of Balasubramanian and Shorey.

In 1986, Shorey proved an interesting fact which I quote now:

(C9.13). *There exists an effectively computable number $C > 0$ such that, if $x > 1$, q is a prime number, $n_1 \equiv n_2 \pmod{q}$ and $\frac{x^{n_1} - 1}{x - 1}$, $\frac{x^{n_2} - 1}{x - 1}$ are q^{th} powers, then $x, q, n_1, n_2 < C$. In* particular, since 1 is a qth power, if $n \equiv 1 \pmod{q}$ and $\frac{x^n-1}{x-1}$ is a qth power, then $n, x, q < C$. Thus, if q is a prime and $x \geq C$, then $\#\{n > 1 | \frac{x^n-1}{x-1}$ is a qth power $\} \leq q - 1$.

Let $C_1 = \sum_{x<C} C(x)$ (where $C(x)$ was defined in (C9.13)). For every prime q, let $S = \{n > 1|$ there exists $x > 1$ such that $\frac{x^n-1}{x-1}$ is a qth power $\}$. Then $\#(S) < q + C_1$. Indeed, let $S_x = \{x > 1 | \frac{x^n}{x-1}$ is a qth power$\}$; then $\#(S_x) \leq C(x)$. By the above remark, $\#(\bigcup_{x \geq C} S_x) \leq q - 1$. Also

$$\# \bigcup_{x<C} S_x \leq \sum_{x<C} S_x \leq \sum_{x<C} C(x) = C_1.$$

Hence
$$\#S \leq \#(\bigcup_{x \geq C} S_x) + \#(\bigcup_{x<C} S_n) < q + C_1.$$

III. The Equation $X^U - Y^V = 1$

Now I consider the Catalan equation

$$X^U - Y^V = 1.$$

The problem is, I recall, to determine the quadruples of natural numbers (x, y, m, n) with $x, y \geq 1$, $m, n \geq 2$, such that $x^m - y^n = 1$. In variance with the equations considered before, neither x, y nor m, n are kept fixed.

10. The Theorem of Tijdeman

All of the results developed up to now in this book do not suffice to conclude that there exist only finitely many quadruples of integers (x, y, m, n) with $x \geq 1$, $y \geq 1$, $m \geq 2$, $n \geq 2$, such that $x^m - y^n = 1$.

The Theorem of Tijdeman

This was finally established by Tijdeman in 1976, using the lower bound for linear forms in logarithms, indicated in (C1.8).

To begin, I prove:

(C10.1). Lemma

1) If $|a| \leq \dfrac{1}{2}$, then $|\log(1+a)| < 2|a|$.

2) If $0 < a \leq \dfrac{1}{2}$ and $0 < c$, then
$$|\log[(1-a)^c + a^c]| \leq 2ac.$$

3) If $0 < a \leq \dfrac{1}{2}$ and $0 < c$, then
$$|\log[(1+a)^c - a^c]| \leq 2ac.$$

Proof:

1) $$|\log(1+a)| = \left|a\left(\sum_{n=1}^{\infty}(-1)^{n-1}\dfrac{a^{n-1}}{n}\right)\right|$$
$$\leq |a|\sum_{n=1}^{\infty}\dfrac{|a|^{n-1}}{n} \leq |a|\sum_{n=1}^{\infty}\dfrac{1}{2^{n-1}} = 2|a|.$$

2) It suffices to show that
$$\left|\log[(1-a)^c + a^c]\right| \leq c\left|\log(1-a)\right|,$$

that is,
$$c\log(1-a) \leq \log[(1-a)^c + a^c] \leq -c\log(1-a),$$

or equivalently,
$$(1-a)^c \leq (1-a)^c + a^c \leq (1-a)^{-c}.$$

One inequality is obvious. It suffices to show that $(1-a)^{2c} + a^c(1-a)^c \leq 1$. For this purpose, consider the function $f(x) = (1-x)^{2c} - x^c(1-x)^c$, defined for $0 \leq x \leq 1$. Its derivative is $f'(x) = -2c(1-x)^{2c-1} - c(1-x)^{c-1}x^c + c(1-x)^c x^{c-1}$. Now I show that $f'(x) \leq 0$ for $0 \leq x \leq 1$.

Indeed, $f'(x) = -2c(1-x)^{2c-1} - c(1-x)^{c-1}x^c + c(1-x)^c x^{c-1}$. I wish to prove that $c(1-x)^c x^{c-1} \leq 2c(1-x)^{2c-1} + c(1-x)^{c-1}x^c$, or equivalently

$$(1-x)x^{c-1} \leq 2(1-x)^c + x^c.$$

If $0 \leq x \leq \frac{1}{2} \leq 1-x \leq 1$, then

$$(1-x)x^{c-1} \leq (1-x)^c < 2(1-x)^c + x^c.$$

If $0 \leq 1-x \leq \frac{1}{2} \leq x \leq 1$, then similarly

$$(1-x)x^{c-1} \leq x^c < 2(1-x)^c + x^c.$$

Therefore, $f(x)$ is a monotone non-increasing function. Thus, $1 = f(0) \geq f(a) = (1-a)^{2c} + (1-a)^c a^c$, for $0 \leq a < \frac{1}{2}$.

3) It suffices to show that

$$\left| log[(1+c)^c - a^c] \right| \leq c \log(1+a),$$

that is,

$$-c\log(1+a) \leq \log[(1+a)^c - a^c] \leq c\log(1+a),$$

or equivalently

$$(1+a)^{-c} \leq (1+a)^c - a^c \leq (1+a)^c.$$

One inequality is trivial and so is the other, because

$$(1+a)^{2c} - a^c(1+a)^c = (1+a)^c[(1+a)^c - a^c] \geq 1. \quad\blacksquare$$

The Theorem of Tijdeman

Now I prove Tijdeman's main theorem:

(C10.2). *There exists a number $C > 0$, which is effectively computable, such that if x, y, m, n are positive integers, $m, n \geq 2$ and $x^m - y^n = 1$, then $x^m, y^n < C$.*

Proof:

1°). If p, q are primes, let $E_{(p,q)}$ be the set of pairs (x, y) of positive integers, such that $x^p - y^q = 1$.

By (C2.3), if $E_{(p,q)} \neq \emptyset$ there exists $C(p, q) > 0$, such that if $(x, y) \in E_{(p,q)}$ then $x^p, y^q < C(p, q)$.

It suffices to show that there exists $C' > 0$ such that if $E_{(p,q)} \neq \emptyset$, then $p, q < C'$. Then $C = \max\{C(p,q) | p, q < C'\}$ satisfies the required condition. Indeed, if $m, n \geq 2$, $x, y \geq 1$ and $y^m - y^n = 1$, let p, q be primes such that $p|m$, $q|n$. Let x', y' be integers such that $x^m = x'^p$, $y^n = y'^q$, so $x'^p - y'^q = 1$. Hence $(x', y') \in E_{(p,q)}$ so $y^n < x^m = x'^p < C(p, q) \leq C$.

2°). After this preamble, I begin the proof of the existence of the constant C. Let $x, y \geq 1$, p, q be primes and $x^p - y^q = 1$. Letting $\epsilon = \pm 1$ and writing $x^p - y^q = \epsilon$, I may assume that $q < p$.

I shall show that

$$q < 2^{44} \times 3^{10} (\log p)^3.$$

By (B2.7)

$$\begin{cases} x - \epsilon = p^{-1} r^q & \text{where } p | r \\ y + \epsilon = q^{-1} s^p & \text{where } q | s. \end{cases}$$

Note that $x, y > 1$, hence $r \neq 0$, $s \neq 0$. The original equation is rewritten as

$$(p^{-1} r^q + \epsilon)^p - (q^{-1} s^p - \epsilon)^q = \epsilon. \tag{10.1}$$

Consider the linear form in logarithms

$$\Lambda_1 = q \log q - p \log p + pq \log \frac{r}{s}, \tag{10.2}$$

so
$$\Lambda_1 = \log \frac{(p^{-1}r^q)^p}{(q^{-1}s^p)^q} = \log \frac{(x-\epsilon)^p}{(y+\epsilon)^q} \neq 0,$$

because
$$\max\left\{(x-1)^p, (y-1)^q\right\} < x^p = y^q + \epsilon < \min\left\{(x+1)^p, (y+1)^q\right\}.$$

Note that
$$\frac{(x-\epsilon)^p}{(y+\epsilon)^q} = \frac{1}{\left(1+\frac{\epsilon}{x-\epsilon}\right)^p}\left[\left(\frac{\epsilon}{y+\epsilon}\right)^q + \left(1-\frac{\epsilon}{y+\epsilon}\right)^q\right].$$

I show that
$$|\Lambda_1| \leq \frac{4p^2}{t^q}, \qquad (10.3)$$

where $t = \min\{r, s\}$. Indeed, by Lemma (C10.1)

$$|\Lambda_1| \leq \left|\log\left(1+\frac{\epsilon}{x-\epsilon}\right)^p\right| + \left|\log\left[\left(\frac{\epsilon}{y+\epsilon}\right)^q + \left(1-\frac{\epsilon}{y+\epsilon}\right)^q\right]\right|$$
$$\leq \frac{2p}{p^{-1}r^q} + \frac{2}{q^{-1}s^p} = \frac{2p^2}{r^q} + \frac{2q^2}{s^q} < \frac{4p^2}{t^q}$$

(note that by (B1.1), $x > 2$, $y > 2$).

The assertion (2°) is true if $q \leq 10(\log p)$. So, I shall assume that $q > 10(\log p)$. Then

$$t^{\frac{q}{2}} > t^{5\log p} = p^{5\log t} \geq p^{5\log 2} \geq 4p^2.$$

Therefore
$$|\Lambda_1| \leq \frac{1}{t^{q/2}}, \qquad (10.4)$$

that is,
$$|\Lambda_1| \leq \exp\left(-\frac{q}{2}\log t\right). \qquad (10.5)$$

The Theorem of Tijdeman

Before computing a lower bound for $|\Lambda_1|$, I shall show that

$$\max\{r, s\} \leq t^4. \tag{10.6}$$

From

$$\left| \frac{q \log q}{p} - \frac{p \log p}{q} + \log \frac{r}{s} \right| \leq \frac{1}{pqt^{q/2}},$$

it follows that

$$\left| \log \frac{r}{s} \right| \leq \frac{1}{pqt^{q/2}} + \left| \frac{q \log q}{p} - \frac{p \log p}{q} \right|$$

$$\leq \frac{1}{pqt^{q/2}} + \frac{q \log q}{p} + \frac{p \log p}{q} \leq 1 + \frac{2 \log p}{q} < 2,$$

because $10 \log p < q$. Thus $\frac{1}{e^2} < \frac{r}{s} < e^2$, so either $2 \leq s < r < se^2 < s^4$ or $2 \leq r < s < re^2 < r^4$. In both cases, $\max\{r, s\} < t^4$. I shall now apply (C1.8) to the linear form Λ_1. Accordingly

$$|\Lambda_1| > \exp(-D_1 A \log B)$$

where $n = 3$, $d = 1$, $D_1 = (16dn)^{2(n+2)} = 48^{10} = 2^{40} \times 3^{10}$, $b_1 = q$, $b_2 = p$, $b_3 = pq$. So $B = \max\{|b_1|, |b_2|, |b_3|\} \leq p^2$, $A_1 = q$, $A_2 = p$, $A_3 = max\{r, s\}$ and

$$A = (\log A_1)(\log A_2)(\log A_3) \leq (\log p)^2 (\log t^4).$$

Thus

$$|\Lambda_1| > \exp(-2^{40} \times 3^{10} (\log p)^2 (\log t^4)(\log p^2)). \tag{10.7}$$

Comparing with (10.5),

$$\frac{q}{2} \log t < 2^{40} \times 3^{10} \times 8 (\log p)^3 (\log t),$$

hence

$$q < 2^{44} \times 3^{10} (\log p)^3. \tag{10.8}$$

3°). Now I show that there exists $C > 0$ such that if p, q are primes and $E_{(p,q)} \neq \emptyset$, then $C(p,q) < C$.

Let, as before, $x^p - y^q = \epsilon$, with $\epsilon = \pm 1$, $q < p$, $x, y \geq 1$.
Since
$$\frac{(p^{-1}r^q + \epsilon)^p}{(q^{-1}s^p)^q} = \frac{x^p}{(y+\epsilon)^q} = \frac{y^q + \epsilon}{(y+\epsilon)^q} \neq 1,$$
then the linear form in logarithms
$$\Lambda_2 = q \log q + p \log\left(\frac{p^{-1}r^q + \epsilon}{s^q}\right) \neq 0. \qquad (10.9)$$

I shall give an upper bound for $|\Lambda_2|$. First, I remark that
$$\frac{(p^{-1}r^q + \epsilon)^p}{(q^{-1}s^p)^q} = \left(1 - \frac{\epsilon}{q^{-1}s^p}\right)^q + \left(\frac{\epsilon}{q^{-1}s^p}\right)^p, \qquad (10.10)$$
and by Lemma (C10.1),
$$|\Lambda_2| = \left|\log\left[\left(1 - \frac{\epsilon}{q^{-1}s^p}\right)^q\right]\right| \leq \frac{2q}{q^{-1}s^p} = \frac{2q^2}{s^p}. \qquad (10.11)$$

If $p \geq 22$, then $2q^2 < 2p^2 < 2^{p/2} \leq s^{p/2}$, hence $|\Lambda_2| \leq \frac{1}{s^{p/2}}$, so
$$|\Lambda_2| \leq \exp\left(-\frac{p}{2}\log s\right). \qquad (10.12)$$

To determine a lower bound for $|\Lambda_2|$, I apply (C1.8), with $n = 2, d = 1$. Also $b_1 = q$, $b_2 = p$, thus $B = p$; next $A_1 = q$ and $A_2 = \max\{p^{-1}r^q + \epsilon, s^q\}$. It is necessary to estimate A_2. From (10.11)
$$\left(\frac{p^{-1}r^q + \epsilon}{s^q}\right)^p = q^{-q}\left[\left(\frac{\epsilon}{q^{-1}s^p}\right)^q + \left(1 - \frac{\epsilon}{q^{-1}s^p}\right)^q\right],$$
hence
$$\left|\log\left(\frac{p^{-1}r^q + \epsilon}{s^q}\right)\right| \leq \frac{q \log q}{p} + \frac{2q^2}{ps^p} < \frac{q \log p}{p} + \frac{1}{p}$$
$$< \frac{2^{44} \times 3^{10}(\log p)^4 + 1}{p}.$$

The Theorem of Tijdeman

So there exists $M > 0$, which is effectively computable, and may be taken as $M > 22$, such that $2^{44} \times 3^{10}(\log p)^4 + 1 < b$ for every $p > M$. Thus, for $p > M$,

$$\left|\log\left(\frac{p^{-1}r^q + \epsilon}{s^q}\right)\right| < 1.$$

Thus
$$\frac{1}{e} < \frac{p^{-1}r^q + \epsilon}{s^q} < e.$$

So $p^{-1}r^q + \epsilon < s^q e < s^{2q}$, whenever $p > M$. This yields $A_2 < s^{2q}$.

Applying (C1.8), with $p > M$:

$$|\Lambda_2| \geq \exp(-20^{40}(\log p)(\log q)(\log s^{2q})).$$

But
$$(\log p)(\log q)(\log s^{2q}) < 2q(\log p)^2(\log s)$$
$$< 2^{45} \times 3^{10}(\log p)^5(\log s).$$

Hence
$$|\Lambda_2| \geq \exp(-2^{85} \times 3^{10}(\log p)^5(\log s)). \qquad (10.13)$$

Comparing with (10.12),

$$\frac{p}{2}\log s < 2^{85} \times 3^{10}(\log p)^5(\log s),$$

hence
$$p < 2^{86} \times 3^{10}(\log p)^5,$$

which holds for $p > M$. Thus, there exists an effectively computable constant C'' such that $p < C''$. Therefore $q < p < C' = \max\{M, C''\}$, and this concludes the proof. ∎

The effective proof of Tijdeman's theorem has been made explicit. Assume that p, q are odd primes; $x, y \geq 1$ and $x^p - y^q = 1$. Langevin follows the original proof of Tijdeman and showed (1976a, 1976b) that $p, q < e^{245}$ and that $x, y < \exp\exp\exp\exp 730$. Such an enormous number is hard to conceive.

With the recent improvement in the bounds for linear forms in logarithms by Mignotte and Waldschmidt (1990, 1992) and Blass, Glass et al. (1990), it has been possible to lower substantially the upper bounds for p, q.

On the one hand, Glass, Meronk, Okada and Steiner (1991, still unpublished) showed that $\max\{p, q\} < 1.7 \times 10^{28}$ and $\min\{p, q\} < 5.43 \times 10^{19}$.

On the other hand, Mignotte (1992a, still unpublished), obtained somewhat better results: $\max\{p, q\} < 1.06 \times 10^{26}$, $\min\{p, q\} < 1.31 \times 10^{18}$.

Moreover, if $p \equiv 3 \pmod{4}$, then $p < 1.23 \times 10^{18}$ and $q < 2.48 \times 10^{24}$. In the above papers, the method of linear forms in logarithms was combined with the recent criteria of Inkeri (1990) and Aaltonen and Inkeri (1991). Glass et al. proved: If $x^p - y^q = \pm 1$, with $x, y \geq 1$, $2 < q < p$, then $q > 53$ except for the following cases (which they could not rule out):

$(q, p) = (17, 46021)$, $(19, 137)$, $(23, 2481757)$,

$(23, 13703077)$, $(31, 2806861)$, $(41, 1025273)$,

$(41, 138200401)$, $(53, 97)$, $(53, 4889)$.

Mignotte (1992a) studied explicitly $X^5 - Y^7 = \pm 1$ and showed that it has no solution (with $x, y \geq 1$). He also showed (1991b) that $x^p - y^{19} = \pm 1$ and $x^{97} - y^{53} = \pm 1$ have only trivial solutions, thus settling cases not dealt with by Glass et al.

In his paper of 1964, already quoted, Hyyrö also considered some special cases of the equation $X^U - Y^V = 1$, and he obtained the following results:

(**C10.3**). *If $a \geq 2$, $n \geq 5$ are integers, the equation $|a^U - y^n| = 1$ has at most one solution in positive integers (u, y).*

(C10.4). *If $a \geq 2$ is an integer, then the equation $|a^U - Y^V| = 1$ has at most ν distinct solutions in integers $u, v \geq \nu$, $y \geq 1$, where ν denotes the number of distinct prime factors of a.*

It should be noted that if a is a prime power, then Gérono's result (B1.1) is better.

Waldschmidt has applied (1990b) his latest inequalities for linear forms in logarithms to the equation $X^m - Y^n = k$.

(C10.5). *Let $W = 1.37 \times 10^{12}$. Assume that $k \geq 1$, m, n, $x \geq 1$, $y \geq 2$, and that $x^m > k^2$. If $0 < x^m - y^n \leq k$, then $m < W \log y$ and $n < W \log x$.*

It should be noted that W is a number which does not depend on k, a new strong feature of the above inequalities which must be satisfied by large powers with difference at most equal to k.

Consider now the sequence of all proper powers, with arbitrary exponents greater than 1:

$$S : z_1 < z_2 < \ldots < z_i < z_{i+1} < \ldots$$

The study of this sequence amounts to the study of the equations $X^U - Y^V = k$ for all integers $k \geq 1$. According to Tijdeman's theorem (C10.2)

$$\liminf_{i \to \infty} (z_{i+1} - z_i) \geq 2.$$

The main conjecture about the sequence of powers is already in Pillai's papers (1936, 1945) (see also Landau, 1959):

Conjecture of Pillai. *If $k \geq 2$, there exist at most finitely many quadruples (x, y, m, n) of natural numbers, with $m \geq 2$, $n \geq 2$, such that $x^m - y^n = k$.*

This conjecture may be restated as follows:

If $z_1 < z_2 < z_3 \ldots$ is the sequence of all natural numbers which are proper powers, then $\lim_{i \to \infty} (z_{i+1} - z_i) = \infty$.

Despite the efforts of numerous mathematicians, this conjecture has yet to be proved. It is interesting to notice the following relationship with the weaker conjecture of Hall (see §C2). Nair showed in 1978 that if the weaker conjecture of Hall (about gaps between squares and cubes) is true, then so is Pillai's conjecture.

11. A Density Result

If $k \neq 1$, not very much is known about the exponential polynomials diophantine equation $X^U - Y^V = k$.

In this section, I show a density theorem concerning the family of equations $X^m - Y^n = k$, for all choices of exponents (m, n); see Ribenboim (1986).

I begin with a lemma:

(C11.1). Lemma. *Let P be a set of $t \geq 1$ prime numbers and $q = \prod_{p \in P} p$. Let $N \geq 1$, $r \geq 1$, and $S_{P,r,N} = \left\{ (n_1, \ldots, n_r) \mid 1 \leq n_i \leq N \text{ and there exists } p \in P \text{ such that } p \mid n_i \text{ for each } i = 1, \ldots, r \right\}$. Then:*

$$N^r \geq \#(S_{P,r,N}) \geq N^r \left\{ 1 - \prod_{p \in P} \left(1 - \frac{1}{p^r}\right) \right\} - 2^t \left\{ (1+N)^r - N^r \right\}.$$

Proof: The first inequality is obvious. Let Σ' denote sums extended over tuples of distinct primes belonging to P.

Then:

$$\#(S_{P,r,N}) = \Sigma' \left[\frac{N}{p}\right]^r - \Sigma' \left[\frac{N}{pp'}\right]^r + \Sigma' \left[\frac{N}{pp'p''}\right]^r - \cdots$$

$$= -\sum_{\substack{d \mid q \\ d \neq 1}} \left[\frac{N}{d}\right]^r \mu(d) = N^r - \sum_{d \mid q} \left[\frac{N}{d}\right]^r \mu(d)$$

$$= N^r \left\{ 1 - \sum_{d \mid q} \frac{\mu(d)}{d^r} \right\} + \sum_{d \mid q} \left\{ \left(\frac{N}{d}\right)^r - \left[\frac{N}{d}\right]^r \right\} \mu(d).$$

A Density Result

But

$$\left|\sum_{d|q}\left\{\left(\frac{N}{d}\right)^r - \left[\frac{N}{d}\right]^r\right\}\mu(d)\right| = \left|\sum_{d|q}\left\{\left(\frac{N}{d} - \left[\frac{N}{d}\right]\right)^r\right.\right.$$
$$+ \binom{r}{1}\left(\frac{N}{d} - \left[\frac{N}{d}\right]\right)^{r-1}\left[\frac{N}{d}\right]$$
$$+ \binom{r}{2}\left(\frac{N}{d} - \left[\frac{N}{d}\right]\right)^{r-2}\left[\frac{N}{d}\right]^2 + \ldots$$
$$\left.\left.+ \binom{r}{r-1}\left(\frac{N}{d} - \left[\frac{N}{d}\right]\right)\left[\frac{N}{d}\right]^{r-1}\right\}\mu(d)\right|$$
$$\leq \sum_{d|q} 1 + \binom{r}{1}N\sum_{d|q}\frac{1}{d} + \binom{r}{2}N^2\sum_{d|q}\frac{1}{d^2} + \ldots$$
$$\ldots + \binom{r}{r-1}N^{r-1}\sum_{d|q}\frac{1}{d^{r-1}}$$
$$\leq \left(\sum_{d|q} 1\right)\left\{1 + \binom{r}{1}N + \binom{r}{2}N^2\right.$$
$$\left.+ \ldots + \binom{r}{r-1}N^{r-1}\right\}$$
$$= 2^t\left\{(1+N)^r - N^r\right\}.$$

Also
$$\sum_{d|q}\frac{\mu(d)}{d^r} = \prod_{p\in P}\left(1 - \frac{1}{p^r}\right),$$

so

$$\#S_{P,r,N}) \geq N^r\left\{1 - \prod_{p\in P}\left(1 - \frac{1}{p^r}\right)\right\} - 2^t\left[(1+N)^r - N^r\right]. \blacksquare$$

Let $S^*_{P,r,N} = \left\{(n_1,\ldots,n_r) \mid 1 \leq n_i \leq N \text{ and for every } i = 1,\ldots,r \text{ there exists } p_i \in P \text{ such that } p_i | n_i\right\}$.

Then $(n_1, \ldots, n_r) \in S^*_{P,r,N}$ if and only if $n_i \in S_{P,1,N}$ for every $i = 1, \ldots, r$. Thus

$$\#(S^*_{P,r,N}) = \left[\#(S_{P,1,N})\right]^r.$$

Let a, b, k be non-zero integers; k may be assumed to be positive. For any integers $m, n \geq 1$, consider the diophantine equation

$$E_{(m,n)}: \quad aX^m - bY^n = k. \tag{11.1}$$

Let $Z(E_{(m,n)})$ be the set of non-trivial solutions of $E_{(m,n)}$, that is, the set of $(x, y) \in \mathbb{Z}^2$, such that $x, y \notin \{0, 1, -1\}$ and $ax^m - by^n = k$.

By (C2.1), if $m, n \geq 2$ and $\max\{m, n\} \geq 3$ then the set $Z(E_{(m,n)})$ is finite.

(C11.2). Lemma. *If $m, n \geq 2$ and $\max\{m, n\} \geq 3$, there exists $h = h_{(m,n)} \geq 1$ such that if $u, v \geq h$ then $Z(E_{(um,vn)}) = \emptyset$.*

Proof: If $Z(E_{(m,n)}) = \emptyset$, let $h = 1$. If

$$Z(E_{(m,n)}) = \{(x_1, y_1), \ldots, (x_s, y_s)\}, \text{ let } h = \max_{1 \leq i \leq s}\{|x_i|, |y_i|\}.$$

Now, if $u, v \geq h$ and $(x, y) \in Z(E_{(um,vn)})$, then $ax^{um} - by^{vn} = k$ so $(x^u, y^v) \in Z(E_{(m,n)})$; therefore $|x|^u, |y|^v \leq h \leq u, v$ hence $x, y \in \{0, 1, -1\}$, which is contrary to the hypothesis. ∎

(C11.3). *If $N \geq 1$ let $D_n = \{(m, n) | 1 \leq m, n \leq N \text{ and } Z(E_{(m,n)}) = \emptyset\}$. I show:*

$$\lim_{N \to \infty} \frac{\#(D_N)}{N^2} = 1.$$

Proof: Let $p_1 = 2 < p_2 = 3 < p_3 \ldots$ be the sequence of prime numbers. For $t \geq 2$, let $P_t = \{p_2, p_3, \ldots, p_t\}$. For $N \geq 1$, $t \geq 2$

A Density Result

let $S_{t,N}^* = S_{P_t,2,N}^* = \{(m,n) \in \mathbb{Z}^2 | \ 1 \leq m, n \leq N$ and there exist primes $p_i, p_j \in P_t$ such that $p_i | m$ and $p_j | n\}$.

For every $(p_i, p_j) \in P_t^2$, consider the integer $h_{(p_i,p_j)}$, defined in Lemma (C11.2), and let $M_t = p_t \max\{h_{(p_1,p_j)} / (p_i, p_j) \in P_t^2\}$.

For $N > M_t$, let

$$S'_{t,N} = \{(m,n) \in S_{t,N}^* | \ M_t < m, n\},$$
$$T_{t,N} = \{(m,n) \in \mathbb{Z}^2 | \ 1 \leq m, n \leq N \text{ and } m \leq M_t \text{ or } n \leq M_t\}.$$

Clearly, $\#(T_{t,N}) = 2M_t N - M_t^2 < 2M_t N$ and $S_{t,N}^* \subseteq S'_{t,N} \cup T_{t,N}$, hence

$$\#(S_{t,N}^*) \leq \#(S'_{t,N}) + \#(T_{t,N}) \leq \#(S'_{t,N}) + 2M_t N.$$

Now I show that $S'_{t,N} \subseteq D_N$. Indeed, let $(m,n) \in S'_{t,N}$, so $M_t < m, n \leq N$, and there exist $(p_i, p_j) \in P_t^2$, such that $p_i | m$ and $p_j | n$. So there exist integers $u, v \geq 1$, such that $p_i h_{(p_i,p_j)} \leq M_t < m = p_i u$ and $p_j h_{(p_i,p_j)} \leq M_t < n = p_j^v$, hence $h_{(p_i,p_j)} < u, v$. By Lemma (C11.2), $Z(E_{(up_i, vp_j)}) = Z(E_{(m,n)}) = \emptyset$, so $(m,n) \in D_n$.

Hence

$$\#(S_{t,N}^*) = \left[\#(S_{P_t,1,N})\right]^2 \geq \left[N\left\{1 - \prod_{j=2}^{t}\left(1 - \frac{1}{p_j}\right)\right\} - 2^{t-1}\right]^2$$
$$= \left[N\left\{1 - 2\prod_{j=1}^{t}\left(1 - \frac{1}{p_j}\right)\right\} - 2^{t-1}\right]^2.$$

Thus

$$\frac{\#(D_N)}{N^2} \geq \left[\left\{1 - 2\prod_{j=1}^{t}\left(1 - \frac{1}{p_j}\right)\right\} - \frac{2^{t-1}}{N}\right]^2 - \frac{2M_t}{N}$$

$$\geq 1 - 2\left[2\prod_{j=1}^{t}\left(1 - \frac{1}{p_j}\right) + \frac{2^{t-1}}{N}\right] - \frac{2M_t}{N}$$

$$= 1 - 4\prod_{j=1}^{t}\left(1 - \frac{1}{p_j}\right) - \frac{2^t + 2M_t}{N}.$$

Since $\prod_p \left(1 - \frac{1}{p}\right) = 0$, given $\epsilon > 0$ there exists t such that $\prod_{j=1}^{t}\left(1 - \frac{1}{p_j}\right) < \frac{\epsilon}{8}$. With this choice of t, let N_0 be such that $N_0 > M_t$ and $\frac{2^t + 2M_t}{N_0} < \frac{\epsilon}{2}$. Finally, if $N \geq N_0$, then $1 \geq \frac{\#(D_N)}{N^2} \geq 1 - \epsilon$, and this concludes the proof. ∎

The above result may be spelled out as follows:

> *The asymptotic density of the set of exponents (m,n), for which the equation $X^m - Y^n = k$ has only trivial solutions, is equal to 1.*

This method has been used by Granville and Heath-Brown, independently in 1985, to show that the density of the set of exponents $n > 2$, for which Fermat's equation has only trivial solutions in integers, is equal to 1.

It should be stressed that, given k, the method indicated does not allow us to assert whether, for a given pair of exponents (m,n), the equation $X^m - Y^n = k$ has non-trivial solutions.

Part X

Appendix 1
Catalan's Equation in Other Domains

Catalan's equations has also been considered in domains other than $\mathbb{Z}$. In this appendix, I shall give some of these results.

(A). Catalan's Equation over Number Fields

In 1986, Brindza, Györy and Tijdeman studied Catalan's equation over a number field. Let K be an algebraic number field of degree n over $\mathbb{Q}$, and let A denote the ring of algebraic integers of K. For each $x \in K$, let $\lceil x \rceil$ denote the maximum of the absolute values of the conjugates of x.

Extending the theorem of Tijdeman (C10.2), Brindza, Györy and Tijdeman proved:

(X1.1). *Given a number field K, there exists an effectively computable number $C > 0$ (depending on K), such that if m, n are natural numbers, $m \geq 2$, $n \geq 2$ and $mn > 4$, and $x, y \in A$ are not roots of unity, which satisfy the equation*

$$x^m - y^n = 1,$$

then

$$\max\left\{\lceil x \rceil, \lceil y \rceil, m, n\right\} < C.$$

It is not difficult to observe that the restrictions on x, y, m, n are necessary. The method of proof involves Baker's linear forms in logarithms, but as the authors stress, the proof is not just a straightforward extension of Tijdeman's proof.

(B) Catalan's Equation over Fields $K(t)$ and Domains $K[t]$

The study of Catalan's equation over a polynomial domain, or a field of rational fractions, is (surprisingly!) related to Fermat's equation.

Accordingly, I begin by proving the following result of Greenleaf (1969), which reinforced previous results by Liouville (1879), Korkine (1880), Shanks (1962); see also Ribenboim (1979).

(X1.2). *Let $n > 2$, and let K be a field of characteristic not dividing n. If $f, g, h \in K[X]$, with $\gcd(f, g, h) = 1$, and $f^n + g^n = h^n$, then f, g, h have degree 0.*

Proof: There is no loss of generality to assume that K is algebraically closed. Indeed, let $\bar{K}$ be an algebraic closure of K. If it is shown that $f, g, h \in \bar{K}$, then $f, g, h \in \bar{K} \cap K(x) = K$. Suppose that there exists a triple (f, g, h) of polynomials, with $\gcd(f, g, h) = 1$, $f^n + g^n = h^n$, and the maximum d of the degrees of f, g, h is not 0. Among all such triples, choose one with d minimal. It is clear that f, g, h are pairwise relatively prime. K contains a primitive n^{th} root of 1, because it is algebraically closed and its characteristic does not divide n. Hence

$$g^n = h^n - f^n = \prod_{j=0}^{n-1}(h - \zeta^j f), \tag{1.1}$$

where ζ is a primitive n^{th} root of 1.

Since $K[X]$ is a unique factorization domain and the polynomials $h - \zeta^j f \in K[X]$ are pairwise relatively prime, then each one is an nth power,

$$h - \zeta^j f = g_j^n, \quad \text{with } g_j \in K[X].$$

The three distinct polynomials $h - f$, $h - \zeta f$, $h - \zeta^2 f$ belong to the K-subspace of $K[X]$ spanned by h, f, hence they are linearly dependent. Thus, there exist $c_0, c_1, c_2 \in K$ such that

$$c_0(h - f) + c_1(h - \zeta f) + c_2(h - \zeta^2 f) = 0,$$

and c_0, c_1, c_2 are not all equal to 0. Actually, c_0, c_1, c_2 must all be different from 0, because f, g, h are pairwise relatively prime.

Let $d_i \in K$ be such that $c_i = d_i^n$ (for $i = 0, 1, 2$). Then

$$(d_0 g_0)^n + (d_1 g_1)^n + (d_2 g_2)^n = 0,$$

and n deg $(d_i g_i) \leq \max\{\deg(h), \deg(g)\} \leq d$. Thus, $\max\{\deg(d_i g_i) | i = 0, 1, 2\} < d$, and this is a contradiction. ∎

The next result is an extension of Nathanson's theorem for Catalan's equation over $K[X]$ or $K(X)$, given in 1974. For this generalization see Ribenboim (1984).

(X1.3). *Let $n \geq 3$, and let K be a field of characteristic not dividing n. Let $P \in K[X]$ be a polynomial having degree $m \geq 2$ with distinct roots. If $f, g, \in K(X)$, $m > 2$, or if $m = 2$, then $n \neq 4$, and $P(f) = g^n$, then $f, g \in K$.*

Proof: As in the preceding proof, there is no loss of generality to assume that K algebraically closed, because if $\bar{K}$ is an algebraic closure of K, then $f, g \in \bar{K} \cap K(X) = K$.

Next, I observe that if $P(f) = g^n$ and $f \in K$, then $g^n = P(f) = c \in K$. Since K is algebraically closed, there exists $d \in K$ such that $d^n = c = g^n$. Hence $g \in K$, because K contains the nth roots of 1.

Similarly, if $g = c \in K$, let $Q(X) = P(X) - c^n \in K[X]$. Then f is a root of the polynomial $Q(X)$, thus $f \in K$, since K is algebraically closed.

So it may be assumed that $f, g \notin K$. Let

$$P(X) = a_0 X^m + a_1 X^{m-1} + \cdots + a_m \qquad (1.2)$$
$$= a_0 \prod_{i=1}^{m} (X - r_i)$$

with $a_0, a_1, \ldots, a_m, r_1, \ldots, r_m \in K$, $a_0 \neq 0$. By hypothesis $r_1, \ldots, r_m$ are distinct. Let $f = \frac{f_1}{f_0}$, $g = \frac{g_1}{g_0}$ with

$$f_0, f_1, g_0, g_1 \in K[X], \quad gcd(f_0, f_1) = 1, \; gcd(g_0, g_1) = 1,$$

then

$$g_1^n f_0^m = a_0 f_1^m + a_1 f_1^{m-1} f_0 + \cdots + a_{m-1} f_1 f_0^{m-1} + a_m f_0^m) g_0^n.$$

Since $gcd(f_0, a_0 f_1^m + \cdots + a_m f_0^m) = 1$, it follows that $g_0^n = h f_0^m$ with $h \in K[X]$. From $gcd(g_0, g_1) = 1$, it follows that

$$a_0 f_1^m + a_1 f_1^{m-1} f_0 + \cdots + a_{m-1} f_1 f_0^{m-1} + a_m f_0^m = h' g_1^n,$$

with $h' \in K[X]$. Hence $hh' = 1$, and therefore $h, h' \in K$. Let $b, e, e' \in K$ be such that $a_0 = b^n$, $h = e^n$ and $h' = e'^n$. Then

$$(b^{-1} e' g_1)^n = \prod_{i=1}^{m} (f_1 - r_i f_0). \qquad (1.3)$$

Since the elements $r_1, \ldots, r_m$ are distinct, and $gcd(f_0, f_1) = 1$, then the polynomials $f_1 - r_i f_0$ ($i = 1, \ldots, m$) are pairwise relative prime. From (1.3) it follows that

$$f_1 - r_i f_0 = k_i^n$$

(for $i = 1, \ldots, m$), where $k_i \in K[X]$.
Now assume that $m \geq 3$.

Catalan's Equation in Other Domains 317

The three polynomials $f_1 - r_1 f_0$, $f_1 - r_2 f_0$, $f_1 - r_3 f_0$ are in the K-subspace of $K[X]$ generated by f_0, f_1, so they are linearly dependent. Hence there exist $b_i K$ ($i = 1, 2, 3$), not all equal to 0 such that

$$b_1(f_1 - r_1 f_0) + b_2(f_1 - r_2 f_0) + b_3(f_1 - r_3 f_0) = 0.$$

Actually, b_1, b_2, b_3 are all different from 0, because $gcd(f_0, f_1) = 1$.

Let $c_i \in K$ be such that $c_i^n = b_i$ ($i = 1, 2, 3$). Then

$$(c_1 k_1)^n + (c_2 k_2)^n + (c_3 k_3)^n = 0.$$

By (X1.2), $c_i k_i \in K$ ($i = 1, 2, 3$), so $f_1 - r_i f_0 \in K$ (for $i = 1, 2, 3$). This implies that $(r_1 - r_2) f_0 \in K$, so $f_0, f_1 \in K$, hence $f \in K$, which has been excluded.

Now assume that $m = 2, n \neq 4$, so $g_0^n = h f_0^2$, $h = e^n$, hence $f_0^2 = (e^{-1} g_0)^n$.

If n is odd, then necessarily $f_0 = \ell^n$ for some $\ell \in K[X]$. If n is even, say $n = 2n'$, then $f_0 = \ell^{n'}$ where $\ell = \pm e^{-1} g_0$. Thus, in both cases, $f_0 = \ell^\nu$ where $\ell \in K[X]$, $\nu \geq 3$. From (1.3), as before,

$$f_1 - r_1 f_0 = k_1^\nu, \qquad f_1 - r_2 f_0 = k_2^\nu$$

where $k_1, k_2 \in K[X]$. Hence $(r_1 - r_2) f_0 = k_2^\nu - k_1^\nu$. If $r_1 - r_2 = s^\nu$, then $(s\ell)^\nu + k_1^\nu + (\rho k_2)^\nu = 0$, where $\rho^\nu = -1$. Once again, by (X1.2), $k_1, k_2, \ell \in K$ and this implies that $f \in K$, already excluded. ∎

Taking $P(X) = X^m - 1$ yields Nathanson's result on Catalan's equation, in a slightly stronger form.

(C) Catalan's Equation over Function Fields of Projective Varieties

Silverman considered in 1982 the equation $aX^m + bY^n = c$ over the function field K of a projective variety.
He proved:

(X1.4). *Let K_0 be any field, let V be a projective non-singular variety over K_0 and let K be the function field of V. Given $a, b, c \in K$, $a, b, c \neq 0$, there exist only finitely many pairs of integers $m, n \geq 2$, such that:*
  i) *If char $K_0 = p \neq 0$, then $p \nmid mn$.*
  ii) *There exists $f, g \in K \backslash K_0$ such that $af^m + bg^n = c$.*

Furthermore, for each pair (m, n) as above, there are only finitely many pairs (f, g) with $f, g \in K \backslash K_0$ satisfying $af^m + bg^n = c$, except in the following cases:

1) $\frac{a}{c}$ is an mth power in K and $\frac{b}{c}$ is an nth power in K.

2) If $(m, n) \in \{(2, 2), (2, 3), (2, 4), (3, 2), (3, 3), (4.2)\}$, then the equation defines a curve of genus 0 or 1 over K_0.

In the above cases (1), (2), there may be infinitely many solutions in $K \backslash K_0$.

Appendix 2
Powerful Numbers

The study of powerful numbers is quite interesting. One may formulate problems about these numbers which are similar to those for powers. In this short appendix, I shall evoke briefly these problems, giving bibliographic references. The interested readers may complete their information directly from the sources.

The first paper dealing with these numbers is by Erdös and Szekeres (1935).

The name "powerful numbers" seems to have been coined by Golomb (1970).

If $k \geq 2$, a natural number n is said to be k-*powerful* if it is of the form $n = \prod_{i=1}^{r} p_i^{e_i}$, where $r \geq 1$, $p_1, \ldots, p_r$ are distinct prime numbers and $e_i \geq k$ (for $i = 1, \ldots, r$).

A 2-powerful number is simply called a *powerful number*.

I denote by W_k the set of k-powerful numbers and I write $W = W_2$. Clearly

$$W = W_2 \supset W_3 \supset \cdots \supset W_{12} \supset \cdots,$$

and each proper power is a powerful number. Moreover, n is k-powerful if and only if $n = a_0^k a_1^{k+1} \cdots a_{k-1}^{2k-1}$, where a_i are integers (not necessarily relatively prime), $a_i \geq 1$. In particular, the powerful numbers are those of the form $a_0^2 a_1^3$; moreover, a_1 may be taken to be square-free.

The main problems concerning powerful numbers are of the following three types:

A) Distribution of powerful numbers
B) Additive problems
C) Difference problems

These questions are also meaningful for the set of proper powers—where they may become trivial (like the distribution of powers) or very difficult and still unsolved (like Fermat's problem and similar questions). Differences of powers, especially consecutive powers, have been the focus of this book. For a survey of these questions, see Ribenboim (1988).

(A) Distribution of Powerful Numbers

The aim is to estimate the number $w_k(x)$ of elements in the set
$$W_k(x) = \{n \in W_k | 1 \leq n \leq x\}, \tag{2.1}$$
where $x \geq 1$, $k \geq 2$.

Already in 1935, Erdös and Szekeres gave the first result about $w_2(x)$:
$$w_2(x) = \frac{\zeta\left(\frac{3}{2}\right)}{\zeta(3)} x^{\frac{1}{2}} + O(x^{\frac{1}{3}}), \text{ as } x \to \infty. \tag{2.2}$$

where $\zeta(s)$ is the Riemann zeta function; see also Bateman (1954) and Golomb (1970).

To describe the more recent results, I introduce the zeta function associated to the sequence of k-powerful numbers. Let
$$j_k(n) = \begin{cases} 1 & \text{if } n \text{ is } k\text{-powerful} \\ 0 & \text{otherwise.} \end{cases}$$

The series $\sum_{n=1}^{\infty} \frac{j_k(n)}{n^s}$ is convergent for $\text{Re}(s) > \frac{1}{k}$ and defines a function $F_k(s)$. This function admits the following Euler product representation
$$F_k(s) = \prod_p \left(1 + \frac{\frac{1}{p^{ks}}}{1 - \frac{1}{p^s}}\right) = \prod_p \left(1 + \frac{1}{p^{(k-1)s}(p^s - 1)}\right), \tag{2.3}$$

which is valid for $\text{Re}(s) > \frac{1}{k}$.

With well-known methods, Ivić and Shiu have shown in 1982:

(X2.1). $w_k(x) = \gamma_{0,k} x^{\frac{1}{k}} + \gamma_{1,k} x^{\frac{1}{k+1}} + \cdots + \gamma_{k-1,k} x^{\frac{1}{2k-1}} + \Delta_k(x)$,
where $\gamma_{i,u}$ is the residue at $\frac{1}{k+i}$ of $\frac{F_k(s)}{s}$.

Explicitly,
$$\gamma_{i,k} = C_{k+i,k} \frac{\Phi_k\left(\frac{1}{k+i}\right)}{\zeta\left(\frac{2k+2}{k+i}\right)}, \tag{2.4}$$

where
$$C_{k+i,k} = \prod_{\substack{j=k \\ j \neq k+i}}^{2k-1} \zeta\left(\frac{j}{k+i}\right), \tag{2.5}$$

$\Phi_2(s) = 1$, and if $k > 2$, then $\Phi_k(s)$ has a Dirichlet series with abscissa of absolute convergence $\frac{1}{2k+3}$, and $\Delta_k(x)$ its error term.

Erdös and Szekeres had already considered this error term and showed that

$$\Delta_k(x) = O\left(x^{\frac{1}{k+1}}\right) \quad \text{as } x \to \infty. \tag{2.6}$$

Better estimates of the error have been obtained since. Let

$$\rho_k = \inf\{\rho > 0 | \Delta_k(x) = O(x^\rho)\}.$$

Bateman and Grosswald showed in 1958 that $\rho_2 \leq \frac{1}{6}$ and $\rho_3 \leq \frac{7}{46}$.

Sharper results are due to Ivić and Shun:

$$\rho_2 \leq 0.128 < \frac{1}{6}, \quad \rho_3 \leq 0.128 < \frac{7}{46},$$
$$\rho_4 \leq 0.1189, \quad \rho_5 \leq \frac{1}{10}, \quad \rho_6 \leq \frac{1}{12}, \quad \rho_7 \leq \frac{1}{14}, \text{ etc.}$$

I refer also to the work on Krätzel (1972) on this matter.

It is conjectured that, for every k:

$$\Delta_k(x) = O\left(x^{\frac{1}{2k}}\right) \quad \text{for } x \to \infty. \tag{2.7}$$

More specifically, taking $k = 2$:

$$w_2(x) = \frac{\zeta\left(\frac{3}{2}\right)}{\zeta(3)} x^{\frac{1}{2}} + \frac{\zeta\left(\frac{2}{3}\right)}{\zeta(2)} x^{\frac{1}{3}} + \Delta_2(x), \tag{2.8}$$

with $\Delta_2(x) = O(x^{\frac{1}{6}})$, as $x \to \infty$.

(B) Additive Problems

If $h \geq 2$, $k \geq 2$, I shall use the following notation:

$$\sum hW_k = \left\{ \sum_{i=1}^{h} n_i \mid \text{each } n_i \in W_k \cup \{0\} \right\},$$

$$\sum hW_k(x) = \left\{ n \in \sum hW_k \mid n \leq x \right\} \quad (\text{for } x \geq 1).$$

The additive problems concern the comparison of the sets $\sum hW_k$ with the set of natural numbers, the distribution of the sets $\sum hW_k$ and similar questions.

The distribution of $\sum 2W_2$ was treated by Erdös in 1975:

(X2.2). $\#\sum 2W_2(x) = o\left(\frac{x}{(\log x)^\alpha}\right)$ (as $x \to \infty$), where $0 < \alpha < \frac{1}{2}$.

In particular, $\#\sum 2W_2(x) = o(x)$, so there exist infinitely many natural numbers which are not the sum of two powerful numbers.

Odoni showed in 1981 that there is no constant $c > 0$ such that

$$\#\sum 2W_2(x) \sim \frac{Cx}{(\log x)^{1/2}} \quad (\text{as } x \to \infty).$$

The following result was conjectured by Erdös and Ivić in the 1970's and proved by Heath-Brown (1985).

(X2.3). *There is an effectively computable number n_0, such that if $n \geq n_0$, then n is the sum of at most three powerful numbers.*

The only known exceptions up to 32000 are 7, 15, 23, 87, 111, 119. Mollin and Walsh conjectured in 1986 that there are no others.

The following problem concerning 3-powerful numbers remains open:

Do there exist infinitely many natural numbers which are not sums of three 3-powerful numbers. Probably yes.

(C) Difference Problems

The problems of this kind are the following.

Problem D1. Given $k \geq 2$, to determine which numbers N are of the form $N = n_1 - n_2$ where, $n_1, n_2 \in W_k$. Such an expression of N is called a *representation as difference of k-powerful numbers*, or simply a k-powerful representation. When $k = 2$, I just say a powerful representation. If $gcd(n_1 n_2) = 1$, the representation is called *primitive*; if n_1 or n_2 is a proper power, the representation is called *degenerate*.

Problem D2. Given $k \geq 2, N \geq 1$, to determine the set, or just the number of representations (primitive or not, degenerate or not) of N as a difference of k-powerful numbers. In the same vein is the following problem:

Problem D3. Given integers $N_1, N_2 \geq 1$, to determine if there exist k-powerful numbers n_1, n_2, n_3 such that

$$n_2 - n_1 = N_1 \quad \text{and} \quad n_3 - n_2 = N_2.$$

In such a case, to study the possible triples of such numbers.

One may also think of similar problems with several differences $N_1, N_2, \ldots, N_r \geq 1$ given in advance, but as I shall indicate, problem D3 in its simplest formulation is unsolved and certainly very difficult.

I begin discussing problems D1 and D2. The first remark, due to Mahler, also shows that these questions bear a close relationship to the equations $X^2 - DY^2 = C$.

Thus, Mahler said: since the equation $X^2 - 8Y^2 = 1$ has infinitely many solutions in integers (x, y), and since the number $8y^2$ is powerful, then 1 admits infinitely many degenerate (primitive) powerful representations.

In 1976, Walker showed that 1 also has an infinite number of non-degenerate powerful (primitive) representations.

In 1981, Sentance showed that 2 has infinitely many primitive degenerate powerful representations, the smallest being:

$$2 = 27 - 25 = 70227 - 70225 = 189750627 - 189750625.$$

More recently, putting together the results in various papers, published independently and almost simultaneously by McDaniel, Mollin and Walsh and Vanden Eyden, it has been established that:

(**X2.4**). *Every natural number has infinitely many primitive degenerate powerful representations and also infinitely many primitive non-degenerate powerful representations.*

Moreover, there is an algorithm to determine such representations.

For a survey of the above results, see also Mollin (1987).

It has been asked by Erdös whether consecutive powerful numbers may be obtained other than as solutions of appropriate equations $EX^2 - DY^2 = 1$.

Concerning the distribution of pairs of consecutive powerful numbers, there are several conjectures by Erdös (1976).

Distribution of Powerful Numbers

First (?) Erdös' conjecture[*]:

$\{n \mid n \text{ and } n+1 \text{ are powerful}, n \leq x\} < (\log x)^c$, where $c > 0$ is a constant.

It is not even yet proved that $c'x^{\frac{1}{3}}$ is an upper bound (with a constant $c' > 0$).

Second Erdös' conjecture:

There do not exist two consecutive 3-powerful numbers.

This may be rephrased as follows: for every $n \geq 1$ there exists a prime p such that p^3 does not divide $n(n+1)$. It is interesting to note that the only known examples of consecutive integers, such that one is 2-powerful and the other is 3-powerful, are $(8, 9)$ and $(12167, 12168)$.

A related conjecture is the following:

Third Erdös' conjecture:

Let $a_1 < a_2 < a_3 < \cdots$ be the sequence of powerful numbers. There exist constants $c > 0$, $c' > 0$, such that for every sufficiently large m,

$$a_{m+1} - a_m > cm^{c'}.$$

In particular,

$$\lim(a_{m+1} - a_m) = \infty.$$

Most likely, these problems are hard to settle.

Now I consider problem D3 in its simplest form, which concerns three consecutive powerful numbers.

With his admirable wisdom, Erdös conjectured:

---

[*] No one can state which was Erdös' first conjecture—I would not be surprised if it was his first meaningful sentence, as a child ...

Fourth Erdös' conjecture:

There do not exist three consecutive powerful numbers.

This goes of course beyond the fact, proved by Makowski (1962), that there do not exist three consecutive perfect powers.

Once again, calculations would only serve to find three consecutive powerful numbers, if some exist.

It is very intriguing that this conjecture has a connection with Fermat's last theorem as I proceed to explain: I shall indicate several statements, of which only the last one is a proven theorem, the others being conjectures.

(E) *There do not exist three consecutive powerful numbers.*

(E') *If x is even, then $x^{2k} - 1$ is not powerful (for every $k \geq 2$).*

(W) *There exist infinitely many primes p, such that $2^{p-1} \not\equiv 1 \pmod{p^2}$.*

(T) *There exist infinitely many primes p, such that the first case of Fermat's last theorem is true for this exponent p; that is, if x, y, z are non-zero integers, not multiples of p, then $x^p + y^p \neq z^p$.*

(T) is the theorem of Adleman, Heath-Brown and Fouvry (1985). Its proof is difficult and requires, among other facts, fine new results in sieve theory.

The statements (E), (E'), (W) are conjectured to be true, but have not yet been proved.

Wieferich proved in 1909 the striking theorem:

If p is odd prime, and there exist non-zero integers x, y, z, not multiples of p such that $x^p + y^p = z^p$, then $2^{p-1} \equiv 1 \pmod{p^2}$.

Therefore, Wieferich's theorem means that $\{p|\ p \text{ odd prime}, 2^{p-1} \not\equiv 1 \pmod{p^2}\} \subseteq \{p|\ p \text{ odd prime, the first case of Fermat's last theorem is true for } p\}$. Thus, the conjecture (W) implies the theorem (T). So, a proof of (W), together with Wieferich's theorem, would provide a new proof, perhaps simpler, of theorem (T).

It is almost immediate that the implication $(E) \Rightarrow (E')$ is true.

Indeed, if x is even, $k \geq 2$ and $x^{2k} - 1$ is powerful, then since $x^{2k} - 1 = (x^k - 1)(x^k + 1)$ and $gcd(x^k - 1, x^k + 1) = 1$, necessarily both $x^k - 1$, $x^k + 1$ are powerful; for $k \geq 2$, x^k is also powerful, so there would exist three consecutive powerful numbers.

The missing link between these statements, namely the implication $(E') \Rightarrow (W)$, was proved by Granville in (1986); the proof is elementary.

As a consequence, it shows the existence of a rather unexpected connection between powerful numbers and Fermat's last theorem.

Numerical evidence shows that (W) is very plausible. Indeed, the only primes $p < 6 \times 10^9$ such that $2^{p-1} \equiv 1 \pmod{p^2}$ are $p = 1093$ and $p = 3511$. Thus, the values of $q_p(2) \bmod p$, for different primes p, are independent of each other. Hence, for any $N > 1$,

$$\text{Probability } (q_p(2) \equiv 0 \pmod{p}), \text{ for some } p \leq N)$$
$$= \sum_{p \leq N} \text{Probability } (q_p(2) \equiv 0 \pmod{p}))$$
$$= \sum_{p \leq N} \frac{1}{p} = \log \log N + O(1)$$

(the last sum is well known). As N tends to infinity, this shows that S is (heuristically) an infinite set. By the prime number theorem $\pi(N) \sim \frac{N}{\log N}$, hence simple heuristic considerations also indicate that the set

$$S' = \{p \text{ odd prime} | 2^{p-1} \not\equiv 1 \pmod{p^2}\}$$

is infinite. Indeed, let

$$S = \{p \text{ odd prime} | 2^{p-1} \equiv 1 \pmod{p^2}\}.$$

For any $N > 1$, let $S(N) = \{p \in S | p \leq N\}$ and

$$S'(N) = \{n \in S' | p \leq N\}.$$

Denote $q_p(2) = \frac{2^{p-1}-1}{p}$; according to Fermat's little theorem, $q_p(2)$ is an integer. For lack of more knowledge, it may be *heuristically assumed that the residue classes $q_p(2)$ mod p are randomly distributed*;

$$\frac{\#S(N)}{\pi(N)} \sim \frac{\log\log N + O(1)}{\frac{N}{\log N}} = 0.$$

It follows that $\frac{S'(N)}{\pi(N)} \sim 1$, which means not only that S' is (heuristically) an infinite set, but even that (heuristically) most primes are in S'.

As a matter of fact, there is also a density result (not based on any heuristic assumption) supporting these conjectures.

For $a \geq 2$, $m \geq 2$, consider the equation

$$(E_{a,m}): \quad (2X)^{2m} - 1 = a^3 Y^2. \tag{2.9}$$

For each $a \geq 2$, let

$Q_a = \{m \geq 2 | \text{ the equation } (E_{a,m}) \text{ has no solution in integers}\}$,

and let, for each $m \geq 2$,

$R_m = \{a \geq 2 | \text{ the equation } (E_{a,m}) \text{ has no solution in integers}\}$.

Of course, conjecture (E') is the same as stating that for every $a \geq 2$, $Q_a = \{m | m \geq 2\}$, or also, for every $m \geq 2$, $R_m = \{a | a \geq 2\}$.

Using the theorem of Schinzel and Tijdeman (C1.6), it is clear that I have proved (Ribenboim, 1992):

For every $a \geq 0$, the complement of the set Q_a is finite.

(X2.5). *For every $m \geq 2$, the complement of the set R_m, relative to the sequence of squares, is equal to 1.*

The concepts of uniform density and uniform density relative to the sequence of squares are fully explained in the above mentioned paper.

In their study of Erdös' conjecture (E), Mollin and Walsh showed (1986a):

(X2.6). *If there exist three consecutive powerful numbers, then there exists a square-free integer $D > 0$, $D \equiv 7 \pmod{8}$ with the following property: there exists a unit $x_k + y_k\sqrt{D} = (x + y\sqrt{D})^k$ of $\mathbb{Q}(\sqrt{D})$ [where $x + y\sqrt{D}$ denotes the fundamental unit], such that $k \geq 1$, is odd and*

a) x_k *is an even powerful number, and*

b) y_k *is odd and D divides y_k.*

Mollin and Walsh also noted that if D does not divide y, then the smallest odd k such that conditions (a), (b) above are satisfied, must be very large.

This remark leads to the difficult question of knowing whether D divides y. In this connection and to illustrate the difficulty of the problem, there are two conjectures.

Conjecture of Ankeny, Artin and Chowla (1952):

If p is a prime, $p \equiv 1 \pmod 4$, and $\frac{x+y\sqrt{p}}{2}$ is the fundamental unit of $\mathbb{Q}(\sqrt{p})$, then p does not divide y.

Conjecture of Mordell (1960):

If p is a prime, $p \equiv 3 \pmod 4$, and $x + y\sqrt{p}$ is the fundamental unit of $\mathbb{Q}(\sqrt{p})$, then p does not divide y.

Needless to say, these conjectures have not yet been proved to be true.

Extending previous work by R. Soleng (unpublished manuscript), Stephens and Williams (1988) checked the conjecture of Ankeny, Artin and Chowla up to 10^9. On the other hand, the conjecture of Mordell was checked by Beach, Williams and Zarnke (1971) for $p \leq 7679299$.

In these calculations, some composite values of D were found, for which D divides y, the smallest such D being 46.

BIBLIOGRAPHY

b.1288, d. 1344 LEVI BEN GERSON See Dickson, L. E., *History of The Theory of Numbers*, Vol. II, p. 731. Carnegie Institution, Washington, 1920. Reprinted by Chelsea Publ. Co., New York, 1971.

1640 FERMAT, P. Lettre à Mersenne (Mai, 1640). *Oeuvres*, Vol. II, 194–195. Publiées par les soins de MM. Paul Tannery et Charles Henry. Gauthier-Villars, Paris, 1894.

1643 FERMAT, P. Lettre à St. Martin (Mai 31, 1643). *Oeuvres*, Vol. II, 258-260. Publiées par les soins de MM. Paul Tannery et Charles Hanry. Gauthier-Villars, Paris, 1894.

1657 FRÉNICLE DE BESSY Solutio duorum problematum circa numeros cubos et quadratos (1657). Bibliothèque Nationale de Paris.

1659 FERMAT, P. Lettre à Carcavi (Septembre 1659). *Oeuvres*, Vol. II, p. 441. Publiées par les soins de MM. Paul Tannery et Charles Henry. Gauthier-Villars, Paris, 1896.

— FERMAT, P. Observations sur Diophante. *Oeuvres*, Vol. III, p. 269. Publiées par les soins de MM. Paul Tannery et Charles Henry. Gauthier-Villars, Paris, 1896.

— FERMAT, P. Commercium Epistolicum. *Oeuvres*, Vol. III, 403–602. Gauthier-Villars, Paris, 1896.

1676 FRÉNICLE DE BESSY *Traité des Triangles Rectangles en Nombres*, Vol. I, Paris, 1676. Reprinted in Mém. Acad. Royale Sci. de Paris 5, 1729, 1666–1699.

1732 EULER, L. Observatione de theoremate quodam Fermatiano aliisque ad numeros primos spectantibus. Comm. Acad. Sci. Petrop. 6 (1732/1733) (1738), 103–107. Reprinted in *Opera Omnia*, Ser. I, Vol. II, Commentationes Arithmeticae I, 1-5. B. G. Teubner, Leipzig, 1915.

1737 EULER, L. De fractionibus continuis dissertatio. Comm. Acad. Sci. Petr., 9 (1737), 1744, 98–137. Reprinted in *Opera Omnia*, Ser. I, Vol. XIV, *Commentationes Analyticae*, 187–215. B. G. Teubner, Leipzig, 1924.

1738 EULER, L. Theorematum quorundam arithmeticorum demonstrationes. Comm. Acad. Sci. Petrop. 10 (1738) (1747), 125–146.Reprinted in *Opera Omnia*, Ser. I, Vol. II, *Commentationes Arithmeticae* I, 38–58. B. G. Teubner, Basel, 1915.

1747 EULER, L. Theoremata circa divisores numerorum. Novi Comm. Acad. Sci. Petrop. 1 (1747/8), 20–48. Reprinted in *Opera Omnia*, Ser. I, Vol. I, *Commentationes Arithmeticae* I, 62–85. B. G. Teubner, Basel, 1915.

1770 EULER, L. *Volständige Anleitung zur Algebra* (2 volumes). Royal Acad. Sci., Sankt Petersburg, 1770. English translation by J. Hewlett. Longman, Orme & Co., London, 1840. Reprinted by Springer-Verlag, New York, 1984.

1770 LAGRANGE, J. L. Additions au mémoire sur la résolution des équations numériques. Mém. Acad. Royale Sci.

Belles-Lettres de Berlin, 24 (1770). Reprinted in *Oeuvres*, Vol. II, 581–652. Gauthier-Villars, Paris, 1868. Publiées par les soins de M. J. A. Serret.

1777 LAGRANGE, J. L. Sur quelques problèmes de l'analyse de Diophante. Nouveaux Mém. Acad. Sci. Belles Lettres, Berlin (1777). Reprinted in *Oeuvres*, Vol. IV, 377–398. Publiées par les soins de M. J. A. Serret. Gauthier-Villars Paris, 1868.

1798 LAGRANGE, J. L. Addition aux *Eléments d'Algeèbre* d'Euler – Analyse Indéterminée. Reprinted in *Oeuvres*, Vol., VII, 3-180. Publiées par les soins de M. J. A. Serret. Gauthier-Villars, Paris, 1877.

1801 GAUSS, C. F. *Disquisitiones Arithmeticae*. Translation into English by A. A. Clarke. Yale Univ. Press, New Haven, 1966.

1830 LEGENDRE, A. M. *Théorie des Nombres*, Vol. II (3e édition). Firmin Didot, Paris, 1830. Reprinted by A. Blanchard, Paris 1955.

1844 CATALAN, E. Note extraite d'une lettre adressée à l'éditeur. J. f. d. reine u. angew. Math. 27 (1844), 192.

1850 LEBESGUE, V. A. Sur l'impossibilité en numbres entiers de l'équation $x^m = y^2 + 1$. Nouv. Ann. de Math. 9 (1850), 178–181.

1853 LEBESGUE, V. A. Résolution des équations biquadratiques (1)(2) $x^2 = x^4 \pm 2^m y^4$, (3) $z^2 = 2^m x^4 - y^4$, (4)(5) $2^m z^2 = x^4 \pm y^4$. J. Math. Pures et Appl. 18 (1853), 73–86.

1870/1 GERONO, G. C. Note sur la résolution en nombres entiers et positifs de l'équation $x^m = y^n + 1$. Nouv. Ann. de Math. (2) 9 (1870), 469–471, and 10 (1871), 204- -206.

1872 BACHMANN, P. *Zahlentheorie*, Vol. III. *Die Lehre von der Kreistheilung.* B. G. Teubner, Leipzig, 1872. Reprinted by Johnson Reprint Corp., New York, 1968.

1876 MORET-BLANC Réponse à la question 1175. Nouv. Ann. de Math. (2) 15 (1876), 44–46.

1879 LIOUVILLE, R. Sur l'impossibilité de la relation algébrique $X^n + Y^n + Z^n = 0$. C. R. Acad. Sci. Paris 87 (1879), 1108–1110.

1880 KORKINE, A. Sur l'impossibilité de la relation algébrique $X^n + Y^n + Z^n = 0$. C. R. Acad. Sci. Paris 90 (1880), 303–304.

1883 GENOCCHI, A. Démonstration d'un théorème de Fermat. Nouv. Ann. de Math. (3) 2 (1883), 306–310.

1885 CATALAN, E. Quelques théorèmes empiriques. (Mélanges Mathématiques, XV). Mém. Soc. Royale Sci. de Liège (2) 12 (1885), 42–43.

1885 HEATH, T. L. *Diophantus of Alexandria, A Study in the History of Greek Algebra.* Cambridge Univ. Press. Cambridge, 1885. Reprinted by Dover, New York, 1964.

1885 MANSION, P. Discours sur les travaux mathématiques de M. Eugène Charles Catalan. Mém. Soc. Royale Sci. de Liège (2) 12 (1885), 1–38.

1886 BANG, A. S. Taltheoretiske Untersogelser. Tidskrift f. Math. (5) 4 (1886), 70-80 and 130–137.

1892 MATHEWS, G. B. *Theory of Numbers.* Deighton Bell, Cambridge, 1892. Reprinted by Chelsea Publ. Co., New York, 1961.

1892 ZSIGMONDY, K. Zur Theorie der Potenzreste. Monatsh. f. Math. 3 (1892), 265–284.

1894 GRAVÉ, D. Question 377. Interm. des Math. 1 (1894), 228.

1895 STØRMER, C. Solution complète en nombres entiers de l'équation m arctan $\frac{1}{x} + n$ arctan $\frac{1}{y} = k\frac{\pi}{4}$. Christiania Vidensk. Selsk. Skrifter, Ser. I, 1895.

1896 STØRMER, C. Sur l'application de la théorie des nombres entiers complexes à la solution en nombres rationnels $x_1, x_2, \ldots, x_n, c_1, c_2, \ldots, c_n, k$ de l'équation c_1 arctg $\frac{1}{x_1}$ + c_2 arctg $\frac{1}{x_2} + \cdots + c_n$ arctg $\frac{1}{x_n} = k\pi/4$. Archiv for Math. og Naturvidenskab 19, N^O. 3 (1896), 95 pages.

1897 STØRMER, C. Quelques théorèmes sur l'équation de Pell $x^2 - Dy^2 = \pm 1$ et leurs applications. Christiania Vidensk. Selskab Skrifter, (I) 1897, N^O. 2, 48 pages.

1899 STØRMER, C. Solution complète en nombres entiers de l'équation m arc tang $\frac{1}{x} + n$ arc tang$\frac{1}{y} = k\frac{\pi}{4}$. Bull. Soc. Math. France 27 (1899), 160–170.

1904 BIRKHOFF, G. D. & VANDIVER, H. S. On the integral divisors of $a^n - b^n$. Annals of Math. (2) 5 (1904), 173–180.

1905 DICKSON, L. E. On the cyclotomic function. Amer. Math. Monthly 12 (1905), 86-89. Reprinted in *Collected Math. Papers* (edited by A. A. Albert), Vol. 3, 136–139. Chelsea Publ., New York, 1975.

1908 STØRMER, C. Solution d'un problème curieux qu'on rencontre dans la théorie élémentaire des logarithmes. Nyt Tidskrift f. Mat. (Copenhagen) B 19, 1908, 1–7.

1908a THUE, A. Bemerkungen über gewisse Näherungsbrüche algebraischer Zahlen. Christiania Vidensk. Selskab Skrifter, (I) 1908, N^O. 3, 31 pages. Reprinted in *Selected Mathematical Papers*, 159-180. Universitetsforlaget, Oslo, 1982.

1908b THUE, A. Om en generel i store hele tal ulösbar ligning. Christiania Vidensk. Selskab Skrifter, (I), 1908, N^O. 7, 15 pages. Reprinted in *Selected Mathematical Papers*, 219–231. Universitetsforlaget, Oslo, 1982.

1909 CARMICHAEL, R. D. Problem 155 (proposed and solved by R. D. Carmichael). Amer. Math. Monthly 16 (1909), 38–39.

1909 THUE, A. Über Annäherungswerte algebraische Zahlen. J. f. d. reine u. angew. Math. 135 (1909), 284–305. Reprinted in *Selected Mathematical Papers*, 232–253. Universitetsforlaget, Oslo, 1982.

1909 WIEFERICH, A. Zum letzten Fermat'schen Theorem. J. f. d. reine u. angew. Math. 136 (1909), 293–302.

1912 SCHUR, I. Über die Existenz unendlich vieler Primzahlen in einiger speziellen arithmetischen Progressionen. Sitzungsber. Berliner Math. Ges. 11, 1912, 40–50.

1913 CARMICHAEL, R. D. On the numerical factors of arithmetical forms $\alpha^n \pm \beta^n$. Annals of Math. 15 (1913), 30–70.

1913 MORDELL, L. J. The diophantine equation $y^2 - k = x^3$. Proc. London Math. Soc. 13 (1913), 60–80.

1913 PERRON, O. *Die Lehre von den Kettenbrüche*. B. G. Teubner, Leipzig, 1913, (2nd edition 1929). Reprinted by Chelsea Publ. C., New York, 1950.

1917 THUE, A. Über die Unlösbarkeit der Gleichung $ax^2 + bx + c = dy^n$ in großen ganzen Zahlen x und y. Archiv f. Math. og Naturvidenskab 34, N° 16 (1917), 4 pages. Reprinted in *Selected Mathematical Papers*, 561–564. Universitetsforlaget, Oslo, 1982.

1918 PÓLYA, G. Zur arithmetischen Untersuchung der Polynome. Math. Z. 1 (1918), 143–148.

1919 BACHMANN, P. *Das Fermatproblem in seiner bisherigen Entwicklung*. W. de Gruyter, Berlin, 1919. Reprinted by Springer-Verlag, Berlin, 1976.

1920 DICKSON, L. E. *History of the Theory of Numbers*, Vol. II. Carnegie Institution, Washington, 1920. Reprinted by Chelsea Publ. Co., New York, 1971.

1920 LANDAU, E. & OSTROWSKI, A. On the diophantine equation $ay^2 + by + c = dx^n$. Proc. London Math. Soc. (2) 19, (1920), 276–280.

1920 NAGELL, T. Note sur l'équation indéterminée $\frac{x^n-1}{x-1} = y^q$. Norske Mat. Tidsskrift 2 (1920), 75–78.

1921a NAGELL, T. Des équations indéterminées $x^2 + x + 1 = y^n$ et $x^2 + x + 1 = 3y^n$. Norsk Mat. Forenings Skrifter Ser. I (1921), N°. 2, 14 pages.

1921b NAGELL, T. Sur l'équation indéterminée $\frac{x^n-1}{x-1} = y^2$. Norsk Mat. Forenings Skrifter, Ser. I, (1921), N°. 3, 17 pages.

1921c NAGELL, T. Sur l'impossibilité de l'équation indéterminée $z^p + 1 = y^2$. Norsk Mat. Forenings Skrifter, Ser. I, 1921, N°. 4, 10 pages.

1921 SIEGEL, C. L. Über den Thueschen Sátz. Vidensk. Selskab Skrifter (Kristiania) Ser. I 16 (1921). Reprinted in *Gesammelte Abhandlungen*, Vol. I, 103–112. Springer-Verlag, Berlin, 1966.

1922a DELONE, B. N. On the number of representations of a number by a cubic binary form with negative determinant. Izv. Akad. Nauk SSSR (6) 16, (1922), 253–272. German translation in Math. Z. 31 (1930), 1–26.

1922b DELONE, B. N. Solution of the indeterminate equation $X^3 q + Y^3 = 1$ (in Russian). Izv. Akad. Nauk SSSR (6) 16 (1922), 273–280. German translation in Math. Z. 28 (1928), 1–9.

1922 NAGELL, T. Résultats nouveaux de l'analyse indéterminée, I. Norsk Mat. Forenings Skrifter, Ser. I (1922), N°. 8, 19 pages.

1923 FRANKLIN, P. Problem 2927. Amer. Math. Monthly 30 (1923), 81.

1924 NAGELL, T. Über die rationale Punkte auf einigen kubischen Kurven. Tôhoku Math. J. 24 (1924), 48–53.

1925a NAGELL, T. Solution complète de quelques équations cubiques à deux indéterminées. J. Math. Pures et Appl. (9) 4 (1925), 209–270.

1925b NAGELL, T. Uber einige kubische Gleichungen mit zwei Unbestimmten. Math. Z. 24 (1925), 422–447.

1926 TARTAKOWSKY, W. Auflösung der Gleichung $x^4 - 2y^4 = 1$. Bull. Acad. Sci. URSS 20 (1926), 301–324.

1929 SIEGEL, C. L. Über einige Anwendungen diophantischer Approximationen. Abh. Preussischen Akad. d. Wiss. zu Berlin, Phys. Math. Kl. I (1929). Reprinted in *Gersammelte Abhandlungen*, Vol. I, 209–266. Springer-Verlag, Berlin, 1966.

1930 FUETER, R. Über kubische diophantische Gleichungen. Comm. Math. Helv. 2 (1930), 69–89.

1931 PILLAI, S. S. On the inequality $0 < a^x - b^y \leq n$. Indian J. Math. Soc. 19 (1931), 1–11.

1932 SELBERG, S. The diophantine equation $x^4 = 1 + y^n$, with $n > 1$, $|x| > 1$ is impossible in integers. Norsk Mat. Tidsskrift 14 (1932), 79-=-80.

1933 MAHLER, K. Über den grössten Primteiler der Polynome. Archiv f. Math. og Naturvidenskab B 41 N^O. 1, (1933), 8 pages.

1933 SKOLEM, T. Einige Sätze über gewisse Reihenentwicklung und exponentiale Bezlehungen mit Anwendung auf diophantische Gleichungen. Chr. Michelsens Inst., Oslo (1933), 61 pages.

1934 MAHLER, K. Eine arithmetische Eigenschaft der kubischen Binärformen. Nieuw Arch. Wiskunde, Groningen (2) 18, N°. 4, (1934), 1–9.

1934 NAGELL, T. Sur une équation diophantienne à deux indéterminées. Det Kong. Norske Vidensk. Selskab Forhandliger, Trondhejm, 1934, N°. 38, 136–139.

1934 SKOLEM, T. Ein Verfahren zur Behandlung gewisser exponentialer Gleichungen und diophantischer Gleichungen. 8te Skand. Mat. Kongr. Förh., Stockholm (1934), 163–188.

1935 ERDÖS, P. & SZEKERES, S. Über die Anzahl der Abelschen Gruppen gegebener Ordnung und über ein verwandtes Zahlentheoretisches Problem. Acta Scient. Math. Szeged 7 (1935), 95 102.

1935 MAHLER, K. Üben den grössten Primteiler spezieller Polynome zweiten Grades. Ark. f. Math. og Naturv. 41, N°. 6, (1935), 26 pages.

1935 NIEWIADOMSKI, R. Zur Fermatschen Vermutung. Prace Mat. Fiz. 42 (1935), 1-10.

1935 SCHEPEL, D. Over de vergeliking van Pell. Nieuw Arch. v. Wisk. (2) 18 (1935), N°. 3, 1–30.

1935 SKOLEM, T. Einige Sätze über p-adische Potenzreihen mit Anwendung auf gewisse exponentielle Gleichungen. Math. Annalen 111 (1935), 399–424.

1936 HERSCHFELD, A. The equation $2^x - 3^y = d$. Bull. Amer. Math. Soc. 42 (1936), 231–234.

1936 PILLAI, S. S. On $a^x - b^y = c$. J. Indian Math. Soc. (New Series), 2, 1936, 119–122, and 215.

1937 SIEGEL, C. L. Die Gleichung $ax^n - by^n = c$. Math. Ann. 114 (1937), 57–68. Reprinted in *Gesammelte Abhandlungen*, Vol. II, 8–19. Springer-Verlag, Berlin, 1966.

1938 LEHMER, D. H. On arc cotangent relations for π. Amer. Math. Monthly 45 (1938), 657–664. Reprinted in *Selected*

Papers, Vol. III, 1074-1081 (edited by D. McCarthy). Ch. Babbage Res. Centre, Winnipeg, Man., 1981.

1938a LJUNGGREN, W. Einige Eigenschaften der Einheitenreeller quadratischer und rein biquadratischer Zahlkörper mit Anwendung auf die Lösung emer Klasse unbestimmter Gleichungen vierten Grades. Det Norske Vidensk. Akad. Oslo Skrifter (I) N° 12 (1936), 73 pages.

1938b LJUNGGREN, W. Über die unbestimmte Gleichung $Ax^2 - By^4 = C$. Archiv f. Math. og Naturvidenskab B 41, N° 10 (1938), 18 pages.

1939 OBLÁTH, R. Über die Zahl $x^2 - 1$. Mathematica A en B, Leyden, Der. B, 8 (1939), 161–172.

1940 DELONE, B. N. & FADDEEV, D. K. *The Theory of Irrationalities of the Third Degree.* Publ. Steklov Institute, Moscow, 1940. English translation by E. Lehmer & S. A. Walker. American Math. Soc., Providence, 1964.

1940 GEL'FOND, A. O. Sur la divisibilité de la différence des puissances de deux nombres entiers par une puissance d'un idéal premier (in Russian). Matem. Sbornik 7 (49) (1940), 7–24.

1940 OBLÁTH, R. Az $x^2 - 1$ Számokról (On the numbers $x^2 - 1$). Mat. és Fiz. Lapok 47 (1940), 58–77.

1941 HAUSMANN, B. A. Problem E444 (proposed by Harry Goheen). Amer. Math. Monthly 48 (1941), 482.

1941 OBLÁTH, R. Sobre ecuaciones diofánticas imposibles de la forma $x^m + 1 = y^n$. Rev. Mat. Hisp. Amer. IV 1 (1941), 122–140.

1942a LJUNGGREN, W. Einige Bemerkungen über die Darstellung ganzer Zahlen durch binäre kubishe Formen mit positiver Diskriminante. Acta Math. 75 (1942), 1–21.

1942b LJUNGGREN, W. Zur Theorie der Gleichung $x^2 + 1 = Dy^4$. Avh. Norsk. Vid. Akad. Oslo (1942), 1–27.

1942c LJUNGGREN, W. Über die Gleichung $x^4 - Dy^2 = 1$. Archv f. Math og Naturvidenskab B 45, N° 6 (1942), 61–70.

1943 LJUNGGREN, W. New propositions about the indeterminate equation $\frac{x^n-1}{x-1} = y^q$ (in Norwegian). Norske Mat. Tidsskrift 25 (1943), 17–20.

1944 HOFMANN, J. E. Neues über Fermats zahlentheoretische Herausforderung von 1657. Abh. Prenβischen Akad. d. Wiss. Berlin N°. 9 (1944), 52 pages.

1944 PILLAI, S. S. On $a^X - b^Y = b^y \pm a^x$. J. Indian Math. Soc. 8 (1944), 10–13.

1944 SKOLEM, T. Utvidelse av et par setninger av C. Størmer. Norsk Matematik Tidsskrift, 26 (1944), 85–95.

1945 PILLAI, S. S. On the equation $2^x - 3^y = 2^X + 3^Y$. Bull. Calcutta Math. Soc. 37 (1945), 18–20.

1945a SKOLEM, T. A method to solve the exponential equation $A_1^{x_1} \ldots A_m^{x_m} - B_i^{y_1} \ldots B_n^{y_n} = C$ (in Norwegian). Norske Mat. Tidsskrift, 27, 1945, 37–51.

1945b SKOLEM, T. A theorem on the equation $\xi^2 - \delta\eta^2 = 1$, where δ, ξ, η are integers in an imaginary quadratic field. Norske Vidensk. Akad. Oslo Avh. 1 (1945), 15 pages.

1945c SKOLEM, T. On certain exponential equations. Det Kong. Norske Vidensk. Selskab Forh. 18, N°. 18 (1945), 71–74.

1947 DYSON, F. J. The approximation to algebraic numbers by rationals. Acta Math. 79 (1947), 225–240.

1947 MORDELL, L. J. On some diophantine equations with no rational solutions. Archiv Math. Natur. 49 (1947), No.6, 143–150.

1950 KANOLD, H. J. Sätze über Kreisteilungspolynome und ihre Anwendungen auf einige zahlentheoretische Probleme, II. J. f. d. reine u. angew. Math. 187 (1950), 355–366.

1951 LJUNGGREN, W. On the diophantine equation $x^2 + 4 = Ay^4$. Det Kong. Norske Vidensk. Selskab Forhandliger 24, N° 18 (1951), 82–84.

1952 ANKENY, N. C., ARTIN, E. & CHOWLA, S. The class number of real quadratic number fields. Annals Math. 56 (1952), 479–493.

1952 GLODEN, A. Sur un problème de Catalan. Mathesis 61 (1952), 302–303.

1952 HEMER, O. On the diophantine equation $y^2 - k = x^3$. Doctoral Dissertation, Uppsala, 1952.

1952 LeVEQUE, W. J. On the equation $a^x - b^y = 1$. Amer. J. Math. 74 (1952), 325–331.

1952 STOLT, B. On the Diophantine equation $u^2 - Dv^2 = \pm 4N$, Parts I, II. Arkiv f. Mat. 2 (1952), 1–23 and 251–268.

1953 CASSELS, J. W. S. On the equation $a^x - b^y = 1$. Amer. J. Math. 75 (1953), 159–162.

1953 GLODEN, A. Histoire du "Problème de Catalan." Actes du 7e Congrès Intern. d'Histoire des Sciences, Jerusalém, (1953), 316–319.

1953 LJUNGGREN, W. On an improvement of a theorem of T. Nagell concerning the diophantine equation $Ax^3 + By^3 = C$. Math. Scand. 1 (1953), 297–309.

1953 MAHLER, K. On the greatest prime factor of $ax^m + by^n$. Nieuw Arch. Wisk. (3) 1 (1953), 113–122.

1953 NAGELL, T. On a special class of Diophantine equations of the second degree. Arkiv f. Mat., 3, 1953, 51–65.

1954 BATEMAN, P. T. Solution of problem 4459. Amer. Math. Monthly 61 (1954), 477–479.

1954 DOMAR, Y. On the diophantine equation $|Ax^n - By^n| = 1$, $n \geq 5$. Math. Scand., 2 (1954), 29–32.

1954 HEMER, O. Notes on the diophantine equation $y^2 - k = x^3$. Ark. f. Mat. 3 (1954), 67–77.

1954 OBLÁTH, R. Über die Gleichung $x^m + 1 = y^n$. Ann. Polon. Math. 1 (1954), 73–76.

1954 STOLT, B. On the Diophantine equation $u^2 - Dv^2 = \pm 4N$, Part III. Arkiv f. Mat. 3 (1954), 117–132.

1955 ARTIN, E. The orders of the linear group. Comm. Pure & Appl. Math. 8 (1955), 355–365. Reprinted in *Collected Papers*, (edited by S. Lang & J. Tate), 387–397. Addison-Wesley, Reading, Mass., 1965.

1955 DAVENPORT, H. & ROTH, K. F. Rational approximation to algebraic numbers. Mathematika 2 (1955), 160–167.

1955 NAGELL, T. Contribution to the theory of a category of diophantine equations of the second degree with two unknowns. Nova Acta Reg. Soc. Scient. Upsaliensis, (IV), 16 N°. 2 (1955), 37 pages.

1955 ROTH, K. F. Rational approximations to algebraic numbers. Mathematika 2 (1955), 1–20 and 168.

1956 HAMPEL, R. On the solution in natural numbers of the equation $x^m - y^n = 1$. Ann. Polon. Math. 3 (1956), 1–4.

1956 LEVEQUE, W. J. *Topics in Number Theory*. Vol. II. Addison-Wesley, Reading, Mass., 1956.

1956 OBLÁTH, R. Une propriété des puissances parfaites. Mathesis 65 (1956), 356–364.

1956 ROTKIEWICZ, A. Sur l'équation $x^z - y^t = a^t$ où $|x-y| = a$. Ann. Polon. Math. 3 (1956), 7–8.

1956 SCHINZEL, A. Sur l'équation $x^z - y^t = 1$ où $|x - y| = 1$. Ann. Polon. Math. 3 (1956), 5–6.

1957 WALL, C. T. C. A theorem on prime powers. Eureka 19 (1957), 10–11.

1958 BATEMAN, P. T. & GROSSWALD, E. On a theorem of Erdös & Szekeres. Illinois J. Math. 2 (1958), 88–98.

1958 NAGELL, T. Sur une classe d'équations exponentielles. Arkiv f. Mat. 3 (1958), 569–581.

1958 SCHINZEL, A. & SIERPIŃSKI, W. Sur certaines hypothèses concernant les nombres premiers. Acta Arithm. 4 (1958), 185-208.

1959 LANDAU, E. *Diophantische Gleichungen mit endlich vielen Lösungen* (new edition by A. Walfisz). V. E. B. Deutscher Verlag d. Wiss., Berlin, 1959.

1960 CASSELS, J. W. S. On a class of exponential equations. Arkiv f. Mat. 4 (1960), 231–233.

1960 GEL'FOND, A. O. *Transcendental and Algebraic Numbers*. Dover, New York, 1960.

1960 HAMPEL, R. O zagadnieniu Catalana (On the problem of Catalan). Roczniki Polsk. Towarzystwa Matem. I (Prace Matem.), 4 (1960), 11–19.

1960 KO, CHAO. On the diophantine equation $x^2 = y^n + 1$. Acta Sci. Natur. Univ. Szechuan 2 (1960), 57–64.

1960 MORDELL, L. J. On a Pellian equation conjecture. Acta Arithm. 6 (1960), 137–144.

1960 ROTKIEWICZ, A. Sur le problème de Catalan. Elem. d. Math. 15 (1960), 121–124.

1960 SIERPIŃSKI, W. On some unsolved problems of arithmetics. Scripta Math. 25 (1960), 125–136.

1960 WRENCH, Jr., J. W. The evolution of extended decimal approximations to π. Math. Teacher 53 (1960), 644–650.

1961 CASSELS, J. W. S. On the equation $a^x - b^y = 1$, II. Proc. Cambridge Phil. Soc. 56 (1961), 97–103.

1961 INKERI, K. & HYYRÖ, S. On the congruence $3^{p-1} \equiv 1$ (mod p^2) and the diophantine equation $x^2 - 1 = y^p$. Ann. Univ. Turku, Ser. AI, N°. 50 (1961), 2 pages.

1961 ROTKIEWICZ, A. Sur le problème de Catalan, II. Elem. d. Math. 16 (1961), 25–27.

1962 MĄKOWSKI, A. Three consecutive integers cannot be powers. Colloq. Math. 9 (1962), 297.

1962 SHANKS, D. *Solved and Unsolved Problems in Number Theory*, Vol. 1. Spartan Books, Washington, 1942. 3rd edition by Chelsea Publ., New York 1985.

1963 FERENTINOU-NICOLACOPOULOU, J. Une propriété des diviseurs du nombre $r^{r^m} + 1$. Applications au dernier théorème de Fermat. Bull. Soc. Math. Grèce 4 (1964), 121–126.

1963 GUT, M. Abschätzungen für fie Klassenzahlen der quadratischen Körper. Acta Arith. 8 (1963), 113–122.

1963 HYYRÖ, S. On the Catalan problem (in Finnish). Arkhimedes (1963), No. 1, 53–54 (see Math. Reviews, 28, 1964, #62).

1964 COHN, J. H. E. Square Fibonacci numbers, etc. Fibonacci Q. 2 (1964), 109–113.

1964a HYYRÖ, S. Über das Catalansche Problem. Ann. Univ. Turku, Ser. AI, N°. 79 (1964), 8 pages.

1964b HYYRÖ, S. Über die Gleichung $ax^n - by^n = c$ und das Catalansche Problem. Annales Acad. Sci. Fennicae, Ser. AI, N°. 355 (1964), 50 pages.

1964 INKERI, K. On Catalan's problem. Acta Arithm. 9 (1964), 285–290.

1964 INKERI, K. & HYYRÖ, S. Über die Anzahl der Lösungen einiger diophantischer Gleichungen. Ann. Univ. Turku, Ser. AI, N°. 78 (1964), 7 pages.

1964 KO, CHAO. On the diphantine equation $x^2 = y^n + 1$. Scientia Sinica (Notes) 14 (1964), 457–460.

1964 LEVEQUE, W. J. On the equation $y^m = f(x)$. Acta Arithm. 9 (1964), 209–219.

1964 MORDELL, L. J. The Diophantine equation $y^2 = Dx^4+1$. J. London Math. Soc. 39 (1964), 161–164.

1964 NAGELL, T. *Introduction to Number Theory.* Chelsea Publ. Co., New York, 1964.

1964 RIESEL, H. Note on the congruence $a^{p-1} \equiv 1 \pmod{p^2}$. Math. Comp. 18 (1964), 149–150.

1965 BIRCH, B. J., CHOWLA, S., HALL Jr., M. & SCHINZEL, A. On the difference $x^3 - y^2$. Norske Videnskaber Selskab Forhandliger (Trondheim) 38 (1965), 65–69.

1965 COHN, J. H. E. Lucas and Fibonacci numbers and some diophantine equations. Proc. Glasgow Math. Assoc. 7 (1965), 24–29.

1965 ROTKIEWICZ, A. Sur les nombres de Mersenne dépourvus de diviseurs carrés et sur les nombres naturels n tels que $n^2|2^n - 2$. Matematicky Vesnik (Beograd) (2) 17 (1965), 78–80.

1966 BOREVICH, Z. I. & SHAFAREVICH, I. R. *Number Theory.* Academic Press, New York, 1966.

1966 COHN, J. H. E. Eight diophantine equations. Proc. London Math. Soc. (3) 16 (1966) 153–166 and (3) 17 (1967), 381.

1966 LJUNGGREN, W. Some remarks on the diophantine equations $x^2 - Dy^4 = 1$ and $x^4 - Dy^2 = 1$. J. London Math. Soc. 41 (1966), 542–544.

1966 MORDELL, L.J. The infinity of rational solutions of $y^2 = x^3 + k$. J. London Math. Soc. 41 (1966), 523–525.

1967 BUMBY, R. T. The diophantine equation $3x^4 - 2y^2 = 1$. Math. Scand. 21 (1967), 144–148.

1967a COHN, J. H. E. The diophantine equation $y^2 = Dx^4 + 1$. J. London Math. Soc. 40 (1967), 475–476.

1967b COHN, J. H. E. Five diophantine equations. Math. Scand. 21 (1967), 61–70.

1967 LJUNGGREN, W. On the diophantine equation $Ax^4 - By^2 = C$ ($C = 1, 4$). Math. Scand. 21 (1967), 149-197-158.

1967 SURYANARAYANA, D. Certain diophantine equations. Math. Student 35 (1967), 197–199.

1967 WALKER, D. T. On the diophantine equation $mX^2 - nY^2 = \pm 1$. Amer. Math. Monthly 74 (1967), 504–513.

1967 WARREN, L. & BRAY, H. On the square-freeness of Fermat and Mersenne numbers. Pacific J. Math. 22 (1967), 563–564.

1968 COHN, J. H. E. Some quartic diophantine equations. Pacific J. Math. 26 (1968), 233–243.

1968 MORDELL, L. J. The diophantine equation $y^2 = Dx^4 + 1$. Nb. Th. Coll. János Boliay Math. Soc., Debrecen (1968), 141–145. North-Holland, Amsterdam, 1970.

1969 GREENLEAF, N. On Fermat's equation in $C(t)$. Amer. Math. Monthly 76 (1969), 808–809.

1969 HOFMANN, J. E. Sobre un problema de Fermat enteoría de números. Rev. Mat. Hisp.-Amer. (4) 29 (1969), 13–50.

1969 MORDELL, L. J. *Diophantine Equations*. Academic Press, New York, 1969.

1970 GOLOMB, S. W. Powerful numbers. Amer. Math. Monthly 77 (1970), 848–852.

1970 SURYANARAYANA, D. Problems in the theory of numbers. Bull. Amer. Math. Soc., 76, 1970, 977.

1971 BEACH, B. D., WILLIAMS, H. C. & ZARNKE, C. R. Computer results on units and quadratic and cubic fields. Proc. 25th. Summer Meeting Can. Math. Congress, Lakehead University (1971), 609–648.

1971 BRILLHART, J., TONASCIA, J. & WEINBERGER, P. On the Fermat quotient. In *Computers in Number Theory* (edited by A. O. L. Atkin & B. J. Birch), 213–222. Academic Press, New York, 1971.

1971 EDGAR, H. On a theorem of Suryanarayna. Proc. Washington State Univ. Conf. Number Theory, Pullman (1971), 52–54.

1971 HALL Jr., M. The diophantine equation $x^3 - y^2 = k$. In *Computers in Number Theory* (edited by A. O. L. Atkin and B. J. Birch), 173-198. Academic Press, Boston, 1971.

1972 INKERI, K. On the diophantine equation $a(x^n - 1)/(x - 1) = y^m$. Acta Arithm. 21 (1972), 299–311.

1972 KRÄTZEL, E. Zahlen k-ter Art. Amer. J. Math. 94 (1972), 309–328.

1972 RIBENBOIM, P. *Algebraic Numbers*. Wiley-Interscience, New York, 1972.

1973 BAKER, A. A sharpening of the bounds for linear forms in logarithms, II. Acta Arithm. 24 (1973), 33–36.

1973 LONDON, H. & FINKELSTEIN, R. (alias STEINER, R.). *On Mordell's equation $y^2 - k = x^3$*. Bowling Green State Univ. Press, Bowling Green, OH, 1973.

1973 MOHANTY, S. P. A note on Mordell's equation $x^2 = y^3 + k$. Proc. Amer. Math. Soc. 39 (1973), 645–646.

1973 STARK, H. M. Effective estimates of solutions of some diophantine equations. Acta Arithm. 24 (1973), 251–259.

1973 TIJDEMAN, R. On integers with many small prime factors. Compositio Math. 26 (1973), 319–330.

1974 HERING, C. Transitive linear groups and linear groups which contain irreducible subgroups of prime order. Geom. Ded. 2 (1974), 425–460.

1974 LANGEVIN, M. Sur la fonction plus grand facteur premier. Sém. Delange-Pisot-Poitou, 16e. année (1974/5), No. G3/7/10/12/22, 22 pages.

1974 NARKIEWICZ, W. *Elementary and Analytic Theory of Algebraic Numbers*. Polish Scientific Publishers, Warszawa, 1974.

1974 NATHANSON, M. B. Catalan's equation in $K(t)$. Amer. Math. Monthly 81 (1974), 371–373.

1974 STOLARSKY, K. B. *Algebraic Numbers and Diophantine Approximation*. M. Dekker, New York, 1974.

1974 TANG SHEN CHI. An alternative proof of diophantine equation $x^p - z^2 = 1$, with trivial solution $x = 1, z = 0$. Rev. Mat. Hisp.-Amer. (4) 34 (1974), 260–262.

1975 BAKER, A. *Transcendental Number Theory*. Cambridge University Press, Cambridge, 1975.

1975 ERDÖS, P. Problems and results on consecutive integers. Eureka 38 (1975), 3–8.

1975 LANG, S. La conjecture de Catalan, d'après R. Tijdeman Acta Arithm. 29 (1976), No. 2, 197-209). Sém. Delange-Pisot-Poitou, 17e année, (1975/76), exposé 29, 9 pages. Institut Henri Poincaré, Paris, 1976.

1975 LANGEVIN, M. Méthodes élémentaires en vue du théorème de Sylvester. Sém. Delange-Pisot-Poitou, 17e. année, (1975/6), N°. G2, 9 pages. Institut Henri Poincaré, Paris, 1976.

1976 CHEIN, E. Z. A note on the equation $x^2 = y^q + 1$. Proc. Amer. Math. Soc. 56 (1976), 83–84.

1976 CHEIN, E. Z. Some remarks on the exponential diophantine equation. Notices Amer. Math. Soc. 26 (1976), A-426.

1976 ERDÖS, P. Problems and results on consecutive integers. Publ. Math. Debrecen 23 (1976), 271–282.

1976a LANGEVIN, M. Quelques applications de nouveaux résultats de van der Poorten. Sém. Delange-Pisot-Poitou, 17e année (1975/76), N°. G 12, 11 pages. Institut Henri Poincaré, Paris, 1976.

1976b LANGEVIN, M. Bornes effectives pour l'équation de Catalan (d'après R. Tijdeman). Unpublished manuscript, 1976.

1976 SALAMIN, E. Computation of π using arithmetic geometric mean. Math. Comp. 30 (1976), 565–570.

1976 SHOREY, T. N. & TIJDEMAN, R. New applications of diophantine approximation to diophantine equations. Math. Scand. 39 (1976), 5–18.

1976 TIJDEMAN, R. On the equation of Catalan. Acta Arithm. 29 (1976), 197-209.

1976 WALKER, D. T. Consecutive integer pairs of powerful numbers and related diophantine equations. Fibonacci Q. 11 (1976), 111–116.

1977 BAKER, A. The theory of linear forms in logarithms. In *Transcendence Theory: Advances and Applications* (edited by A. Baker and D. W. Masser), 1-27. Academic Press, New York, 1977.

1977 McCALLUM, W. *Consecutive Perfect Powers.* Honours Project, Univ. New South Wales, Sydney, 1977, 54 pages (unpublished manuscript).

1977 WEIL, A. Fermat et l'équation de Pell. Prismata (W. Hartner Festschrift), 441-448. Fr. Steiner Verlag, Wiesbaden, 1977. Reprinted in *Oeuvres Scientifiques*, Vol. III, 413–420. Springer-Verlag, New York, 1979.

1978 DUGAC, P. L'Équation de Catalan. Paris, 1978, 19 pages (unpublished manuscript).

1978 NAIR, M. A note on the equation $x^2 - y^3 = k$. Quart. J. Math. Oxford (2) 29 (1978), 483–487.

1978 PERLIS, R. On the class numbers of arithmetically equivalent fields. J. Nb. Th. 10 (1978), 489–459.

1979a RIBENBOIM, P. *13 Lectures on Fermat's Last Theorem.* Springer-Verlag, New York, 1979.

1979b RIBENBOIM, P. On the square factors of the numbers of Fermat and Ferentinou-Nicolacopoulou. Bull. Soc. Math. Grèce 20 (1979), 81–92.

1980 BALASUBRAMANIAN, R. & SHOREY, T. N. On the equation $a(x^m - 1)/(x - 1) = b(y^n - 1)/(y - 1)$. Math. Scand. 46 (1980), 177–182.

1980 — Numéro Spécial π, Supplément au *Petit Archimède* 64–65, (1980), 289 pages.

1981 LÜNEBURG, H. Ein einfacher Beweis für den Satz von Zsigmondy über primitive Primteiler von $A^n - 1$. In *Geometries and Groups*, (edited by M. Aigner & D. Jungnickel). Lect. Notes in Math. #893, 219–222. Springer-Verlag, New York, 1981.

1981 ODONI, R. W. K. On a problem of Erdös on sums of two squarefull numbers. Acta Arith. 39 (1981), 145–162.

1982 EVERTSE, J. H. On the equation $aX^n - bY^n = c$. Comp. Math. 47 (1982), 289–315.

1982 HUA, L. K. *Introduction to Number Theory*. Springer-Verlag, New York, 1982.

1982 IVIĆ, A. & SHIU, P. The distribution of powerful integers. Illinois J. Math. 26 (1982), 576–590.

1982 SENTANCE, W. A. Occurrences of consecutive odd powerful numbers. Amer. Math. Monthly 88 (1981) 272–274.

1982 SILVERMAN, J. H. The Catalan equation over function fields. Trans. Amer. Math. Soc. 273 (1982), 201–205.

1983 EVERTSE, J. H. *Upper Bounds for the Numbers of Solutions of Diophantine Equations*. Thesis, Amsterdam, 1983.

1983a GURALNICK, R. Subgroup of prime power index in a simple group. J. Alg. 81 (1983), 304–311.

1983b GURALNICK, R. Subgroups inducing the same permutation representation. J. Alg. 81 (1983), 312–319.

1984a RIBENBOIM, P. Consecutive powers. Expo. Math. 2 (1984), 193–221.

1984b RIBENBOIM, P. Remarks on existentially closed fields and diophantine equations. Rend. Sem. Mat. Univ. Padova 71 (1984), 229–237.

1984 SHOREY, T. N. On the equation $a(x^m-1)/x-1 = b(y^n-1)/(y-1)$, (II). Hardy-Ramanijan J. 7 (1984), 1–10.

1985 ADLEMAN, L. M. & HEATH-BROWN, D. R. The first case of Fermat's last theorem. Invent. Math. 79 (1985), 409–416.

1985 EDGAR, H. Problems and some results concerning the diophantine equation $1 + a + a^2 + \cdots + a^{x-1} = p^y$. Rocky Mountain J. Math. 15 (1985), 325–327.

1985 ESTES, D., GURALNICK, R., SCHACHER, M., STRAUS, E. Equations in prime powers. Pacific J. Math. 118 (1985), 359–367.

1985 FOUVRY, E. Théorème de Brun-Titchmarsh: applications au théorème de Fermat. Invent. Math. 79 (1985), 383–407.

1985 GOLDSTEIN, B. R. *The Astronomy of Levi ben Gerson*. Springer-Verlag, New York, 1985.

1985 GRANVILLE, A. J. The set of exponents for which Fermat's last theorem is true has density one. C. R. Math. Rep. Acad. Sci. Canada 7 (1985), 55–60.

1985 HEATH-BROWN, D. R. Fermat's last theorem for "almost all" exponents. Bull. London Math. Soc. 17 (1985), 15–16.

1986 BORWEIN, J. M. & BORWEIN, P. B. *Pi and the AGM*. J. Wiley & Sons, New York, 1986.

1986 BRINDZA, B., GYÖRY, K. & TIJDEMAN, R. On the Catalan equation over algebraic number fields. J. f. d. reine u. angew. Math. 367 (1986), 90–102.

1986 GRANVILLE, A. Powerful numbers and Fermat's last theorem. C. R. Math. Rep. Acad. Sci. Canada 8 (1986), 215–218.

1986 McDANIEL, W. L. Représentation comme la différence de nombres puissants non-carrés. C. R. Math. Rep. Acad. Aci. Canada 8 (1986), 53–57.

1986a MOLLIN, R. A. & WALSH, P. G. A note on powerful numbers, quadratic fields and the Pellian. C. R. Math. Rep. Acad. Sci. Canada 8 (1986), 109–114.

1986b MOLLIN, R. A. & WALSH, P. G. On powerful numbers. Intern. J. Math. & Math. Sci. 9 (1986), 801–806.

1986a RIBENBOIM, P. A note on Catalan's equation. J. Nb. Th. 24 (1986), 245–248.

1986b RIBENBOIM, P. Some fundamental methods in the theory of diophantine equations. In *Aspects of Mathematics and Its Applications* (edited by J. A. Barroso), 635–663. Elsevier Sci. Publ., Amsterdam, 1986.

1986 SHOREY, T. N. On the equation $z^q = (x^n - 1)/(x - 1)$. Indag. Math. A 89 (3) (1986), 345–351.

1986 SHOREY, T. N. & TIJDEMAN, R. *Exponential Diophantine Equations.* Cambridge Univ. Press, Cambridge, 1986.

1987 BLASS, J., GLASS, A. M. W., MERONK, D. B. & STEINER, R. P. Practical solutions to Thue equations over the rational integers. Preprint, 1987, Bowling Green State Univ.

1987 DE WEGER, B. M. M. *Algorithms for Diophantine Equations.* Thesis, Univ. Twente, 1987. Reprinted in C. W. I. Tract No. 65 Amsterdam, 1989.

1987 McDANIEL, W. L. Representations of every integer as the difference of nonsquare powerful numbers. Port. Math. 44 (1987) 69–75.

1987 MOLLIN, R. A. The power of powerful numbers. Intern. J. Math. & Math. Sci. 10 (1987), 125–130.

1987 MOLLIN, R. A. & WALSH, P. G. On nonsquare powerful numbers. Fibonacci Q. 25 (1987), 34–37.

1987 ROTKIEWICZ, A. Note on the diophantine equation $1 + x + x^2 + \cdots + x^n = y^m$. Elem. d. Math. 42 (1987), 76.

1988 BAILEY, D. H. The computation of π to $29,360,000$ decimal digits using Borwein's quartically convergent algorithm. Math. Comp. 50 (1988), 283–296.

1988 CASTELLANOS, D. The ubiquitous π. Math. Mag. 61 (1988), 67–98 and 148–163.

1988 HEATH-BROWN, D. R. Ternary quadratic forms and sums of three square-full numbers. Séminaire de Théorie des Nombres, Paris 1986-87 (edited by C. Goldstein), 137-163. Birkhäuser, Boston, 1988.

1988 MOLLIN, R. A. & WALSH, P. G. Proper differences of non-square powerful numbers. C. R. Math. Rep. Acad. Sci. Canada 10 (1988), 71–76.

1988 PHILIPPON, P. & WALDSCHMIDT, M. Lower bounds for linear forms in logarithms. In *New Advances in Transcendence Theory* (edited by A. Baker), 280–312. Cambridge Univ. Press, Cambridge, 1988.

1988 RIBENBOIM, P. Impuissants devant les puissances. Expo. Math. 6 (1988), 3–28.

1988 STEPHENS, A. J. & WILLIAMS, H. C. Some computation results on a problem concerning powerful numbers. Math. Comp. 50 (1988), 619–632.

1988 WALSH, P. G. *The Pell Equation and Powerful Numbers*. M.Sc. Thesis, Univ. Calgary, Calgary, 1988, 117 pages.

1988 WÜSTHOLZ, G. A new approach to Baker's theorem on linear forms in logarithms, III. In *New Advances in Transcendence Theory* (edited by A. Baker), 399–410. Cambridge Univ. Press, Cambridge, 1988.

1989a TZANAKIS, N. & DE WEGER, B. M. M. On the practical solution of the Thue equation. J. Nb. Th. 31 (1989), 99–132.

1989b TZANAKIS, N. & DE WEGER, B. M. M. Solving a specific Thue-Mahler equation. Memorandum N°. 793, Univ. Twente, 1989.

1990 BLASS, J., GLASS, A. M. W., MANSKI, D. K., MERONK, D. B. & STEINER, R. P. Constants for lower bounds for linear forms in the logarithms of algebraic numbers, I and II. Acta Arithm. 55 (1990), 1–22.

1990 INKERI, K. On Catalan's conjecture. J. Nb. Th. 34 (1990), 142–152.

1990 MIGNOTTE, M. & WALDSCHMIDT, M. Linear forms in two logarithms and Schneider's method, III. Ann. Fac. Sci. Toulouse (V), special issue (1990), 43–75.

1990 RIBENBOIM, P. Gauss and the class number problem. Symposia Gaussiana, I (1990), 3–63. Institutum Gaussianum, Toronto, 1990.

1990a WALDSCHMIDT, M. Nouvelles méthodes pour minorer des combinaisons linéaires de nombres algébriques. Sém. Th. Nombres Bordeaux 3 (1990/1), 129–185.

1990b WALDSCHMIDT, M. Sur l'équation de Pillai et la différence entre deux produits de puissances de nombres entiers. C. R. Math. Rep. Acad. Sci. Canada 12 (1990), 173–178.

1991 AALTONEN, M. & INKERI, K. Catalan's equation $x^p - y^q = 1$. Math. Comp. 56 (1991), 359–370.

1992 BAKER, A. The evolution of transcendence theory. Mitt. Math. Ges. Hamburg 12 (1991, 513-552).

1991 GLASS, A. M. W., MERONK, D. B., OKADA, T. & STEINER, R. P. A small contribution to Catalan's equation. Preprint, to appear.

1992 McDANIEL, W. L. & RIBENBOIM, P. Squares and double squares in Lucas sequences. C. R. Math. Rep. Acad. Sci. Canada 14 (1992), ...

1992a MIGNOTTE, M. Sur l'équation de Catalan. Comptes Rendus Acad. Sci. Paris 314, Sér. 1, (1992), 165-168.

1992b MIGNOTTE, M. Sur l'équation de Catalan, II. Preprint, to appear.

1992 RIBENBOIM, P. Density results on parametrized families of diophantine equations with finitely many solutions. Preprint, to appear.

1992a WALDSCHMIDT, M. Minoration de combinaisons linéaires de logarithmes de nombres algébriques, Can. J. Math., to appear.

1992b WALDSCHMIDT, M. Linear independence of logarithms of algebraic numbers. Mat. Science Lect. Notes, Madras, 1992.

1993 BAKER, & WÜSTHOLTZ, G. Logarithmic forms and group varieties. Journal f. d. reine u. angew. Math. (1993), to appear.

Index of Names

Aaltonenen, M., 225, 234, 306
Adleman, A. M., 326
Alkarkhi, 60
Allison, D., 282
Ankeny, N. C., 329
Archimedes, 60
Artin, E., 19, 329

Bachet, C. G., 61, 144
Bachmann, P., 76
Bailey, D. H., 135
Baker, A., 7, 245–247, 250, 313
Balasubramanian, R., 121, 297
Bang, A. S., 19
Bateman, P. T., 320, 321
Baudhayana, 60
Beach, B. D., 330
Bháscara Achárya, 60
Birch, B. J., 249

Birkhoff, G. D., 19
Blass, J., 192, 245, 306
Bond, R., 120
Borevich, Z. I., 234
Borwein, J. M., 132, 135
Borwein, P. M., 132, 135
Bouyer, M., 135
Brahmegupta, 60
Brillhart, J., 234
Brindza, B., 313
Brouncker, W., 60, 61
Brûlard de St. Martin, 144
Bumby, R. T., 196

Carmichael, R. D., 19, 202
Cassels, J. W. S., 6, 78, 202, 204–206, 212, 213, 216, 219, 233, 255, 257, 258, 271, 282, 294
Castellanos, D., 132

Catalan, E., 5–6, 89, 94, 202, 214
Chein, E. Z., 6, 89, 92, 93, 279
Chowla, S., 249, 329
Chudnovsky, D., 136
Chudnovsky, G., 136
Clausen, T., 134
Clark, D., 234
Cohn, J. H. E., 193, 194, 196–199
Crelle, A. L., 6

Dase, Z., 134
Davenport, H., 251, 254
Dedekind, R., 121
Delone, B. N., 181, 182
Dickson, L. E., 5, 19, 60
Digby, D., 61
Diophantus, 32, 60, 128, 145
Dirichlet, C. G. L., 187
Domar, Y., 296
Dugac, P. 351
Dyson, F. J., 243

Edgar, H., 121
Erdös, P., 319–326, 329
Estes, D., 96, 121
Euler, L., 5, 11, 43, 60, 67, 69, 75, 76, 80, 134, 147, 149, 151, 152, 156, 166, 282
Evertse, J. H. 183, 296

Faddeev, D. K., 181
Faltings, G., 244
Ferentinov-Nicolacopoulou, J., 213
Fermat, P., 5, 34, 40, 60, 61, 80, 89, 128, 144–147, 152, 164, 166, 326
Finkelstein, R., *see* Steiner, R. P.
Fouvry, E., 326
Franklin, P., 124

Frénicle de Bessy, 60, 61, 89
Fueter, R., 249, 250

Galois, E., 60
Gauss, C. F., 23, 43, 135
Gel'fond, A. O., 243, 294
Genocchi, A., 40
Gérono, G. C., 201, 202, 214, 307
Glass, A. M. W., 192, 245, 306
Gloden, A., 342
Goheen, H., 202
Goldstein, B. R., 124
Golomb, S. W., 319, 320
Goormaghtigh, R., 121
Granville, A., 312, 327
Gravé, D., 128, 132, 135–138, 143
Greenleaf, N., 314
Gregory, J., 133
Grosswald, E., 321
Guilloud, J., 135
Guralnick, R., 96, 121
Gut, M., 221
Györy, K., 313

Hall, M., Jr., 249, 308
Hampel, R., 204, 214, 215
Hausmann, B. A., 202
Heath, T. L., 60
Heath-Brown, D. R., 312, 323, 326
Hemer, O., 249
Hering, C., 19
Herschfeld, A., 125, 127, 271, 274
Hofmann, J. E., 60, 164
Hua, Loo Keng, 55
Hutton, C., 133, 134, 143
Hyyrö, S., 7, 91, 92, 212, 216, 244, 251, 253, 255, 256, 261, 263, 265, 266, 306

Index of Names

Inkeri, K., 6, 43, 91, 92, 119, 120, 219–222, 225, 229, 231, 234, 244, 306
Ivic, A., 321, 323

Jacobi, C. G. J., 196

Kanada, Y., 135, 136
Kanold, H.-J., 19
Keller, W., 234
Klingenstierna, S., 135
Kloss, K. E., 234
Ko, Chao, 6, 91–93, 95, 112
Korkine, A., 314
Krätzel, E., 321
Kummer, E. E., 29–31

Lagny, F. de, 134
Lagrange, J. L., 26, 55, 59–61, 63, 128, 141, 147, 152, 166, 173, 177, 192
Landau, E., 193, 248, 307
Lang, S., 350
Langevin, M., 287, 306
Lebesgue, V. A., 6, 77, 78, 80, 95, 118, 140, 152, 166, 192
Legendre, A. M., 49, 52, 60, 61, 76, 77
Lehman, W., 134
Lehmer, D. H., 91, 234
Leibniz, G. W., 133
Leo Hebraeus, 5
LeVeque, W. J., 7, 52, 213, 244, 271, 280–282
Levi Ben Gerson, 5, 124
Liouville, J., 243
Liouville, R., 314
Ljunggren, W., 96, 110, 111, 117, 128, 141, 142, 177, 178, 182, 184, 187, 192–196, 198, 296
London, H., 249

Lubelski, S., 91
Lucas, E., 194
Lüneburg, H., 19

Machin, J., 134, 143
Mahler, K., 80, 88, 89, 181, 192, 248, 287, 324
Makowski, A., 6, 213, 326
Mancion, P., 334
Manski, D. K., 245
McCallum, W., 52, 53
McDaniel, W. L., 198, 324
Mcronk, D. B., 192, 245
Mignotte, M., 239, 306\
Möbius, A., 15
Mollin, R. A., 323, 324, 329, 330
Mordell, L. J., 49, 61, 198, 244, 249, 250, 329, 330
Moret-Blanc, M., 94, 203

Nagell, T., 6, 65, 76, 88, 91, 92, 96, 105, 110, 111, 117, 119, 182, 215, 231, 271, 272, 293
Nair, M., 249, 308
Narkiewicz, W., 220
Nathanson, M. B., 315, 317

Obláth, R., 91, 92, 120, 204, 214, 215
Odoni, R. W. K., 322
Okada, T., 305, 306
Ostrowski, A., 193, 248

Pell, J., 60
Perlis, R., 121
Perron, O., 55
Philippon, P., 245
Pillai, S. S., 125, 271, 272, 279, 280, 307, 308

Pólya, G., 272, 272, 284

Ramanujan, S., 136
Ribenboim, P., 61, 179, 198, 213, 215, 242, 308, 314, 315, 320, 328
Riesel, H., 263, 265
Roth, K. F., 243, 251, 254
Rotkiewicz, A., 95, 120, 204, 214, 216

Salamin, E., 135
Sato, D., 136
Schacher, M., 96, 121
Schepel, D., 66
Schinzel, A., 204, 244, 249, 328
Schur, I., 284
Selberg, S., 6, 89–91
Sentance, W. A., 324
Shafarevich, I. R., 234
Shanks, D., 135, 314
Shanks, W., 34
Shiu, P., 321
Shorey, T. N., 121, 242, 287, 297
Siegel, C. L., 7, 124, 183, 243
Sierpinski, W., 213
Silverman, J. H., 318
Skandalis, G., 95
Skolem, T., 183, 184, 190, 193, 271, 292, 293
Soleng, R., 329
Stark, H. M., 248
Steiner, R. P., 192, 245, 249, 306
Stephens, A. J., 329
Stolt, B., 65
Størmer, C., 6, 80, 81, 85–90, 92, 93, 118, 128, 129, 135, 136, 141, 143, 144, 192, 271, 284, 287–290, 292

Straus, E., 96, 121
Suryanarayana, D., 121
Szekeres, G., 319–321

Tamura, Y., 135
Tang, Shen Chi, 78
Tartakowsky, W., 195
Theon of Smyrna, 60
Thue, A., 91, 181, 192, 193, 242, 243, 248, 270–272, 284, 286
Tijdeman, R., 7, 121, 242, 244, 245, 287, 296–299, 301, 306, 307, 313, 314, 328
Tonascia, J., 234
Tzanakis, N., 192

Vanden Eyden, C., 324
Vandiver, H. S., 19, 91
Vega, G. von, 134, 143
Vitry, P. de, 5

Waldschmidt, M., 245, 306, 307
Walker, D. T., 65, 80, 324
Wall, C. T. C., 202
Wallis, J., 60, 61
Walsh, P. G., 65, 323, 324, 329
Weger, B. M. M. de, 192
Weil, A., 61, 250
Weinberger, P., 234
Wieferich, A., 326, 327
Williams, H. C., 329, 330
Wrench, J. W., Jr., 132, 135
Wüstholz, G., 245, 246

Zarnke, C. R., 330
Zsigmondy, K., 19

Subject Index

Associated elements, 28

Binary cubic forms, 177

Catalan's conjecture, 6
Class number
   of cyclotomic field, 30, 220
   of imaginary quadratic field, 220
Computation of π, 132ff
Congruence modulo, 28
Conjecture
   of Ankeny, Artin, and Chowla, 329
   of Erdös, 325, 326
   of Hall, 249
   of Pillai, 207–308
Continued fraction, 55–60
   convergents of, 58
   partial quotients of, 56
   period, pre-period of, 59

purely periodic, 59
Cyclotomic integers, 27
Cyclotomic fields, 27
   units of, 30
Cyclotomic polynomials, 14

Equivalent ideals, 30

Fermat's equation, 31, 34ff, 45ff, 80
Fermat's little theorem, 29
Fermat's numbers, 80
Fermat's problem, 34, 144ff
Fermat's quotients, 221
Fibonacci numbers, 194
Fractional ideal, 28

Grave's problem, 132ff

Ideal class group, 30

Infinite descent, 34
Integral ideal, 28

Linear forms in logarithms, 245ff
Linear recurring sequences of second order, 197
Lucas numbers, 194

Möbius function, 15

Non-singular curve, 245
Numbers of Ferentinou-Nicolacopoulou, p213

p-adic valuation, 10
Powerful number, 319ff
Prime ideal, 29

Primitive factor, 16
Principal fractional ideal, 28
Property of Størmer, 81
Pythagorean equation, 31ff

Real quadratic field, 61ff
   units of, 61
Regular pair, 88
Riemann zeta function, 320

Singular pair, 88
Størmer, property of, 81

Twin primes, 226

Units of cyclotomic fields, 30
Units of real quadratic fields, 61